SATELLITE COMMUNICATIONS

SATELLITE COMMUNICATIONS

Timothy Pratt
Charles W. Bostian

Department of Electrical Engineering
Virginia Polytechnic Institute and State University

JOHN WILEY & SONS
New York · Chichester · Brisbane · Toronto · Singapore

Library of Congress Cataloging in Publication Data:

Pratt, Timothy.
    Satellite communications.

    Includes bibliographies and index.
    1. Artificial satellites in telecommunication.
I. Bostian, Charles W. II. Title.
TK5104.P725   1986   621.38′0422   85-13986
ISBN 0-471-87837-5

TIMOTHY PRATT is a professor in the Electrical Engineering Department at Virginia Polytechnic Institute and State University, where he has been a faculty member since 1981. He received the B.Sc. and Ph.D. degrees in electrical engineering from the University of Birmingham, England and subsequently taught there for ten years after spending five years in industry. He is a senior member of the IEEE and a member of the IEE (London).

CHARLES W. BOSTIAN is Clayton Ayre Professor of Electrical Engineering at Virginia Polytechnic Institute and State University, where he has been a faculty member since 1969. His primary research interests are in the areas of satellite communications and radio wave propagation. He is a coauthor of the Wiley text *Solid State Radio Engineering*, published in 1980. Professor Bostian received his B.S., M.S., and Ph.D. degrees in electrical engineering from North Carolina State University and is a senior member of the IEEE.

PREFACE

Communications satellites play a major role in telephone transmission, television and radio program distribution, computer communications, maritime navigation, and military command and control. The current strong demand for electrical engineers originates from all segments of the industry, ranging from satellite and earth station manufacturers to lessors of satellite channels. This book was written as a text for a one-semester senior or beginning graduate level course to provide these engineers with an appropriate background in satellite technology, link design, and operations. We hope that graduate engineers will also find it useful as a reference.

Satellite communications engineering combines such diverse topics as radio wave propagation, antennas, orbital mechanics, modulation, detection, coding, and radio electronics. Each of these is a major field of study, and each has its own extensive literature. In preparing this text we emphasized the material from these areas that is important to satellite communications and derived those equations that a beginning engineer might reasonably be expected to know and to understand. We have tried to make our coverage as practical as possible, stressing those techniques that are or soon will be in use.

We assume that our readers will have completed the usual undergraduate courses in physics (to provide a background for orbital mechanics), electromagnetic fields (to understand wavelength, power density, and electromagnetic radiation), and communications theory (for a background in modulation, detection, noise, and spectra). Although some familiarity with microwaves and radio electronics would be useful, these are not required.

Although later chapters build upon material covered in earlier ones, we have made each chapter reasonably self-contained for practicing engineers who are seeking information about a particular topic. To help with this process we have provided an end-of-book glossary and a list of symbols.

Chapter 1 introduces satellite communications and provides a brief history of the subject. Chapter 2 is a detailed discussion of orbital mechanics and orbital considerations in satellite launches and satellite link design. So far as we are aware, it is the first treatment of the subject for electrical engineers that covers everything from Newton's laws of motion to look angles. Chapter 3 presents a review of those aspects of spacecraft design and organization that are important in communications satellites. Chapter 4 derives the equations needed for calculating the carrier-to-noise ratio (C/N) of a link and discusses techniques for achieving a specified (C/N). Chapter 5 reviews digital and analog modulation and multiplexing techniques and provides the necessary information for determining the (C/N) requirement that links designed by the procedure of Chapter 4 must meet. Chapter 6 discusses multiple access, and Chapter 7 presents error detection and correction, applications of coding theory that are particularly important to satellite systems. Chapter 8 presents propagation effects and their impact on system design; these are severe at frequencies above 10 GHz. Chapter 9 discusses earth station technology, emphasizing antennas. Chapter 10 describes Intelsat (The International Telecommunications Satellite Organization) and Inmarsat (The International Maritime Satellite Organization). Chapter 11 is about television distribution and satellite broadcasting, topics that the general public most strongly associates with satellite communications. The appendix provides background information on how decibels are used in communications engineering and on the effect of bandlimiting on digital radio signals.

Timothy Pratt
Charles W. Bostian

ACKNOWLEDGMENTS

This text originated as class notes for our course in satellite communications at Virginia Polytechnic Institute and State University. Several "generations" of students have tested it, and we appreciate their patience and their skill as proofreaders. In the latter, Mr. Shiao Chiu Lee was particularly helpful.

Several of our colleagues at Virginia Tech read portions of this book and provided invaluable information and criticism. In particular we would like to acknowledge the contributions of Professors Warren L. Stutzman and Daniel B. Hodge. We also received much essential information from friends in industry and government who are too numerous to name here.

Initial word processing of this manuscript was by Cynthia Will Dettmer and Linda White Kipps. All later drafts were prepared by Patricia Wojciechouski, who completed the book via a long-distance telephone hookup to our computer. We appreciate their contributions and their struggles with our handwriting and notation.

Finally we acknowledge with gratitude the support of our wives, Maggie Pratt and Frieda Bostian, who have provided constant encouragement during the writing of this book.

T. P.
C. W. B.

CONTENTS

1

INTRODUCTION

In less than 20 years communications satellites have become the dominant carriers of long distance communications. From the first commercial launch of INTELSAT I[1] (also called EARLY BIRD) on April 6, 1965, the satellite industry has grown until it handles most international telephone traffic, all international and almost all domestic long-distance television program distribution, and a rapidly growing proportion of new domestic voice and data channels. Direct satellite broadcasting will soon begin, and proposals for electronic mail and personal two-way satellite radios are under discussion. Satellites have significantly improved the reliability and the accuracy of aviation and maritime communications and navigation, removing these functions from the high-frequency (HF) portion of the spectrum. The International Telecommunication Satellite Organization (*Intelsat*) has grown at a rate of 20 percent per year since 1965 and, as of 1984, it operated over 35,000 two-way traffic links. United States domestic satellite use is expected to grow at an annual rate of 15 percent until the end of this century [2].

These changes have occurred because the technology is now available to put large *spacecraft* into synchronous orbit where, to an observer on the ground, they remain permanently at the same place in the sky. At an altitude of about 35,870 km (22,291 miles), the satellites can receive, amplify, and retransmit radio signals for most of a hemisphere. Thus, with one relay via a satellite, a single transmitter on the ground can reach nearly half the world. With three relays it can reach all of it.

---

[1] In this text we have followed the conventional practice of printing the names of individual spacecraft and families of spacecraft in capital letters. Individual members of multiple families are generally identified by Roman numerals, for example, INTELSAT III, except for those cases like the COSMOS series, where there have been over 1300 launches [1].

As satellites have grown larger and more powerful, the cost of required terrestrial equipment has fallen. Where once an earth terminal was a multimillion dollar proposition, receive-only stations are now available for under $1500. The result is that satellites are the cheapest way to send information reliably over a long distance, particularly if that information is intended for a large number of receivers, who may not always be in the same places.

1.1 THE ORIGIN OF SATELLITE COMMUNICATIONS

Most authorities credit Arthur C. Clarke, famous British science fiction writer and author of *2001: A Space Odyssey* [3], with originating the idea of a synchronous communications satellite. In 1945 Clarke noted [4] that a satellite in a circular equatorial orbit with a radius of about 42,242 km would have an angular velocity that matched the earth's. Thus it would always remain above the same spot on the ground, and it could receive and relay signals from most of a hemisphere. Three satellites spaced 120 degrees apart could cover the whole world (with some overlap); provided that messages could be relayed between satellites, reliable communication between any two points in the world would be possible.

As is appropriate for a science fiction writer, Clarke had ideas ahead of their time. It was not until the USSR launched SPUTNIK I on October 4, 1957, that rocket technology was available to put a satellite into even a low orbit; synchronous orbit was not achieved until 1963.

1.2 A BRIEF HISTORY OF SATELLITE COMMUNICATIONS [5]

The 1957 launch of SPUTNIK I was followed by the "space race" and a sustained effort by the United States to catch up with the USSR. This was reflected in SCORE (Signal Communicating by Orbiting Relay Equipment) launched by the U.S. Air Force on December 18, 1958. Essentially an Atlas 10B missile with a modified upper stage, SCORE was placed in a low elliptical orbit with a period of 101 min. It broadcast a taped message from President Eisenhower, but it was also the first successful "bent pipe in the sky" satellite repeater. Its normal operating mode was to record an uplink transmission while passing over one earth station and to play it back when requested by another earth station. The maximum message length was 4 min, and the spacecraft's capacity was either one voice channel or seventy 60-words-per-minute teletype channels. SCORE's transponder was a marvel of "quick and dirty" engineering; the receiver was a modified FM pocket pager and the transmitter was a "handy-talkie" with an outboard amplifier to boost transmitter power to 8 W. The uplink frequency was 150 MHz and the downlink frequency was 132 MHz, and the spacecraft carried a tracking beacon at 108 MHz. SCORE's batteries failed after 35 days in orbit.

The first communications satellites to draw widespread popular interest (because on clear nights they were visible to the naked eye) were ECHO I and II, launched by *AT&T* on August 12, 1960, and January 25, 1964. These were orbiting balloons 100 ft in diameter which served as passive reflectors. As such, they had no transponder batteries to run down, and they did not require a strict frequency channeling of uplink signals to accommodate transponder input bands. On the other hand, they operated like radar reflectors and incurred path losses that were proportional to the fourth power of path length rather than to the square of path length as is the case with active satellites. This as well as the available launch vehicles limited the ECHOs to very low orbits with periods of 118 min for ECHO I and 108.8 min for ECHO II. Low orbits meant that an ECHO was in view of two widely separated earth stations for only a few minutes on each pass. Power and antenna requirements were severe; a typical ECHO link from Bell Laboratories in New Jersey to the Jet Propulsion Laboratory in California used 10 kW transmitters at both ends, an 85-ft dish in California, and a 60-ft dish in New Jersey. Typical frequencies were 960 MHz westbound and 2390 MHz eastbound.

The Bell System also developed and launched the first successful broadband real-time transponders in TELSTAR I and II on July 10, 1962, and May 7, 1963. These were also in low orbits; their periods were 158 and 225 min, respectively. The TELSTARs provided 50 MHz of bandwidth for analog FM signals centered at 6389.58 MHz for the uplink and 4169.72 MHz for the downlink. This choice of frequencies set the precedent for 4/6 GHz operation and also for potential interference between satellite and terrestrial links sharing these bands.

In 1963 Congress passed the Communications Satellite Act, establishing the Communications Satellite Corporation (*Comsat*) and barring the Bell System from further direct participation in satellite communications. While we will not go into the many conflicting reasons why this should or should not have been done (the authors have friends who were involved on all sides of the matter!), this caused considerable bitterness in the Bell System, which had invested substantial resources in the ECHO and TELSTAR programs. The Bell engineers involved felt that, once their company proved that communications satellites would work, the opportunity to profit by their investment was taken away and given to someone else. Unhappiness over this situation persisted well into the 1970s and the restriction ultimately was lifted.

The SYNCOM series provided the first successful geosynchronous communications satellites beginning in 1963, less than 20 years after Clarke first conceived the idea. SYNCOM I failed during launch, but SYNCOM II and III were successfully placed in orbit on July 26, 1963, and July 19, 1964. This was a joint NASA-Department of Defense effort, and the satellites used military frequencies of 7.36 GHz for the uplink and 1.815 GHz for the downlink. Using FM or PSK, the transponders could support two carriers at a time for full duplex operation. These spacecraft continued in service until some time after 1965.

The first commercial geosynchronous satellite was INTELSAT I (first called EARLY BIRD), developed by Comsat for Intelsat. Launched April 6, 1965, it remained active until 1969. Routine operation between the United States and Europe began on June 28, 1965, a date that should be recognized as the birthday of commercial satellite communications. The spacecraft had two 25 MHz bandwidth transponders with uplinks centered at 6301 MHz for Europe and 6390 MHz for the United States. U.S. receivers operated with a 4081 MHz center frequency and the European downlink band was centered at 4161 MHz [6]. With this spacecraft the modern era of satellite communications had begun.[2]

1.3 THE CURRENT STATE OF SATELLITE COMMUNICATIONS

The number of operational and planned satellite communications systems is growing so rapidly that it is difficult to summarize. We will describe it by presenting Table 1.1 and Figure 1.1, both from reference 7. These show over 60 North American spacecraft positions in the geostationary arc and over 80 satellite *systems* worldwide. For a complete list of what is in orbit see reference 8.

1.4 AN OVERVIEW OF SATELLITE SYSTEM ENGINEERING

The design of a satellite communications system is a complicated process that involves the interaction of many disciplines. Later chapters of this book will discuss the more important aspects.

The first problem is the satellite itself. It has to be as small and lightweight as possible and use a minimum of energy. Putting a kilogram into synchronous orbit is extremely expensive. Generating electricity to run the spacecraft requires both weight and surface area for the solar cells. Since communications capacity is what earns revenue, the satellite must be capable of carrying as many communications channels as possible. Since launches and satellites are expensive, spacecraft must be capable of functioning without maintenance for many years in a hostile hard-vacuum environment, through severe thermal cycling, and under constant bombardment of radiation, subatomic particles, and occasional micrometeorites. Finally, since new trends in communications technology often occur quickly and unexpectedly, spacecraft must be designed with as much flexibility as possible.

Another consideration is the distance between the spacecraft and the earth. A path length of 35,870 km corresponds to about 90 percent of a trip around the earth at the equator, and a signal has to travel this distance once to get to a synchronous spacecraft and once to return to the ground. The inverse square losses are enormous, and at frequencies above 10 GHz rain losses add to these. On the uplink, powerful transmitters and large antennas can be used, but these are expensive and inconvenient. For the downlink, antenna size and transmitter power are limited by what the satellite can carry and generate electricity for, and the received signals are extremely weak—weaker than those encountered in almost any other kind of communications system. The result is that careful attention must be paid to antenna gain, transmitter efficiency, receiver noise figure, and the like.

---

[2] The reader who is interested in the early history of satellite communications should consult reference 5. An anonymous document first put together as a NASA contractor report, it was prepared at a time that NASA's further role in satellite communications was in doubt. It is an impressive testimonial to the pioneers in the field.

Table 1.1
Operational and Planned Communications Satellite Systems

Program Name	Category Code[a]	Coverage Type[b]	Operational Status[c]	Date of First Operation	Frequency Bands	Operational Satellites
ABC	F	G	IP	1986	Ku	2 planned (+ 1 spare)
AEROSAT	AM	R	IP	198?	C, L, VHF	2 planned
AMERICAN SATELLITE	F	D	UC	1985	C, Ku	3 under construction (incl. spare)
ANIK A (Canada)	F	D	OP	1972	C	1 operational
ANIK B	F	D	OP	1978	C, Ku	1 operational
ANIK C	F	D	OP	1982	Ku	2 operational, 1 under construction
ANIK D	F	D	OP	1982	C	1 operational, 1 under construction
APPLE	F	D	OP	1981	C	1 operational
ARABSAT	E	D	UC	1984	C, L	2 under construction
ASETA (Andean)	F	R	IP	198?	C, Ku	2 or 3 planned
AUSTRALIA	F	D	UC	1985	Ku	3 under construction
AUTOSAT	LM	D	IP	198?	UHF	1 planned
BS-2 (Japan)	B	D	UC	1984	Ku	2 under construction
CBS	B	D	IP	198?	Kc, Ku	4 planned + 1 or 2 spares
CHINA	B	D	IP	198?	C, Ku	2 planned
COMSTAR	F	D	OP	1976	C	4 operational
CS-2 (Japan)	F	D	OP	1983	C, Kc, K	2 operational
DBSC	B	D	IP	1986	Kc, Ku	3 planned + 1 spare
Dominion Video Satellite Network	B	D	IP	198?	Kc, Ku	2 planned + 1 spare
DSCS II	F/MM, Mi	G	OP	1966	X	8 operational
DSCS III	F/MM, Mi	G	OP	1982	X	1 operational
ECS (EUTELSAT)	F	R	OP	1983	Ku	1 operational, 4 more planned

[a] AM = Aeronautical Mobile, B = Broadcast, E = Experimental, F = Fixed, LM = Land Mobile, Mo = Mobile (General), Mi = Military, MM = Maritime Mobile.
[b] G = Global, R = Regional, D = Domestic.
[c] IP = In Planning, OP = Operational, UC = Under Construction.

Source: Wilber L. Pritchard, "The History and Future of Satellite Communications," *IEEE Communications Magazine,* **22,** 22–37 (May 1984) © 1984 IEEE).

(continued)

Table 1.1 (*continued*)

Program Name	Category Code[a]	Coverage Type[b]	Operational Status[c]	Date of First Operation	Frequency Bands	Operational Satellites
EKRAN (USSR)	B	D	OP	1976	C, UHF	3 operational
FLTSATCOM	F/MM, Mi	G	OP	1978	UHF, X	5 operational
FORDSAT	F	D	IP	1987	C, Ku	3 planned (incl. spare)
GALAXY	F	D	OP	1983	C	1 operational, 2 more planned
GALAXY Ku	F	D	IP	1987	Ku	3 planned
GALS (USSR)	F, Mi	G	IP	198?	X	4 planned
GORIZONT (USSR)	F, Mi	R	OP	1978	C, X	4 operational
GRAPHSAT	B	D	IP	198?	Kc, Ku	2 planned + 1 spare
G-STAR	F	D	UC	1984	Ku	3 under construction, 1 more planned
ILHUICAHUA (Mexico)	F	D	UC	1985	C, Ku	2 under construction
INSAT (India)	F, B	D	OP	1983	C, S	1 operational, 1 more planned
INTELSAT IV	F	G	OP	1965	C	4 operational
INTELSAT IV-A	F	G	OP	1976	C	5 operational
INTELSAT V	F, MM	G	OP	1980	C, L, Ku	6 operational, 3 more under construction
INTELSAT V-A	F	G	UC	1984	C, Ku	6 under construction
INTELSAT VI	F	G	UC	1986	C, Ku	5 under construction, up to 11 more planned
ITALSAT	E	D	IP	1987	EHF, Kc	3 planned (incl. preoperational + spare)
LEASAT	Mo, Mi	G	UC	1984	C, VHF	4 under construction
LES	E/Mo, Mi	G	OP	1976	K, UHF	2 operational
LOUTCH (USSR)	F	R	IP/UC	1983	Ku	8 under construction or planned
LUXSAT (Luxembourg)	B	R	IP	1986	Kc, Ku	2 planned (incl. spare)
MARECS	MM	G	OP	1982	C, L	1 operational, 1 more planned
MARISAT	MM	G	OP	1975	C, L, UHF	3 operational
MOLNIYA 1 (USSR)	F	G	OP	1965	C, UHF	Maybe 6 operational
MOLNIYA 3 (USSR)	F	G	OP	1974	C, UHF	Maybe 7 operational
NATO III	F, Mi	G	OP	1979	X	3 operational, 2 more under construction
NORDSAT (Scandinavia)	B	R	IP	198?	Kc, Ku	2 or 3 planned
OLYMPUS (Europe)	B/E	R	UC	1986	Kc, Ku	1 under construction
OTS (Europe)	E, F	R	OP	1978	Ku	1 operational

PALAPA I (Indonesia)	F	D	OP	1976	C	2 operational
PALAPA II (Indonesia)	F	R	OP	1983	C	1 operational, 1 more planned
POSTSAT (West Germany)	F	D	IP	1986	Kc, Ku	3 planned
RADUGA (USSR)	F, Mi	G	OP	1975	C, X	Maybe 5 operational
RAINBOW	F	D	IP	198?	Ku	2 planned + 1 spare
RCA SATCOM	F	D	OP	1975	C	6 operational, 2 more under construction
RCA SATCOM Ku	F	D	UC	1985	Ku	3 under construction
RCA DBS	B	D	IP	1985	Ku	4 planned + 2 spares
SATCOL (Colombia)	F	D	IP	198?	C or Ku	2 or 3 planned
SARIT (Italy)	B	D	IP	1986	Kc, Ku	1 or 2 planned
SATELLITE SYNDICATED SYSTEMS	B	D	IP	198?	Kc, Ku	4 planned + 1 spare
SBS	F	D	OP	1981	Ku	3 operational, 3 more planned
SBTS (Brazil)	F	D	UC	1985	C	Under construction
SKYNET IV (U.K.)	F, Mi	G	UC	1985	X	2 under construction
SPACENET	F	D	UC	1984	C, Ku	3 under construction, 1 spare planned
STC	B	D	UC	198?	Kc, Ku	2 under construction, up to 4 more planned
SYMPHONIE	F, E	R	OP	1974	C	1 operational
TDF	B	D	UC	1985	Kc, Ku	2 under construction
TELECOM	F	R, D	UC	1984	Ku	3 planned (incl. 2 spares)
TELE-X	E, F/B	R, D	UC	1987	K, Kc, Ku	1 under construction
TELESTAR 3	F	D	OP	1983	C	1 operational, 2 more planned
TVSAT (Germany)	B	D	UC	1985	Kc, Ku	1 or 2 under construction
UNISAT	B	D	IP	1986	Kc, Ku	1 planned + 2 spares
USAT	F	D	IP	1985	Ku	2 planned + 1 spare
USSB	B	D	IP	198?	Kc, Ku	4 planned + 1 spare
VOLNA (USSR)	MM	G	IP	198?	L, VHF/UHF	7 planned
WESTAR	F	D	OP	1974	C	5 operational, 3 more planned
WESTAR Ku	F	D	IP	1985	Ku	3 planned
WESTERN UNION DBS	B	D	IP	198?	Kc, Ku	2 planned + 2 spares

[a] AM = Aeronautical Mobile, B = Broadcast, E = Experimental, F = Fixed, LM = Land Mobile, Mo = Mobile (General), Mi = Military, MM = Maritime Mobile.

[b] G = Global, R = Regional, D = Domestic.

[c] IP = In Planning, OP = Operational, UC = Under Construction.

Source: Wilber L. Pritchard, "The History and Future of Satellite Communications," *IEEE Communications Magazine,* **22,** 22–37 (May 1984) © 1984 IEEE).

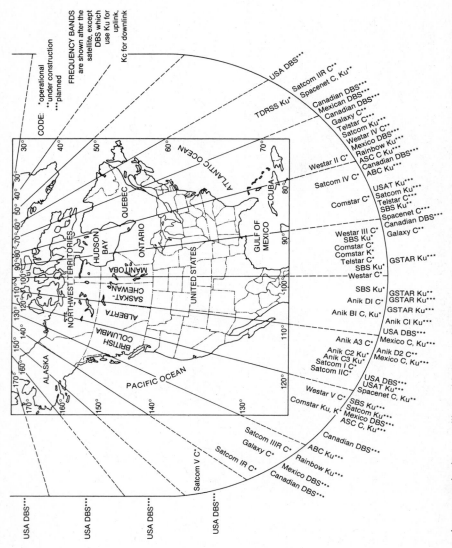

Figure 1.1 North American domestic communications satellite positions as of August 1983. (Reprinted with permission from Wilber L. Pritchard, "The History and Future of Commercial Satellite Communications," *IEEE Communications Magazine,* **22**, 22–37 (May 1984). Copyright © 1984 IEEE.)

Even then, inherent technical and economic limitations of the hardware require a lot of effort in the software aspects of satellite communications. Much work goes into developing modulation and coding schemes for detecting and correcting the transmission errors introduced by noise.

Multiple access is also a problem. One satellite can have many users scattered over a whole country or even over an entire hemisphere. To maximize revenue it must be able to serve a large but changing number simultaneously and efficiently with a minimum of external input or control.

Finally there are the earth stations. They must be cheap enough for users to afford, but powerful and sophisticated enough to communicate efficiently with their satellites. They must also comply with many government licensing and regulatory requirements. They must be able to find their satellites quickly and maintain communication with them through unexpected orbital changes.

Later sections of this book will discuss all of these problems.

1.5 ORGANIZATIONS CITED IN THIS BOOK

In the chapters that follow we refer to a number of organizations that may not be familiar to all readers. Although some of these, like Intelsat, will be described in detail after the text has given sufficient background information, we feel that it would be appropriate to give preliminary definitions at this point.

NASA is of course the National Aeronautics and Space Administration. An agency of the U.S. Government, it is well known for its role in all aspects of space exploration and research. Its European counterpart is the multinational European Space Agency (*ESA*). Unlike NASA, ESA is engaged in the operation of communications and broadcast satellites as well as in space research. The Japanese counterpart to NASA is the National Space Development Agency.

Intelsat is the International Telecommunications Satellite Organization with headquarters in Washington, D.C. An organization of 109 countries (national representatives are called signatories), it maintains and operates the international satellite communications network. The U.S. signatory is Comsat, short for the Communications Satellite Corporation, a private profit making U.S. company. Comsat in turn is owner or has been part-owner of several other companies like Satellite Business Systems (*SBS*) and Comsat General, which compete in the U.S. domestic communications market. Comsat is also the U.S. signatory for Inmarsat, the International Maritime Satellite Organization. An organization of countries similar to Intelsat, Inmarsat provides satellite communications for ships and offshore oil platforms.

The primary regulatory agency for U.S. satellite communications is the Federal Communications Commission (*FCC*). In most other countries this role is played by a Post and Telecommunications Authority (PTT), a government agency or government-owned corporation that is responsible for postal service and almost all other forms of internal communications. Advisory standards are set by the *CCIR* and *CCITT* (the letters come from the French language initials for International

Radio Consultative Committee and for International Telegraph and Telephone Consultative Committee). These are agencies of the International Telecommunications Union (ITU), which in turn is part of the United Nations. See reference 9 for a detailed description of the ITU.

1.6 SUMMARY

Satellites in geostationary orbit remain permanently at the same place in the sky. At an altitude of about 35,870 km, they can receive, process, amplify, and retransmit signals for most of a hemisphere. Since the first communications satellites were launched in the early 1960s, there have been rapid and continued increases in spacecraft capacity and power. These have been accompanied by significant decreases in the cost of satellite channels and sharp increases in the amount of communications traffic carried by satellite.

Satellite systems engineering involves the interaction of many disciplines. The spacecraft themselves must be designed to operate reliably under their own power for years in a hostile hard-vacuum environment. The communications links must be able to function with extremely weak signals and long propagation delays.

REFERENCES

1. M. W. Sherman, *TRW Space Log*, TRW Inc., Redondo Beach, CA, 1983.
2. B. I. Edelson and R. S. Cooper, "Business Use of Satellite Communications," *Science*, **215**, 837–842 (February 1982).
3. Arthur C. Clarke, *2001: A Space Odyssey*, New American Library, New York, 1968.
4. Arthur C. Clarke, "Extraterrestrial Relays," *Wireless World*, **51**, 305–308 (October 1945).
5. *NASA Compendium of Satellite Communications Programs* (Document X-751-73-178), NASA Goddard Space Flight Center, Greenbelt, MD, June 1973.
6. Martin P. Brown, Jr., Ed., *Compendium of Communication and Broadcast Satellites 1958 to 1980*, IEEE Press, New York, 1981.
7. Wilber L. Pritchard, "The History and Future of Commercial Satellite Communications," *IEEE Communications Magazine*, **22**, 22–37 (May 1984).
8. Walter L. Morgan, "Satellite Locations—1984," *Proceedings of the IEEE*, **72**, 1434–1444 (November, 1984).
9. W. H. Bellchambers, J. Francis, E. Hummel, and R. L. Nickelson, "The International Telecommunications Union and Development of Worldwide Telecommunications," *IEEE Communications Magazine*, **22**, 72–83 (May 1984).

2
ORBITAL ASPECTS OF SATELLITE COMMUNICATIONS

2.1 ORBITAL MECHANICS

The Equations of the Orbit

We will begin our development of the equations for a satellite's orbit with the rectangular coordinate system shown in Figure 2.1. The origin is at the center of the earth, and the z axis extends through the north geographic pole. The coordinate system is right-handed; the directions of the x and y axis are otherwise arbitrary. The coordinate system is fixed in space, and the earth rotates on the z axis. We will assume that the center of mass of the earth–satellite system coincides with the center of mass of the earth at the origin.

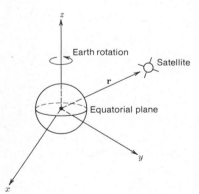

Figure 2.1 The first coordinate system used to describe the earth and a satellite. The x, y, and z axes are fixed by the earth's equatorial plane and north pole. The vector $\mathbf{r}$ locates the moving satellite with respect to the center of the earth.

A satellite of mass m is located at a vector distance $\mathbf{r}$ from the center of the earth. The gravitational force $\mathbf{F}$ on the satellite is given by

$$\mathbf{F} = -\frac{GM_E m \hat{\mathbf{r}}}{r^2} \tag{2.1}$$

where M_E is the mass of the earth and $G = 6.672 \times 10^{-11}$ Nm/kg^2 [1] is the universal gravitational constant. The unit vector in the $\mathbf{r}$ direction is represented by $\hat{\mathbf{r}}$. The product $GM_E = 3.9861352 \times 10^5$ km^3/s^2 is called Kepler's constant, symbol μ. Writing Newton's second law

$$\mathbf{F} = m \frac{d^2 r}{dt^2} \hat{\mathbf{r}} \tag{2.2}$$

and equating the inertial force on the satellite to the gravitational force, we have

$$-\frac{\mu \hat{\mathbf{r}}}{r^2} = \frac{d^2 r}{dt^2} \hat{\mathbf{r}} \tag{2.3}$$

This is a second-order vector linear differential equation, and its solution will involve six undetermined constants called the *orbital elements*. Writing

$$\mathbf{r} = r \hat{\mathbf{r}} \tag{2.4}$$

and rearranging we have

$$\frac{1}{r} \frac{d^2 \mathbf{r}}{dt^2} + \frac{\mu \mathbf{r}}{r^3} = 0 \tag{2.5}$$

Adopting the approach and notation of reference 2, we can show that the motion defined by Eq. (2.5) is confined to a plane. Taking $\mathbf{r} \times$ each term yields

$$\mathbf{r} \times \frac{d^2 \mathbf{r}}{dt^2} = 0 \tag{2.6}$$

Invoking the rule for finding the derivative of a product, it follows that

$$\frac{d}{dt} \left[\mathbf{r} \times \frac{d\mathbf{r}}{dt} \right] = \frac{d\mathbf{r}}{dt} \times \frac{d\mathbf{r}}{dt} + \mathbf{r} \times \frac{d^2 \mathbf{r}}{dt^2} \tag{2.7}$$

The cross product of any vector with itself is zero; hence, (Eq. 2.7) may be rewritten

$$\frac{d}{dt} \left[\mathbf{r} \times \frac{d\mathbf{r}}{dt} \right] = 0 \tag{2.8}$$

This is equivalent to

$$\mathbf{r} \times \frac{d\mathbf{r}}{dt} = \mathbf{h} \tag{2.9}$$

where the constant, $\mathbf{h}$, is the orbital angular momentum of the satellite. But the orbital angular momentum can be a constant only if the orbit lies in a plane. Hence, the problem of satellite motion in three dimensions reduces to the problem of motion in a plane. Note that the orientation of this plane is as yet undetermined.

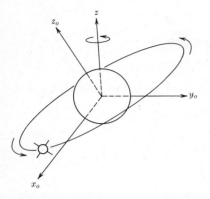

Figure 2.2 The orbital plane coordinate system. The x_o and y_o axes lie in the orbital plane and the z_o axis is perpendicular to it.

While Eq. (2.5) is an appealingly compact form of the equation of motion for a satellite, it is difficult to solve because the second derivative of **r** involves second derivatives of the unit vector $\hat{\mathbf{r}}$. We may avoid this problem by expressing the equation in a rectangular coordinate system where the unit vectors are constant. To do that we will use the orbital plane to define a second rectangular coordinate system (x_o, y_o, z_o). See Figure 2.2. Here the subscript o indicates orbital plane. The first two coordinates lie in the plane and z_o is normal to it. It is not necessary to specify the orientation further at this point. Expressing Eq. (2.5) in terms of these coordinates yields

$$\hat{\mathbf{x}}_o\left(\frac{d^2 x_o}{dt^2}\right) + \hat{\mathbf{y}}_o\left(\frac{d^2 y_o}{dt^2}\right) + \frac{\mu(x_o\hat{\mathbf{x}}_o + y_o\hat{\mathbf{y}}_o)}{(x_o^2 + y_o^2)} = 0 \qquad (2.10)$$

Equation (2.10) is easier to solve if it is expressed in the polar coordinate system (r_o, ϕ_o) of Figure 2.3 in the orbital plane. Using the transformations

$$x_o = r_o \cos \phi_o \qquad (2.11a)$$

$$y_o = r_o \sin \phi_o \qquad (2.11b)$$

$$\hat{\mathbf{x}}_o = \hat{\mathbf{r}}_o \cos \phi_o - \hat{\boldsymbol{\phi}}_o \sin \phi_o \qquad (2.11c)$$

$$\hat{\mathbf{y}}_o = \hat{\boldsymbol{\phi}}_o \cos \phi_o + \hat{\mathbf{r}}_o \sin \phi_o \qquad (2.11d)$$

and equating the $\hat{\mathbf{r}}_o$ components of Eq. (2.10) yields

$$\frac{d^2 r_o}{dt^2} - r_o\left(\frac{d\phi_o}{dt}\right)^2 = -\frac{\mu}{r_o} \qquad (2.12)$$

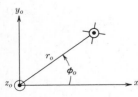

Figure 2.3 Polar coordinates in the orbital plane. In this drawing the orbital plane coincides with the plane of the paper.

Likewise equating the $\hat{\phi}_o$ components we have

$$r_o\left(\frac{d^2\phi_o}{dt^2}\right) + 2\left(\frac{dr_o}{dt}\right)\left(\frac{d\phi_o}{dt}\right) = 0 \tag{2.13}$$

Following closely the approach presented in reference 2 we may rewrite Eq. (2.13) as

$$\frac{1}{r_o}\frac{d}{dt}\left(r_o^2\frac{d\phi_o}{dt}\right) = 0 \tag{2.14}$$

This is equivalent to

$$r_o^2\frac{d\phi_o}{dt} = \text{a constant} = |\mathbf{h}| = h \tag{2.15}$$

where h is the magnitude of the angular momentum vector in Eq. (2.9). Writing

$$r_o\left(\frac{d\phi_o}{dt}\right)^2 = \frac{h^2}{r_o^3} \tag{2.16}$$

and substituting this expression into Eq. (2.12), we have

$$\frac{d^2r_o}{dt^2} - \frac{h^2}{r_o^3} = -\frac{\mu}{r_o} \tag{2.17}$$

In order to find the equation relating r_o and ϕ_o (i.e., the equation of the orbit), we must eliminate time t from Eq. (2.17). To do this we define a new variable u by

$$u = \frac{1}{r_o} \tag{2.18}$$

so that

$$\frac{dr_o}{d\phi_o} = -\frac{1}{u^2}\frac{du}{d\phi_o} \tag{2.19}$$

and using the relationship

$$\frac{dr_o}{dt} = \left(\frac{dr_o}{d\phi_o}\right)\left(\frac{d\phi_o}{dt}\right) = \left(\frac{dr_o}{d\phi_o}\right)\left(\frac{h}{r_o^2}\right) = -h\frac{du}{d\phi_o} \tag{2.20}$$

to transform d^2r_o/dt^2 in Eq. (2.17) to

$$\frac{d^2r_o}{dt^2} = -h^2u^2\left(\frac{d^2u}{d\phi_o^2}\right) \tag{2.21}$$

We may rewrite Eq. (2.17) as

$$\frac{d^2u}{d\phi_o^2} + u = \frac{\mu}{h^2} \tag{2.22}$$

According to reference 2, the solution to this differential equation is

$$u = \frac{\mu}{h^2} + C \cos(\phi_o - \theta_o) \tag{2.23}$$

where C and θ_o are constants to be determined from the boundary conditions. If we express Eq. (2.23) in terms of r_o, we have

$$r_o = \frac{1}{\mu/h^2 + C \cos(\phi_o - \theta_o)}$$

$$= \frac{\left(\dfrac{h^2}{\mu}\right)}{1 + \left(\dfrac{h^2}{\mu}\right) C \cos(\phi_o - \theta_o)}$$

$$= \frac{p}{1 + e \cos(\phi_o - \theta_o)} \tag{2.24}$$

For $e < 1$ this is the equation of an ellipse whose semilatus rectum p is given by

$$p = \frac{h^2}{\mu} \tag{2.25}$$

and whose eccentricity e is $h^2 C/\mu$. That the orbit is an ellipse is Kepler's first law of planetary motion. Under the limiting condition $e = 0$, the orbit is a circle with the earth at its center.

Describing the Orbit

The quantity θ_o in Eq. (2.24) serves to orient the ellipse with respect to the orbital plane axes x_o and y_o. Now that we know the orbit is an ellipse, we can always choose x_o and y_o so that θ_o is zero. For the rest of this discussion we will assume that this has been done, making the equation of the orbit

$$r_o = \frac{p}{1 + e \cos \phi_o} \tag{2.26}$$

The path of the satellite in the orbital plane is shown in Figure 2.4. The lengths a and b of the semimajor and semiminor axis are given by reference 2 as

$$a = \frac{p}{1 - e^2} \tag{2.27}$$

$$b = a(1 - e^2)^{\frac{1}{2}} \tag{2.28}$$

The point in the orbit where the satellite is closest to the earth is called the *perigee* and the point where the satellite is farthest from the earth is called the *apogee*. To make θ_o equal to zero, we have chosen x_o axis so that both the apogee and perigee lie along it.

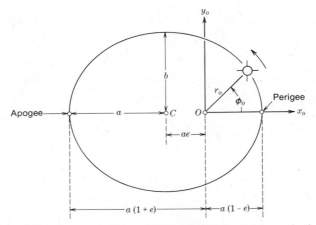

Figure 2.4   The orbit as it appears in the orbital plane. The point O is the center of the earth and the point C is the center of the ellipse. The two centers do not coincide unless the eccentricity e is zero. The dimensions a and b are the semimajor and semiminor axes of the orbital ellipse.

The differential area swept out by the vector $\mathbf{r}_o$ from the origin to the satellite in time dt is given by

$$dA = 0.5r_o^2 \left(\frac{d\phi_o}{dt}\right) dt = 0.5\,h\,dt \tag{2.29}$$

Thus the radius vector to the satellite sweeps out equal areas in equal times. This is Kepler's second law of planetary motion. Following reference 2 we may use it to derive an expression for the orbital period T by equating the area of the ellipse (πab) to the area swept out in one orbital revolution.

$$T = \frac{2\pi a^{\frac{3}{2}}}{\mu^{\frac{1}{2}}} \tag{2.30}$$

or

$$T^2 = \frac{4\pi^2 a^3}{\mu} \tag{2.31}$$

These last two expressions are a mathematical statement of Kepler's third law of planetary motion: the square of the period of revolution is proportional to the cube of the semimajor axis.

We may use Eq. (2.31) to calculate the radius of the *geosynchronous orbit*. Taking the earth's rotational period to be exactly 86,400 s (24 h), substituting this value into the equation, and solving for a, we obtain $a = 42,241.558$ km. Since a geosynchronous satellite must have a constant angular velocity, the orbit must be circular, and the radius of the circle must equal the length, a, of the semimajor axis. Rounded off to the nearest kilometer, then, the orbital radius of a geosynchronous satellite is 42,242 km. A geosynchronous orbit that lies in the earth's equatorial plane (i.e., a geosynchronous orbit with zero inclination) is *geostationary*.

Locating the Satellite in the Orbit

Consider now the problem of locating the satellite in its orbit. The equation of the orbit may be rewritten by combining Eqs. (2.26) and (2.27) to obtain

$$r_o = \frac{a(1 - e^2)}{1 + e \cos \phi_o} \tag{2.32}$$

The angle ϕ_o (see Figure 2.4) is measured from the x_o axis and is called the *true anomaly*. Since we defined the positive x_o axis so that it passes through the perigee, ϕ_o measures the angle from the perigee to the instantaneous position of the satellite. The rectangular coordinates of the satellite are given by

$$x_o = r_o \cos \phi_o \tag{2.33}$$

$$y_o = r_o \sin \phi_o \tag{2.34}$$

The orbital period T is the time required for the satellite to complete one revolution and travel 2π rad. The average angular velocity is thus

$$\eta = \frac{2\pi}{T} = \frac{\mu^{\frac{1}{2}}}{a^{\frac{3}{2}}} = \frac{1}{a}\left(\frac{\mu}{a}\right)^{\frac{1}{2}} \tag{2.35}$$

The time required for a spacecraft moving at this angular velocity to go around any circle is T s. If we enclose the orbit in its *circumscribed circle* of radius a (see Figure 2.5.), then an object going around the circumscribed circle with a

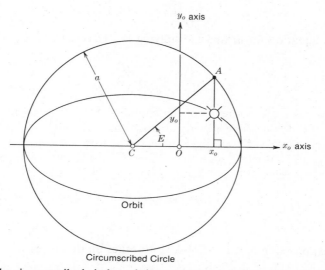

Orbit

Circumscribed Circle

Figure 2.5   The circumscribed circle and the eccentric anomaly E. Point O is the center of the earth and point C is both the center of the orbital ellipse and the center of the circumscribed circle. The satellite location in the orbital plane coordinate system is specified by (x_0, y_0). A vertical line through the satellite intersects the circumscribed circle at point A. The eccentric anomaly E is the angle from the x_o axis to the line joining C and A.

constant angular velocity of η would complete one revolution in exactly the same period of time as the satellite requires to complete one orbital revolution.

To find the equation of the orbit, we eliminated time t from the equations of motion. But to find the spacecraft location at a given time, we must reintroduce t. Continuing to follow closely the approach of reference 2, we relate the linear velocity v to r_o and ϕ_o by

$$v^2 = \left(\frac{dx_o}{dt}\right)^2 + \left(\frac{dy_o}{dt}\right)^2 = \left(\frac{dr_o}{dt}\right)^2 + r_o^2\left(\frac{d\phi_o}{dt}\right)^2 \tag{2.36}$$

It can be shown [2] that

$$v^2 = \left(\frac{\mu}{a}\right)\left(\frac{2a}{r_o} - 1\right) \tag{2.37}$$

From Eqs. (2.16), (2.15), and (2.27)

$$r_0^2\left(\frac{d\phi_0}{dt}\right)^2 = \frac{h^2}{r_0^2} = \frac{\mu p}{r_0^2} = \frac{\mu a(1 - e^2)}{r_0^2} \tag{2.38}$$

Thus Eq. (2.36) becomes

$$\left(\frac{\mu}{a}\right)\left(\frac{2a}{r_o} - 1\right) = \left(\frac{dr_o}{dt}\right)^2 + \left(\frac{\mu a}{r_o^2}\right)(1 - e^2) \tag{2.39}$$

and

$$\frac{dr_o}{dt} = \left\{\left(\frac{\mu}{ar_o^2}\right)[a^2 e^2 - (a - r_o)^2]\right\}^{\frac{1}{2}} \tag{2.40}$$

Solving this equation for dt and multiplying by the mean angular velocity η, we obtain

$$\eta\, dt = \left(\frac{r_o}{a}\right)\frac{dr_o}{[a^2 e^2 - (a - r_o)^2]^{\frac{1}{2}}} \tag{2.41}$$

Now consider the geometry of the circumscribed circle as shown in Figure 2.5. Locate the point where a vertical line drawn through the position of the satellite intersects the circumscribed circle. A line from the center of the ellipse to this point makes an angle E with the x_o axis; E is called the *eccentric anomaly* of the satellite. It is related to the radius r_o by

$$r_o = a(1 - e \cos E) \tag{2.42}$$

Thus

$$a - r_o = ae \cos E \tag{2.43}$$

and when expressed in terms of E, Eq. (2.41) takes on the surprisingly simple form

$$\eta\, dt = (1 - e \cos E)\, dE \tag{2.44}$$

Let t_p be the *time of perigee*. This is simultaneously the time of closest approach to the earth, the time when the satellite is crossing the x_o axis, and the time when E is zero. If we integrate both sides of Eq. (2.44), we then obtain

$$\eta(t - t_p) = E - e \sin E \tag{2.45}$$

The left side of Eq. (2.45) is called the *mean anomaly*, M.

$$M = \eta(t - t_p) = E - e \sin E \qquad (2.46)$$

M is the arc length (in radians) that the satellite would have traversed since the perigee passage if it were moving on the circumscribed circle at the mean angular velocity η.

Provided that we know the time of perigee, t_p, the eccentricity, e, and the length of the semimajor axis, a, we now have the necessary equations to determine the coordinates (r_o, ϕ_o) and (x_o, y_o) of the spacecraft in the orbital plane. Here is the process:

1. Calculate η by Eq. (2.35)
2. Calculate M by Eq. (2.46)
3. Solve Eq. (2.46) for E
4. Find r_o from E using Eq. (2.43)
5. Solve Eq. (2.32) for ϕ_o
6. Use Eqs. (2.33) and (2.34) to calculate x_o and y_o.

Now we must locate the orbital plane with respect to the earth.

Locating the Satellite with Respect to the Earth

At the end of the last section we summarized the process for locating the satellite at the point (x_o, y_o, z_o) in the rectangular coordinate system of the orbital plane. Now we will develop the transformations that permit the satellite to be located from a point on the rotating surface of the earth.

We will begin (in the notation of reference 1) with a fixed rectangular coordinate system (x_i, y_i, z_i) called the *geocentric equatorial coordinate system* whose origin is the center of the earth. This is shown in Figure 2.6. The z_i axis coincides with the earth's axis of rotation and extends through the geographic north pole. The x_i axis points toward a fixed location in space called the *first point of Aries*. This is the direction of a line from the center of the earth through the center of the sun at the vernal equinox (about March 21), the instant when the subsolar

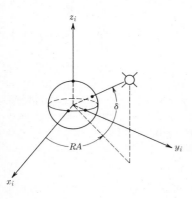

Figure 2.6 The geocentric equatorial system. This differs from the first coordinate system of Figure 2.1 only in that the x_i axis points toward the first point of Aries. An object may be located by its right ascension RA and its declination δ.

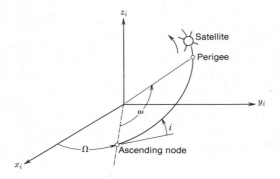

Figure 2.7   Locating the orbit in the geocentric equatorial system. The satellite penetrates the equatorial plane (while moving in the positive z_i direction) at the ascending node. The right ascension of the ascending node is Ω and the inclination i is the angle between the equatorial plane and the orbital plane. Angle ω, measured in the orbital plane, locates the perigee with respect to the equatorial plane.

point crosses the equator north to south [3]. This coordinate system moves through space; it translates as the earth revolves around the sun, but it does not rotate. The x_i direction is always the same, whatever the earth's position. The (x_i, y_i) plane contains the earth's equator and is called the *equatorial plane*.

Angular distance measured eastward in the equatorial plane from the x_i axis is called *right ascension* and given the symbol RA. The two points at which the orbit penetrates the equatorial plane are called nodes; the satellite moves upward through the equatorial plane at the *ascending node* and downward through the equatorial plane at the *descending node*. The *right ascension of the ascending node is called Ω*. The angle that the orbital plane makes with the equatorial plane (the planes intersect at the line joining the nodes) is called the *inclination, i*. Figure 2.7 illustrates these quantities.

The variables Ω and i together locate the orbital plane with respect to the equatorial plane. To locate the orbital coordinate system with respect to the equatorial coordinate system we need ω, the *argument of perigee*. This is the angle measured along the orbit from the ascending node to the perigee.

The satellite coordinates in the orbital plane (x_o, y_o, z_o) are related to the satellite coordinates (x_i, y_i, z_i) by a linear transformation. From reference 2 it is given by[1]

$$
\begin{bmatrix} x_i \\ y_i \\ z_i \end{bmatrix} = \begin{bmatrix} \cos(\Omega)\cos(\omega) & -\cos(\Omega)\sin(\omega) & \sin(\Omega)\sin(i) \\ -\sin(\Omega)\cos(i)\sin(\omega) & -\sin(\Omega)\cos(i)\cos(\omega) & \\ \sin(\Omega)\cos(\omega) & -\sin(\Omega)\sin(\omega) & -\cos(\Omega)\sin(i) \\ +\cos(\Omega)\cos(i)\sin(\omega) & +\sin(\Omega)\cos(i)\cos(\omega) & \\ \sin(i)\sin(\omega) & \sin(i)\cos(\omega) & \cos(i) \end{bmatrix} \begin{bmatrix} x_o \\ y_o \\ z_o \end{bmatrix}
$$

$$(2.47)$$

One more coordinate transformation is needed to locate the satellite with respect to a point on the rotating earth. To do this as shown in Figure 2.8 we define a rotating rectangular system (x_r, y_r, z_r) attached to the earth whose z axis and x-y plane correspond to those of the geocentric equatorial system. The x_r axis passes through the point where the prime geographic meridian crosses the equator.

---

[1] We have corrected an apparent sign error in the (2, 1) term of the matrix.

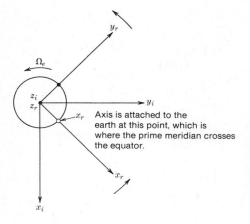

Axis is attached to the earth at this point, which is where the prime meridian crosses the equator.

Figure 2.8 The relationship between the rotating coordinate system (x_r, y_r, z_r) and the geocentric equatorial system (x_i, y_i, z_i). The equatorial plane coincides with plane of the paper. The earth rotates counterclockwise with angular velocity Ω_e. The x_r and y_r axes are attached to the earth and rotate with it. The z_i and z_r axes coincide.

The rotating system turns at angular velocity Ω_e and T_e measures the elapsed time since the x_r axis coincided with the x_i axis. This coincidence occurs once with every rotation of the earth, but it does not occur at the same time every day because of the earth's motion in its orbit around the sun. The coordinates of the satellite in the rotating system are related to the coordinates in the geocentric equatorial system by reference 2

$$\begin{bmatrix} x_r \\ y_r \\ z_r \end{bmatrix} = \begin{bmatrix} \cos(\Omega_e T_e) & \sin(\Omega_e T_e) & 0 \\ -\sin(\Omega_e T_e) & \cos(\Omega_e T_e) & 0 \\ 0 & 0 & 1 \end{bmatrix} \begin{bmatrix} x_i \\ y_i \\ z_i \end{bmatrix} \qquad (2.48)$$

The angle $\Omega_e T_e$ may be conveniently calculated by methods summarized in reference 4. To explain the procedure we must first digress a bit and talk about timekeeping and dates.

Standard time for space operations and most other scientific and engineering purposes is *universal time* (UT); this is essentially mean solar time at the Greenwich Observatory near London, England [4]. Universal time is measured in hours, minutes, and seconds or in fractions of a day. It is 5 h later than Eastern Standard Time, so that 07:00:00 EST is 12:00:00 hours UT.

The civil or calendar day begins at 00:00:00 hours UT, frequently written 0 hours. This of course is also midnight (24:00:00) on the previous day. Astronomers employ a second dating system involving *Julian days* and *Julian dates*. Julian days start at noon UT in a counting system whereby noon on December 31, 1899, was the beginning of Julian day 2415020, usually written 241 5020. These are extensively tabulated in reference 5; to give the reader a convenient reference point, noon UT on December 31, 1984, is the start of Julian day 244 6066. Julian dates can be used to indicate time by appending a decimal fraction; 00:00:00 hours UT on January 1, 1985, is Julian date 244 6066.5.

Returning now to the problem of finding the angle $\Omega_e T_e$ and the procedure outlined in reference 4, the right ascension $\alpha_{g,o}$ of the Greenwich meridian at 0 h UT at Julian date JD is given by

$$\alpha_{g,o} = 99.6909833 + 36000.7689 T_c + 0.00038708 T_c^2 \text{ degrees} \qquad (2.49)$$

where T_c is the elapsed time in Julian centuries between 0 h UT on Julian day JD and noon UT on January 1, 1900. It is calculated by

$$T_c = (JD - 2415020)/36525 \text{ Julian centuries} \qquad (2.50)$$

Since Eq. (2.49) is for 0 h UT and the Julian day begins at the previous noon UT, the JD value used in Eq. (2.50) should be the tabulated value shown in reference 5 for the civil date plus 0.5. The example to follow in Section 2.3 should make this clear.

The value of $\Omega_e T_e$ at any time t expressed in minutes after midnight UT is given by

$$\Omega_e T_e = \alpha_{g,o} + 0.25068447\, t \text{ degrees} \qquad (2.51)$$

Orbital Elements

To specify the absolute (i.e., the inertial) coordinates of the satellite at time t we need to know six quantities. (This was evident earlier when we determined that a satellite's equation of motion is a second-order vector linear differential equation.) These six quantities are called the orbital elements. There is some arbitrariness in exactly what six quantities are chosen. We will adopt a set commonly used in satellite communications: eccentricity (e), semimajor axis (a), time of perigee (t_p), right ascension of ascending node (Ω), inclination (i), and argument of perigee (ω). Frequently the mean anomaly (M) at a given time is substituted for t_p.

2.2 LOOK ANGLE DETERMINATION

The coordinates to which an earth station antenna must be pointed to communicate with a satellite are called the *look angles*. These are most commonly specified as *azimuth* (Az) and *elevation* (El), although other pairs exist. For example, right

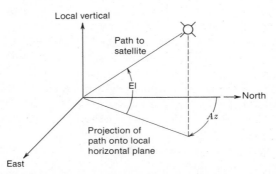

Figure 2.9   The definitions of azimuth (*Az*) and elevation (*El*). Elevation is measured upward from local horizontal and azimuth is measured from north eastward to the projection of the satellite path onto the local horizontal plane.

ascension and declination are standard for radio astronomy antennas. Azimuth is measured eastward from geographic north to the projection of the satellite path on a (locally) horizontal plane at the earth station. Elevation is the angle measured upward from this horizontal plane to the path. Figure 2.9 illustrates these look angles.

The Subsatellite Point

Although look angles may be calculated directly from the satellite coordinates (x_r, y_r, z_r) of the previous section, in practice, for geostationary satellites this is conveniently done using the geographic coordinates of the *subsatellite point* as intermediaries.[2] The subsatellite point is the place where a line drawn from the center of the earth to the satellite passes through the earth's surface. For an ideal geostationary satellite the subsatellite point is on the equator at some fixed longitude and it—rather than six orbital elements—is all that needs to be remembered when calculating the look angles.

We will represent the north latitude of the subsatellite point by L_s and the west longitude by l_s. (The reader should note that other texts may use south latitude or east longitude or both. The choice of a coordinate system for the subsatellite point obviously has some effect on the equations that follow.) In terms of the satellite coordinates (x_r, y_r, z_r) of the rotating system of the last section, the subsatellite latitude L_s in degrees north is given by

$$L_s = 90° - \cos^{-1}\left[\frac{z_r}{(x_r^2 + y_r^2 + z_r^2)^{\frac{1}{2}}}\right] \qquad (2.52)$$

The equation used for calculating the subsatellite longitude l_s depends on the quadrant in which the point (x_r, y_r) lies. Its value in degrees west is given by

$$l_s = \begin{cases} -\tan^{-1}\left(\dfrac{y_r}{x_r}\right), & y_r \geq 0 \quad \text{and} \quad x_r \geq 0 \text{ (first quadrant)} \\[2mm] 180° + \tan^{-1}\left(\dfrac{y_r}{|x_r|}\right), & y_r \geq 0 \quad \text{and} \quad x_r \leq 0 \text{ (second quadrant)} \\[2mm] 90° + \tan^{-1}\left(\dfrac{|x_r|}{|y_r|}\right), & y_r \leq 0 \quad \text{and} \quad x_r \leq 0 \text{ (third quadrant)} \\[2mm] \tan^{-1}\left(\dfrac{|y_r|}{x_r}\right), & y_r \leq 0 \quad \text{and} \quad x_r \geq 0 \text{ (fourth quadrant)} \end{cases} \qquad (2.53)$$

See Figure 2.10

---

[2] For those whose geography is hazy, latitude is measured in degrees of arc north or south from the equator. Longitude is measured in degrees of arc east or west from the "prime meridian," which passes through Greenwich, England. Lines of constant longitude run north and south and are called meridians; all meridians join at the north and south geographic poles. Lines of constant latitude run east and west and are parallel to each other; hence they are called parallels.

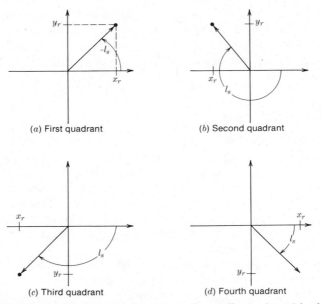

(*a*) First quadrant   (*b*) Second quadrant

(*c*) Third quadrant   (*d*) Fourth quadrant

Figure 2.10   The relationship between the spacecraft coordinates (x_r, y_r) in the rotating co-ordinate system and the latitude l_s of the subsatellite point. The equatorial plane is the plane of the paper and the x_r axis passes through the point where the prime meridian crosses the equator. Positive l_s is measured west from the prime meridian.

Elevation Calculation

Figure 2.11 shows the geometry of the elevation angle calculation. In it $\mathbf{r}_s$ is the vector from the center of the earth to the satellite, $\mathbf{r}_e$ is the vector from the center of the earth to the earth station, and $\mathbf{d}$ is the vector from the earth station to the satellite. These three vectors lie in the same plane and form a triangle. The angle γ measured between $\mathbf{r}_e$ and $\mathbf{r}_s$ is the central angle between the earth station and the satellite, and ψ is the angle (within the triangle) measured from $\mathbf{r}_e$ to $\mathbf{d}$. Defined so that it is nonnegative, γ is related to the earth station north latitude L_e and west longitude l_e and the subsatellite point north latitude L_s and west longitude l_s by

$$\cos(\gamma) = \cos(L_e)\cos(L_s)\cos(l_s - l_e) + \sin(L_e)\sin(L_s) \tag{2.54}$$

The magnitudes of the vectors joining the center of the earth, the satellite, and the earth station are related by the law of cosines. Thus

$$d = r_s\left[1 + \left(\frac{r_e}{r_s}\right)^2 - 2\left(\frac{r_e}{r_s}\right)\cos(\gamma)\right]^{\frac{1}{2}} \tag{2.55}$$

Since the local horizontal plane at the earth station is perpendicular to $\mathbf{r}_e$, the elevation angle El is related to the central angle ψ by

$$El = \psi - 90° \tag{2.56}$$

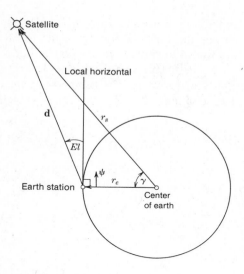

Figure 2.11 The geometry of elevation angle calculation. The plane of the paper is the plane defined by the center of the earth, the satellite, and the earth station. The central angle is γ. The elevation angle El is measured up from local horizontal to the satellite path.

By the law of sines

$$\frac{r_s}{\sin(\psi)} = \frac{d}{\sin(\gamma)} \tag{2.57}$$

Combining the last three equations yields

$$\cos(El) = \frac{r_s \sin(\gamma)}{d}$$

$$= \frac{\sin(\gamma)}{\left[1 + \left(\dfrac{r_e}{r_s}\right)^2 - 2\left(\dfrac{r_e}{r_s}\right)\cos(\gamma)\right]^{\frac{1}{2}}} \tag{2.58}$$

Equations (2.58) and (2.54) permit the elevation angle El to be calculated from the subsatellite and earth station coordinates, the orbital radius r_s, and the earth's radius r_e. A good value for the last is 6370 km.

Azimuth Calculation

Since the earth station, the center of the earth, the satellite, and the subsatellite point all lie in the same plane, the azimuth Az from the earth station to the satellite is the same as the azimuth from the earth station to the subsatellite point. This is more difficult to compute than elevation because the exact geometry involved depends on whether the subsatellite point is east or west of the earth station and on which hemispheres contain the earth station and the subsatellite point. The problem simplifies somewhat for ideal geosynchronous satellites or for cases where the subsatellite point and the earth station are in the same hemisphere; we will treat these situations after exploring the general case.

Our solution for the general case follows the procedure presented in reference 6 for finding the azimuths between any two points A and B on the earth's surface.

Their latitudes in degrees north are L_A and L_B; their longitudes in degrees west are l_A and l_B. Either point A or point B can be the earth station; the other must be the subsatellite point. By definition B is closer to the pole that is nearer to both A and B than is A. Points A and B and that pole form a spherical triangle with polar angle C and angles Y at vertex B and X at vertex A. The geometry of the triangle depends on the location of A and B; see Figure 2.12 for the four possible cases. The polar angle C is given by

$$C = |l_A - l_B| \quad \text{or} \quad |360 - |l_A - l_B|| \text{ degrees} \tag{2.59}$$

whichever makes $C \leq 180$ degrees.

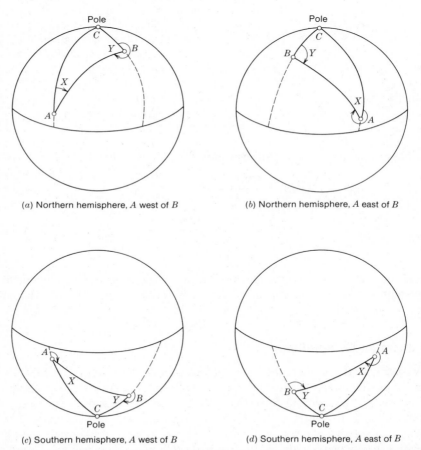

(a) Northern hemisphere, A west of B

(b) Northern hemisphere, A east of B

(c) Southern hemisphere, A west of B

(d) Southern hemisphere, A east of B

Figure 2.12 The geometry of azimuth calculation. The arrows indicate the azimuths from point A to point B and from point B to point A. Point B is either the earth station or the subsatellite point, whichever is closer to the pole. If either the earth station or the subsatellite point or both are in the northern hemisphere, cases (a) or (b) apply. If both points are in the southern hemisphere, cases (c) and (d) apply.

If at least one point is in the northern hemisphere, B must be chosen so that it is closer to the north pole than A, making $L_B > L_A$. Then bearings X and Y may be found from

$$\tan\left[0.5(Y - X)\right] = \frac{\cot(0.5C)\,\sin\left[0.5(L_B - L_A)\right]}{\cos\left[0.5(L_B + L_A)\right]} \tag{2.60}$$

$$\tan\left[0.5(Y + X)\right] = \frac{\cot(0.5C)\,\cos\left[0.5(L_B - L_A)\right]}{\sin\left[0.5(L_B + L_A)\right]} \tag{2.61}$$

$$X = 0.5(Y + X) + 0.5(Y - X) \tag{2.62}$$

$$Y = 0.5(Y + X) - 0.5(Y - X) \tag{2.63}$$

If both points are in the southern hemisphere, then point B must be closer to the south pole making $L_B < L_A$ but $|L_B| > |L_A|$. Then Eqs. (2.60) and (2.61) become

$$\tan\left[0.5(Y - X)\right] = \frac{\cot(0.5C)\,\sin\left[0.5(|L_B| - |L_A|)\right]}{\cos\left[0.5(|L_B| + |L_A|)\right]} \tag{2.64}$$

$$\tan\left[0.5(Y + X)\right] = \frac{\cot(0.5C)\,\cos\left[0.5(|L_B| - |L_A|)\right]}{\sin\left[0.5(|L_B| + |L_A|)\right]} \tag{2.65}$$

The relationship between X, Y, and the azimuth Az depends on the identity of points A and B and on their geographical relationship. It is summarized in Table 2.1.

Table 2.1
Formulas for Calculating Azimuth

At Least One Point in the Northern Hemisphere			
Subsatellite Point	**Earth Station**	**Relation**	**Azimuth in Degrees**
A	B	A west of B	$360 - Y$
B	A	A west of B	X
A	B	B west of A	Y
B	A	B west of A	$360 - X$
Both Points in the Southern Hemisphere			
Subsatellite Point	**Earth Station**	**Relation**	**Azimuth in Degrees**
A	B	A west of B	$180 + Y$
B	A	A west of B	$180 - X$
A	B	B west of A	$180 - Y$
B	A	B west of A	$180 + X$

Specialization to Geostationary Satellites

For most geosynchronous satellites the subsatellite point is at the equator at longitude l_s, and latitude L_s is 0. The geosynchronous radius r_s is 42,242 km. Equation (2.54) for the central angle γ simplifies to

$$\cos(\gamma) = \cos(L_e)\cos(l_s - l_e) \tag{2.66}$$

The distance d from the earth station to the satellite is given by

$$d = 42,242[1.02274 - 0.301596\cos(\gamma)]^{\frac{1}{2}} \text{ km} \tag{2.67}$$

The elevation angle El is given by

$$\cos(El) = \frac{\sin(\gamma)}{[1.02274 - 0.301596\cos(\gamma)]^{\frac{1}{2}}} \tag{2.68}$$

The problem of azimuth calculation simplifies considerably for geostationary satellites.[3] The derivation of the needed equations is based on the spherical triangle with vertices E, S, and G, where E is the earth station, S is the subsatellite point, and G is the point where the meridian (longitude line) of the earth station crosses the equator. Figure 2.13 illustrates all of the possible orientations of this triangle for various locations of the subsatellite point with respect to the earth station. The three sides of the triangle are arcs of length γ, a, and c. The first is the central angle γ given by Eq. (2.64); a and c are related to the coordinates of the earth station and the subsatellite point by

$$a = |l_s - l_e| \tag{2.69}$$

$$c = |L_e - L_s| \tag{2.70}$$

If we call the half-perimeter of the triangle s,

$$s = 0.5(a + c + \gamma) \tag{2.71}$$

then the angle α at the vertex may be obtained from the equation [6, pp. 46–9ff]

$$\tan^2\left(\frac{\alpha}{2}\right) = \frac{\sin(s - \gamma)\sin(s - c)}{\sin(s)\sin(s - a)} \tag{2.72}$$

The azimuth Az is related to α by the equations that appear in Figure 2.13 and in Table 2.2.

Equations (2.69) through (2.72) may be combined in one form, which, although compact, tends to obscure the derivation.

$$\alpha = 2\tan^{-1}\left\{\frac{\sin(s - \gamma)\sin(s - |L_e|)}{\sin(s)\sin(s - |l_e - l_s|)}\right\}^{\frac{1}{2}} \tag{2.73}$$

---

[3] The authors wish to acknowledge the contribution of Warren L. Stutzman [7] in drawing their attention to this convenient approach to azimuth calculation. It is considerably easier to use than most of the techniques that appear in the literature.

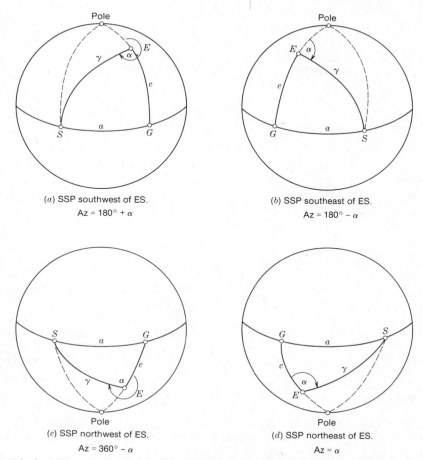

Figure 2.13   The spherical trigonometry of azimuth calculation. In all cases E is the earth station (ES), S is the subsatellite point (SSP), and G is the intersection of the earth station's longitude line with the equator. The arrow indicates the azimuth angle Az.

Table 2.2
Equations for Calculating Azimuth Az from
Spherical Triangle Angle α

Situation	Equation
1. Subsatellite point southwest of earth station	$Az = 180° + \alpha$
2. Subsatellite point southeast of earth station	$Az = 180° - \alpha$
3. Subsatellite point northwest of earth station	$Az = 360° - \alpha$
4. Subsatellite point northeast of earth station	$Az = \alpha$

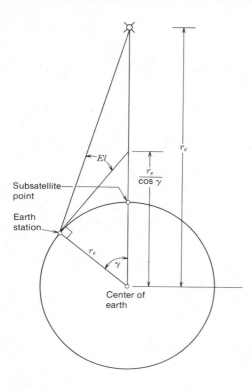

Figure 2.14   The geometry of the visibility problem. The satellite is said to be visible from the earth station if the elevation angle *El* is positive. This requires that the orbital radius r_s be greater than $r_e/\cos \gamma$ where r_e is the earth radius and γ is the central angle.

Visibility

For a satellite to be visible from an earth station its elevation angle *El* must be above some minimum value, which is at least 0°. A positive or zero elevation angle requires (see Figure 2.14).

$$r_s \geq \frac{r_e}{\cos(\gamma)} \tag{2.74}$$

This means that the maximum central angular separation between the earth station and the subsatellite point is limited by

$$\gamma \leq \cos^{-1}\left(\frac{r_e}{r_s}\right) \tag{2.75}$$

For a nominal geosynchronous orbit, the last equation reduces to $\gamma \leq 81.3°$ for visibility. To avoid the propagation problems associated with extremely low elevation angles (see Chapter 8), a smaller angular separation is desirable.

2.3 NUMERICAL EXAMPLE

To illustrate the orbital and look angle calculations described in the previous two sections, we will make some calculations using real data for the CTS spacecraft.

Here are the osculating orbital elements[4] as measured at 00:00:00 UT on December 27, 1978, and received at the authors' earth station:

Symbol	Name	Value
a	semimajor axis	42164.765 km
e	eccentricity	0.001181
i	inclination	0.802°
M	mean anomaly	116.636°
ω	argument of perigee	138.167°
Ω	RA of ascending node	84.178°

While they are not needed for look angle calculations with this set of elements, we will begin by calculating the mean angular velocity η and using it to find the time of perigee t_p. Using Eq. (2.35),

$$\eta = \frac{1}{a}\left(\frac{\mu}{a}\right)^{\frac{1}{2}} = \frac{1}{42164.765}\left(\frac{3.9861352 \times 10^5}{42164.765}\right)^{\frac{1}{2}}$$

$$= 7.29208108 \times 10^{-5} \text{ rad/s}$$

or $4.17805467 \times 10^{-3}$ °/s. Using this result and M in Eq. (2.46) we may solve for $(t - t_p)$, the elapsed time since perigee passage.

$$(t - t_p) = \frac{M}{\eta} = \frac{116.636}{4.17805467 \times 10^{-3}} = 2.79163413 \times 10^4 \text{ s}$$

Thus 7 h 45 min and about 16.3 s have passed since perigee passage. Working backwards from 00:00:00 hours on December 27, the passage immediately preceding this set of orbital elements occurred on December 26 at 16:14:43.66 UT.

To locate the spacecraft in its orbit we must determine the eccentric anomaly, E. Since we know M, we may find E by a trial-and-error solution of Eq. (2.46). To the accuracy of the author's hand calculator, the answer is $E = 116.637056°$. From Eq. (2.42) we can determine the orbital plane radial coordinate r_o; the result is $r_o = 4.218709065 \times 10^4$ km. Substituting this into Eq. (2.32) yields an angular coordinate ϕ_o of $116.5461901°$. Using Eqs. (2.33) and (2.34) to transform the spacecraft orbital plane coordinates to a rectangular system yields $x_o = -1.885421816 \times 10^4$ km and $y_o - 3.773948960 \times 10^4$ km. Transforming these coordinates to the geocentric equatorial system is a tedious process; the intermediate step of evaluating the matrix elements in Eq. (2.47) gives

$$x_i = -0.739037283x_o + 0.673520595y_o$$

$$y_i = -0.673599838x_o - 0.739094880y_o$$

$$z_i = 9.33551890 \times 10^{-3}x_o - 1.04291146 \times 10^{-2}y_o$$

---

[4] The osculating elements describe the orbit that the spacecraft would follow in the absence of perturbing forces. Section 2.4 will discuss orbital perturbations.

Substituting values for x_o and y_o into the above, we find $x_i = 3.93522813 \times 10^4$ km, $y_i = -1.519286524 \times 10^4$ km, and $z_i = -5.69603370 \times 10^2$ km. These hand-calculator values compare favorably with the published official geocentric equatorial coordinates of the spacecraft that were distributed with orbital elements: $x_i = 3.940769666 \times 10^4$ km, $y_i = -1.504868175 \times 10^4$ km, and $z_i = -5.7005836 \times 10^2$ km.

Now we must find the value of $\Omega_e T_e$ to substitute into Eq. (2.48). According to Table I of reference 5, Julian day 244 3843 begins at noon UT on November 30, 1978. To get to 00:00:00 h on December 1 we add 0.5 days, and to get from 00:00:00 h on December 1 to 00:00:00 h on December 27 we add 26. Thus the Julian day number we must use for JD in Eq. (2.50) is 244 3843 + 0.5 + 26 or 244 3869.5. The elapsed time in Julian centuries since noon UT on January 1, 1900, is T_c, which by Eq. (2.50) is $(2443869.5 - 2415020)/36525 = 0.78985626$ Julian centuries. Substituting this into Eq. (2.49) we get $\alpha_{g,o} = \Omega_e T_e = 2.853512400 \times 10^{4\circ}$. Dividing by 360° per revolution, this number is equivalent to 79.2642333 revolutions. Subtracting out the 79 revolutions and multiplying the remainder by 360°, we obtain a final value for $\Omega_e T_e$ of 95.12399880°. Using this value in Eq. (2.48) to determine the spacecraft coordinates in the rotating system, we obtain $x_r = -1.864676149 \times 10^4$, $y_r = -3.783812210 \times 10^4$, and $z_r = -5.696033700 \times 10^2$ km. The subsatellite point lies in the third quadrant, and its longitude l_s is given by $90° + \tan^{-1}|x_r/y_r|$ or $90 + 26.23424233$, which equals 116.2342° W. The published official longitude of the subsatellite point at this time was 116.0230° W; the difference between these two numbers reflects computational accuracy. The subsatellite latitude L_s is $90 - \cos^{-1}[z_r/(x_r^2 + y_r^2 + z_r^2)]^{\frac{1}{2}\circ}$ north; calculating from the rotating coordinates developed in this example, we obtain $L_s = -0.7736°$ N. The official published value was $-0.775°$ N.

Now that we have located the subsatellite point we can compute the look angles from the authors' earth station at longitude $l_e = 80.438°$ W and latitude $L_e = 37.229°$ N. We will do this for our calculated subsatellite coordinates. Starting with the elevation, we find that the central angle γ is $\cos^{-1}[\cos(L_e)\cos(L_s)\cos(l_s - l_e) + \sin(L_e)\sin(L_s)] = 50.387511°$. The orbital radius r_s is $(x_r^2 + y_r^2 + z_r^2)^{\frac{1}{2}}$; substituting in the values calculated above yields $r_s = 42183.23$ km. Using $r_e = 6370$ km, we calculate $\cos(El) = 0.8455$ and $El = 32.28°$. For the azimuth calculations, the angle c of Eq. (2.70) is 35.7962°, $\tan(0.5(Y + X))$ is 9.359795, and $\tan(0.5(Y - X))$ is 1.061419. The argument $0.5(Y + X)$ is 83.9016° and $0.5(Y - X)$ is 46.7066°. These lead to bearing angles of 37.1950° for X and 130.6081° for Y. The azimuth Az is then 229.39°. This completes the computations.

The tedious set of calculations presented in this section illustrates the process that must be followed to determine look angles from orbital elements. The reader who goes through it will appreciate the computer programs that do this automatically.

2.4 ORBITAL PERTURBATIONS

The orbital equations derived in Section 2.1 modeled the earth and the satellite as point masses influenced only by mutual gravitational attraction. Under these conditions a "Keplerian" orbit results; this is an ellipse whose properties are constant with time. In practice, the satellite and the earth respond to many other influences including asymmetry of the earth's gravitational field, the gravitational fields of the sun and the moon, solar radiation pressure, and some other factors like atmospheric drag that are not important in geosynchronous orbits. These interfering forces cause the true orbit to be different from a Keplerian ellipse; if unchecked they would cause the subsatellite point of a nominally geosynchronous satellite to move with time.

Historically much attention has been given to techniques for incorporating these additional perturbing forces into orbit descriptions. Some of the mathematical methods that have been developed and a complete set of references appear in Chapter IV of reference 2. The approach normally adopted for communications satellites is first to derive an *osculating orbit* for some instant of time (the Keplerian orbit that the spacecraft would follow if all perturbing forces were removed at that time) with orbital elements $(a, e, t_p, \Omega, i, \omega)$. The perturbations are assumed to cause the orbital elements to vary with time and the orbit and satellite location at any instant are taken from the osculating orbit calculated with orbital elements corresponding to that time. To visualize the process, assume that the osculating orbital elements at time t_o are $(a_o, e_o, t_{po}, \Omega_o, i_o, \omega_o)$. Then assume that the orbital elements vary linearly with time at constant rates given by $(da/dt, de/dt, \text{etc.})$. The satellite's position at any time t_1 is then calculated from a Keplerian orbit with elements

$$\left(a_o + \frac{da}{dt}(t_1 - t_o), \qquad e_o + \frac{de}{dt}(t_1 - t_o), \text{etc.}\right)$$

This approach is particularly useful in practice because it permits the use of either theoretically calculated derivatives or empirical values based on satellite observation.

As the perturbed orbit is not an ellipse, some care must be taken in defining an orbital period. Since the satellite does not return to the same point in space once per revolution, the quantity most frequently specified is the so-called *anomalistic period*: the elapsed time between successive perigee passages.

Effects of the Earth's Oblateness

Since the earth is not a perfect sphere with a symmetric distribution of mass, its gravitational potential does not have the simple $(1/r)$ dependence assumed by the equations of Section (2.1). The earth's gravitational potential is represented more accurately by an expansion in Legendre polynomials J_n in ascending powers of (earth's radius r_e)/(orbital radius r) [1]. In this expansion the dominant term is $J_2(r_e/r)^2$; its value is called the J_2 coefficient. (Some authors use a different notation for the expansion and call it the J_{22} coefficient.) The effect of this term is to cause an unconstrained geosynchronous satellite to drift toward and circulate around the nearer of two stable points. These correspond to subsatellite longitudes of $105°$ W and $75°$ E, locations called "graveyards" because they collect old satellites whose stationkeeping fuel is exhausted.

Effects of the Sun and Moon

Gravitational attraction by the sun and moon causes the orbital inclination of a geosynchronous satellite to change with time. If not countered by north–south stationkeeping, these forces would increase the orbital inclination from an initial $0°$ at launch to $14.67°$ 26.6 years later. While these numbers are academic in the sense that no commercial satellite has had a 26-year useful lifetime, they indicate the magnitude of the problem. The rate of change varies with the inclination of the moon's orbit, but values of about $0.86°$/year are quoted for the 1970–1980 time period [8].

2.5 ORBIT DETERMINATION

Orbit determination involves making sufficient measurements of the location or location and the velocity of a spacecraft to determine the six orbital elements. This is an old problem in astronomy, and techniques were developed by Laplace, Gauss, and others to calculate an orbit based on the angular position of an object measured at three different times from the same terrestrial location. Three angular position measurements are required because there are six unknowns and each measurement provides two equations. Conceptually these may be thought of as one equation giving the azimuth and another equation giving the elevation as a function of the six unknown orbital elements. While traditional optical astronomy was limited to measuring angular position, radar techniques permit measurement of range and range-rate simultaneously with position, and these reduce the number of observations required at one location to two. Alternatively, accurate time coordination between two or more earth stations permits orbits to be determined from simultaneous position and range measurements at two or more sites.

Reference 1 provides a detailed summary of orbit determination by the methods of Laplace and Gauss. The latter is used for contemporary satellites [9]. The required computations are quite tedious, and since in contrast to look angle calculations they are almost never done by hand, we will not reproduce the necessary equations here. For a particularly good tutorial explanation of Gauss's method using two angular position and range measurements, see reference 10.

2.6 LAUNCHES AND LAUNCH VEHICLES

For a spacecraft to achieve synchronous orbit, it must be accelerated to a velocity of 3070 m/s in a zero-inclination orbit and raised to a distance of 42,242 km from the center of the earth [11]. There are two competing technologies for doing this: expendable launch vehicles (*ELV*) and the space shuttle (*STS*, for "space transportation system"). At present (1984) the situation is fluid, and long-term developments are difficult to predict. U.S. government policy is to phase out ELV launches for its own satellites and to rely exclusively on STS. If this happens a number of private firms would like to pick up the ELV programs and offer ELV launches for sale to nongovernment users. At the same time the European company Arianespace is marketing ELV launches in competition with U.S. industry and with STS. There are also rumors of a forthcoming Japanese entry into the commercial launch market. In this section we will try to summarize the problems of launching a synchronous satellite and to describe the systems that are currently available for solving them, warning the reader that the situation may change drastically in the next few years.

The Mechanics of Launching a Synchronous Satellite

While theoretically a satellite could be placed into geosynchronous orbit in one operation, practical considerations of cost and launch vehicle capability dictate a two- or three-step process, as depicted in Figure 2.15. Most ELV launchers like Delta and Ariane (Figure 2.16) put the satellite in an inclined elliptical orbit called a transfer orbit with an apogee at geosynchronous altitude and a 185 to 370 km (100 to 200 nautical mile) perigee [12]. At the transfer orbit apogee, a rocket engine called the apogee kick motor (AKM) puts the satellite into a circular geosynchronous orbit with (ideally) zero inclination. The AKM is an integral part of the satellite except in some Titan IIIC launches where the last stage of that rocket both places the satellite into transfer orbit and serves as an AKM [13].

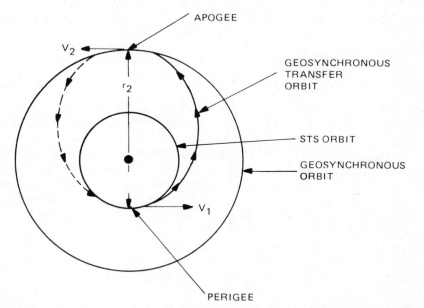

Figure 2.15 The steps in launching a communications satellite. V_1 and V_2 are the velocity increments that move the spacecraft into and out of the transfer orbit. (Reprinted with permission from K. H. Muller, "Launch Vehicles for Commercial Communications Satellites," in *RCA Engineer*, **28**, 72–76 (March–April 1983).)

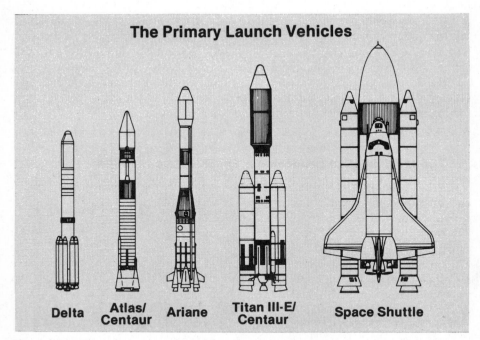

Figure 2.16 The primary launch vehicles. (Reprinted from Kim E. Degnan, "Commercial Launch Vehicle Services," *Satellite Communications*, **7**, 34–36 (September 1983). Reprinted with permission of *Satellite Communications*, 6530 So. Yosemite St., Englewood, Colorado 80111.)

The AKM must be capable of increasing the satellite velocity from 1585 m/s in the transfer orbit to about 3048 m/s in geosynchronous orbit while simultaneously reducing the orbital inclination to zero. The bigger the inclination, the more energy this requires. Rocket engineers express required energy in terms of an equivalent velocity change of the spacecraft in level flight. The minimum inclination that the transfer orbit can have is equal to the latitude of the launching pad. For Cape Canaveral this is 28.3°, and the energy required to overcome this inclination corresponds to a velocity change of 366 m/s. Hence, for a Cape Canaveral launch the AKM must be able to deliver a total velocity change of 1463 + 366 or 1829 m/s. Ariane rockets launched from the Guiana Space Center (French initials CGS) in French Guiana have a transfer orbit inclination of about 5°; thus an Ariane launch requires less energy from the AKM for inclination correction and can deliver a bigger payload to synchronous orbit than a rocket of the same size launched from Florida [13].

The Space Shuttle flies a nominal 296-km circular orbit. Geosynchronous satellites launched by STS are moved from this orbit to a transfer orbit by an additional stage frequently called a PAM (for payload assist module) or a perigee motor. An AKM injects the satellites from transfer orbit into geosynchronous orbit in the same way as with ELV launches.

Approximately 30 min elapse between ELV liftoff and placement of the satellite into the transfer orbit. With STS the time may be much longer, depending on the rest of the Shuttle's mission. After transfer orbit insertion, control passes from the launching agency to the satellite owner, who determines when to inject the spacecraft into geosynchronous orbit. Factors determining the time spent in the transfer orbit include how long it takes to verify the transfer orbit, required visibility of the satellite from earth stations, and the like [13]. RCA SATCOMs, for example, go into geosynchronous orbit at their seventh apogee passage [14].

U.S. Expendable Launch Vehicles

While several types of militarily derived rockets have served as launch vehicles for communications satellites, the work horse of the recent years and the main competitor for STS and Ariane is the McDonald-Douglas Delta. See Figures 2.17 and 2.18. The models currently in use form what is called the 3900 series, developed in a program that was jointly funded by McDonald-Douglas and RCA [14]. The last stage of the Delta is usually a PAM (the same kind of PAM that is used with STS); fitted with a PAM-D, a Delta 3920 rocket can put a 1200-kg payload into transfer orbit at a 1983 cost of $18.5M. With a PAM-D2 the same rocket can deliver 1500 kg to transfer orbit; the PAM-A now under development is expected to carry 2000 to 2500 kg [15]. While NASA plans to stop using Delta in 1986 [12], the rocket may continue to be available commercially.

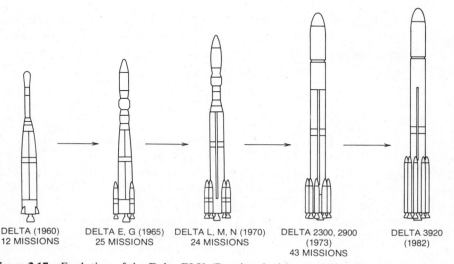

DELTA (1960)
12 MISSIONS

DELTA E, G (1965)
25 MISSIONS

DELTA L, M, N (1970)
24 MISSIONS

DELTA 2300, 2900
(1973)
43 MISSIONS

DELTA 3920
(1982)

Figure 2.17 Evolution of the Delta ELV. (Reprinted with permission from K. H. Muller, "Launch Vehicles for Commercial Communications Satellites," in *RCA Engineer*, **28**, 72–76 (March–April 1983).)

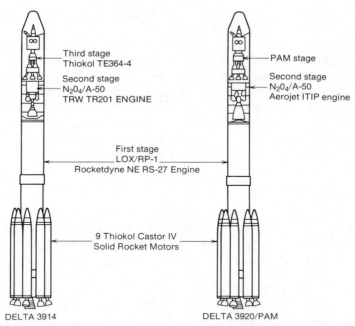

Third stage
Thiokol TE364-4

Second stage
N₂0₄/A-50
TRW TR201 ENGINE

First stage
LOX/RP-1
Rocketdyne NE RS-27 Engine

9 Thiokol Castor IV
Solid Rocket Motors

PAM stage

Second stage
N₂0₄/A-50
Aerojet ITIP engine

DELTA 3914

DELTA 3920/PAM

Figure 2.18   Current (1984) Delta configurations. (Reprinted with permission from K. H. Muller, "Launch Vehicles for Commercial Communications Satellites," in *RCA Engineer*, **28**, 72–76 (March–April 1983).)

The Titan missile can be used for launching very heavy satellites. When used with the Centaur upper stage, it is capable of delivering 4000 kg to geosynchronous orbit. Titans nominally cost $35–40M per launch, but 1983 bids for a Titan launch for INTELSAT VI were as high as $75M [15].

Ariane [16]

Ariane (Figure 2.19) is a European ELV offering commercial launches in competition with STS. Operational versions are the Ariane-2 and the Ariane-3, and a larger Ariane-4 will be available in 1986. Arianes are designed to launch two satellites at the same time from an enclosure called SYLDA (*System de Lancement Double Ariane*), illustrated in Figure 2.20.

On-orbit capabilities of Ariane-2 are a total payload (could be one or two satellites) of 2100 kg and of Ariane-3 are 2580 kg. Ariane-4 is being designed for a maximum on-orbit payload of about 4300 kg, competitive with Titan. Because they are launched from French Guiana, Ariane rockets inject satellites into transfer orbits with inclinations of about 5°, permitting more payload weight to be devoted to communications systems and less to AKM fuel.

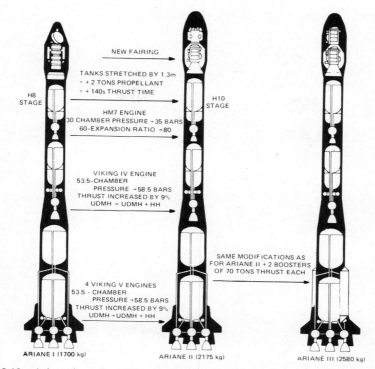

NEW FAIRING

TANKS STRETCHED BY 1.3m
= + 2 TONS PROPELLANT
≈ + 140s THRUST TIME

H8
STAGE

H10
STAGE

HM7 ENGINE
30 CHAMBER PRESSURE →35 BARS
60-EXPANSION RATIO →80

VIKING IV ENGINE
53.5-CHAMBER
PRESSURE →58.5 BARS
THRUST INCREASED BY 9%
UDMH → UDMH + HH

SAME MODIFICATIONS AS
FOR ARIANE II + 2 BOOSTERS
OF 70 TONS THRUST EACH

4 VIKING V ENGINES
53.5 - CHAMBER
PRESSURE →58.5 BARS
THRUST INCREASED BY 9%
UDMH →UDMH + HH

ARIANE I (1700 kg) ARIANE II (2175 kg) ARIANE III (2580 kg)

Figure 2.19 Ariane launchers. (Reprinted with permission from K. H. Muller, "Launch Vehicles for Commercial Communications Satellites," in *RCA Engineer*, **28**, 72–76 (March–April 1983).)

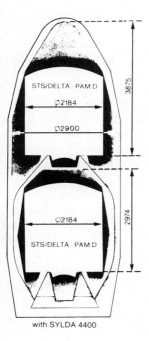

STS/DELTA PAM D
⌀2184
⌀2900

3875

⌀2184
STS/DELTA PAM D

2974

with SYLDA 4400

Figure 2.20 The SYLDA module with which Ariane can launch two satellites at the same time. (Reprinted with permission from K. H. Muller "Launch Vehicles for Commercial Communications Satellites," in *RCA Engineer*, **28**, 72–76 (March–April 1983).)

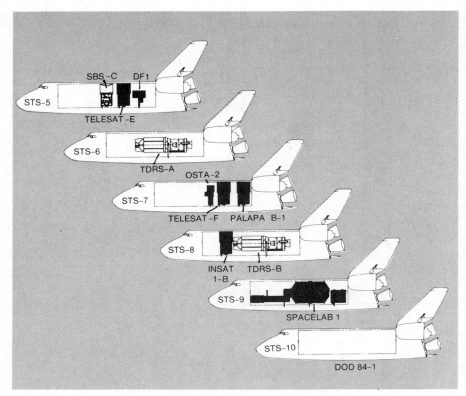

Figure 2.21 Satellites launched by the Space Shuttle. (Reprinted with permission from K. H. Muller, "Launch Vehicles for Commercial Communications Satellites," in *RCA Engineer*, **28**, 72–76 (March–April 1983).)

STS

Figure 2.21 illustrates the Space Shuttle configuration for several of the early STS communications satellite launches. The payload weights that the Shuttle can deliver are limited by the availability of propulsion systems to take satellites from the parking orbit into the transfer orbit. Operating with the PAM-D developed for Delta, STS can put about 1200 kg into geosynchronous orbit [12]. The Shuttle can carry the U.S. Air Force IUS (now Inertial Upper Stage, originally Interim Upper Stage), which is cited as having the capability to deliver 2000 kg [17] to 4000 kg [12]. Military requirements and restrictions may make the IUS unattractive for commercial use, and a civilian alternative called TOS™ is under development. TOS is expected to deliver a 3000-kg payload to synchronous orbit from the Shuttle [17].

Selecting a Launch Vehicle

At the time this text was written, a satellite owner could choose among essentially three launch vehicles: the Delta with its established performance record or the new STS and Ariane systems. While each has its own proponents, no system has yet emerged as technically superior or lowest in cost for all applications. This is reflected in the current industrial approach of designing such satellites as GSTAR, SPACENET, SATELLITE TV SYSTEMS DBS, and ADVANCED SATCOM to be compatible with two or even with all three launchers. GSTAR, for example, can be launched by Delta, by Ariane, and by STS with either PAM-D or PAM-DII [2]. In this section we will discuss the economic and technical trade-offs involved in selecting a launch vehicle and avoid taking a position about which is best.

The true costs of launching a satellite can be estimated accurately only by securing bids for a particular system at a particular time. Some recent estimates from the literature appear in Table 2.3 from reference 12 and Table 2.4 from

Table 2.3
Launch Vehicle Comparison

Comparative Launch Weight Capability		
Launch system	Weight (kg) Transfer orbit	Weight (kg) Geosynchronous orbit[a]
Delta 3920	1270	662
STS PAM-D	1254	660
1/2 Ariane-3 (low)	1140	666[b]
1/2 Ariane-3 (high)	1195	696[b]
STS PAM-DII	1796	943
STS PAM-A	1996	1050
1/2 Ariane-4[c]	1882	1098

[a] Assumes optimum AKM.
[b] Due to location of launch site.
[c] Under development, available 1986.

Comparative Launch Costs	
Launch system	Cost (millions of dollars)
Delta 3910/PAM	36.0
Delta 3920/PAM	41.0
Delta 3924	37.8
STS/PAM	17.5[a]
Ariane	31.0

[a] NASA has increased STS launch costs. Current projections for 1986 would place this value at $28.5M.
Source: Reprinted with permission from K. H. Muller, "Launch Vehicles for Commercial Communication Satellites," in *RCA Engineer,* **28,** 72–76 (March–April, 1983).

Table 2.4

Launch Vehicle Comparison

System	Capacity (kg)	Cost ($1000/kg)	ILC[a]
Shuttle/PAM-D	544	40	1982
Delta 3910/PAM-D	488	41	1981
Ariane-1	953	37	1982
Shuttle/PAM-A	998	40	1983
Atlas G/Centaur	1225	53	1984
Ariane-3	1406	35	1985
Titan 34D/IUS	1860	75	1982
Shuttle/IUS	2268	55	1983
Ariane-4	2359	33	1987
Shuttle/TOS	3084	26	1986
Shuttle/Centaur	5443	31	1986

[a] Initial launch capability (year).

Source: Scott Webster, "TOS: Commercial Launch Vehicles for the 80's," *Satellite Communications,* **7**, pp. 44–50 (September 1983). Reprinted with permission of *Satellite Communications*, 6530 So. Yosemite St., Englewood, Colorado 80111.

reference 17. Note that the two tables are not consistent with each other. Part of the difference may be in what the preparers of the tables included in the launch cost—is it just the vehicle or does it include all of the preparation and support costs inherent in putting up a satellite? Other variances probably result from different assumptions about (1) the true costs of STS after NASA completes the first series of demonstration launches at artificially low prices, (2) the projected entry of private U.S. companies into the launch business, (3) Arianespace's marketing strategy, and so on. There are also differences in insurance costs between launch vehicles. These fluctuate with the industry's recent experiences with launch failures. See reference 18 for a detailed discussion of satellite insurance.

Turning to technical questions, let us first consider STS. Its advantages are that it offers a large cargo bay, presumably lower G forces during launch, and in-space checkout and repair before transfer orbit insertion. One of its minor disadvantages is that a user must provide a stage to put the satellite into transfer orbit, but PAMs are available to do that. A major disadvantage is that launch from a manned vehicle imposes significantly more difficult safety restrictions on the satellite designers than does ELV launch. For example, satellites for STS launch must use nonflammable materials and all pyrotechnics must be triply inhibited [12].

Ariane benefits from its ability to achieve lower-inclination transfer orbits from its more southerly launch site. As an unmanned vehicle, it imposes fewer safety constraints on satellite designers but it obviously cannot offer in-space checkout. An Ariane launch includes transfer orbit injection; no PAM is required.

While older than Ariane and less powerful than some of the Ariane configurations, Delta has the advantage of an established technology. It offers the same manned versus unmanned trade-offs against STS as does Ariane. With a U.S. launch, Delta cannot achieve the low-inclination transfer orbits that Ariane offers. Of course a primary question about Delta is whether or not it will be available after 1986.

Cost and administrative differences between the available launch vehicles may be more significant than technical matters. NASA ELV launches are not fixed-price; NASA provides a "best estimate" cost for a launch contract, but all U.S. government launch expenses must be reimbursed, and a satellite owner may be billed for charges as long as two years after launch [13]. Presumably this same policy will apply to STS once the first experimental flights are over. Delta launches will also be conducted on a "best estimate" basis until private companies offer them (if they offer them) after NASA terminates the Delta program. Ariane has an advantage here; as a private company rather than a government agency Ariane-space offers fixed commercial contracts to users [16]. Arianespace also is able to offer a more flexible launch schedule with more openings than the heavily committed STS.

2.7 ORBITAL EFFECTS IN COMMUNICATIONS SYSTEM PERFORMANCE

Doppler Shift

To a stationary observer, the frequency of a moving radio transmitter varies with the transmitter's velocity. If the true transmitter frequency (i.e., the frequency that the transmitter would put out when at rest) is f_T, the received frequency f_R is higher than f_T when the transmitter is moving toward the receiver and lower than f_T when the transmitter is moving away from the receiver. Mathematically the relationship between the transmitted and received frequencies is

$$\frac{f_R - f_T}{f_T} = \frac{\Delta f}{f_T} = \frac{V_T}{v_p} \tag{2.76}$$

where V_T is the component of the transmitter velocity directed toward the receiver and v_p is the phase velocity of light (3×10^8 m/s in free space). If the transmitter is moving away from the receiver, then V_T is negative. This change in frequency is called *Doppler shift*, the Doppler effect, or more commonly just "Doppler" after the physicist who first studied it in sound waves. See reference 19 for a complete derivation.

Doppler shift is quite pronounced for low-orbit satellites and compensating for it requires frequency tracking in narrowband receivers. Its effects are negligible for geosynchronous satellites.

While most engineers think of Doppler shift as it affects the downlink, Doppler is also present on the uplink. As shown in Figure 2.22, the magnitude of the

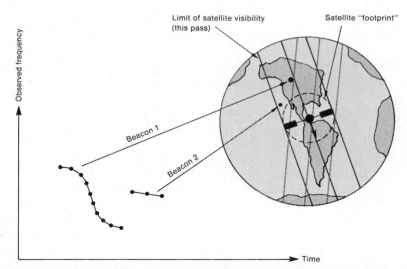

Figure 2.22 Doppler shift observed for two different earth stations. Beacon 1 is close to the center of the satellite's ground track and shows wide Doppler shift and long visibility. Beacon 2 is near the limit of visibility. (Reprinted with permission from Walter C. Scales and Richard Swanson, "Air and Sea Rescue Via Satellite Systems," *IEEE Spectrum*, **21**, No. 3, 48–52 (March 1984). Copyright © 1984 IEEE.)

frequency shift is proportional to the earth station's closeness to the ground "track" of the satellite. Search and rescue satellites use this effect to locate emergency beacons [20].

Range Variation

With the best stationkeeping systems available, the position of a geosynchronous satellite with respect to the earth exhibits a cylic daily variation. While the resulting range uncertainty has negligible effect on the power equations of Chapter 4, the variable round-trip time delay that it could add would require unacceptably large guard times in the time division multiple access (*TDMA*) systems described in Chapter 6. For these reasons stations using TDMA must continually monitor satellite range and adjust their burst timing accordingly.

Eclipse

A satellite is said to be in *eclipse* when the earth prevents sunlight from reaching it, that is, when the satellite is in the shadow of the earth. For geosynchronous satellites, eclipses occur during two periods that begin 23 days before equinox (about March 21 and about September 23; see reference 3) and end 23 days after equinox. Figure 2.23 from reference 21 and Figure 2.24 from refer-

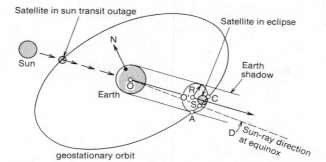

Figure 2.23 Eclipse geometry. (*Source:* J. J. Spilker, Jr., *Digital Communications by Satellite,* copyright © 1977, p. 144. Reprinted by permission of Prentice-Hall, Inc., Englewood Cliffs, NJ.)

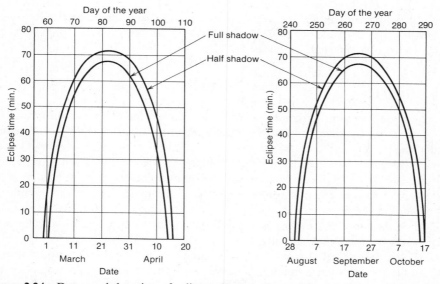

Figure 2.24 Dates and duration of eclipses. (*Source:* James Martin, *Communications Satellite Systems,* copyright © 1978, p. 37. Reprinted by permission of Prentice-Hall, Inc., Englewood Cliffs, NJ.)

ence 22 illustrate the geometry and duration of eclipses. They occur near the equinoxes because these are the times when the sun, the earth, and the spacecraft are nearly in the same plane.

During full eclipse a satellite receives no power from its solar array and it must operate entirely from batteries. This can reduce the available prime power significantly as the spacecraft nears the end of its life, and it may necessitate shutting down some of the transponders during the eclipse period. Spacecraft designers must guard against harmful transients as solar power fluctuates sharply at the beginning and end of an eclipse. Such transients were rumored to have caused the failure of the primary power system of at least one early communications satellite. In addition, passage into and out of the earth's shadow can place severe thermal stress on a satellite. For these reasons satellite failure is probably more likely to occur during eclipse than at any other time following deployment.

Sun-Transit Outage

When the sun passes through the beam of an earth station antenna, the overall receiver noise level will rise significantly and interfere with or prevent normal operation. This effect is predictable and can cause outage for as much as 10 min a day for several days and for about 0.02 percent of an average year [23]. A receiving earth station can do nothing about it except wait for the sun to move out of the main lobe of the antenna. While the time involved may seem insignificant, sun-transit outage is very costly to domestic data users. It always occurs in the daytime when their circuits are carrying traffic, and it forces them to arrange alternative channels.[5]

2.8 SUMMARY

Starting with Newton's second law and the equation for the gravitation force on a satellite, we have shown that the satellite's position is described by a second-order linear vector differential equation. The solution to this equation involves six constants, which are called the orbital elements. The satellite orbit is an ellipse with the earth at one focus; three of the orbital elements describe the shape of the ellipse and the satellite's position on it, and the other three describe the orientation of the ellipse with respect to the rotating earth. For a geostationary orbit, the ellipse degenerates to a circle about the center of the earth lying in the equatorial plane with a radius of 42,242 km.

Given the orbital elements, we may calculate the geographic coordinates of the subsatellite point as a function of time, and from the subsatellite point and the orbital radius we may calculate the azimuth and elevation of the satellite from any point on the earth. These constitute the look angles; they are the coordinates to which an earth station antenna must be pointed to communicate with the satellite. For a geostationary satellite, the look angles and the subsatellite point are constant with time.

At the present time, satellites may be launched by expendable launch vehicles (ELV) or by the Space Shuttle (STS). In both methods, the satellite is placed into a low inclined elliptical orbit called a transfer orbit before it is injected into synchronous orbit by its apogee kick motor (AKM). Shuttle-launched satellites first

---

[5] The authors wish to thank the anonymous reviewer who brought the economic impact of sun-transit outage to our attention.

follow the Shuttle's orbit; then they are put into a transfer orbit by a payload assist module (PAM).

The primary U.S. ELV is the Delta; its European competitor is the Ariane. Although the current U.S. government policy is to phase out the Delta by 1986 in favor of STS, the Ariane family of rockets and the STS are expected to compete with each other for the foreseeable future.

REFERENCES

1. Archie E. Roy, *The Foundations of Astrodynamics*, the Macmillan Company, New York, 1965.
2. *Orbital Flight Handbook Part 1—Basic Techniques and Data* (NASA SP-33 Part 1), National Aeronautics and Space Administration, Washington, D.C., 1963.
3. Lloyd Motz and Anneta Duveen, *Essentials of Astronomy* (2nd ed.), Columbia University Press, New York, 1977.
4. David R. Brooks, *An Introduction to Orbit Dynamics and Its Application to Satellite-Based Earth Monitoring Missions* (NASA Reference Publication 1009), National Aeronautics and Space Administration, Washington, D.C., 1977.
5. *The American Ephemeris and Nautical Almanac*, U.S. Government Printing Office, Washington, D.C. (published annually).
6. *Reference Data for Radio Engineers* (6th ed.), Howard W. Sams and Co., Indianapolis, 1975, pp. 28–9 through 28–11.
7. W. L. Stutzman, *Signal Power Calculation Procedures for Communications Systems*, unpublished manuscript, March 1976.
8. W. C. Isley and K. I. Duck, "Propulsion Requirements for Communications Satellites," in P. L. Bargellini, Ed., *Communications Satellite Technology* (Vol. 33 of *Progress in Astronautics and Aeronautics*), MIT Press, Cambridge, MA, 1972.
9. Arnold A. Saterlee, "Mechanics of Synchronous Orbit Insertion," *Comsat Technical Review*, **2**, 422–435 (Fall 1972).
10. *Space Navigation Handbook* (NAVPERS 92988), U.S. Navy, Washington, D.C., 1962. (U.S. Government Printing Office Item Number O-628762)
11. Rene C. Collette and Bernard L. Herdan, "Design Problems of Spacecraft for Communications Missions," *Proceedings of the IEEE*, **65**, 342–356 (March 1977). [Reprinted in Harry L. VanTrees, Ed., *Satellite Communications*, IEEE Press, New York, 1979, 451–465].
12. K. H. Muller, "Launch Vehicles for Commercial Communications Satellites," *RCA Engineer*, **28**, 72–76 (March-April, 1983).
13. Z. O. Bleviss, "Expendable Launch Vehicles for Synchronous Communication Satellites," in D. Jarett, Ed., *Satellite Communications: Future Systems* (Vol. 54 of *Progress in Astronautics and Aeronautics*), American Institute of Aeronautics and Astronautics, New York, 1977, pp. 319–346.

14. J. H. Schwarze, "Satellite Launching—a Combined Effort," *RCA Engineer*, **28**, 77–82 (March–April, 1983).
15. Kim E. Degnan, "Commercial Launch Vehicle Services," *Satellite Communications*, **7**, pp. 34–36 (September, 1983).
16. Theo Pirard and Kim E. Degnan, "Ariane in 1983: Shuttle Beware," *Satellite Communications*, **7**, pp. 28–32 (September 1983).
17. Scott Webster, "TOS: Commercial Launch Vehicles for the 80's," *Satellite Communications*, **7**, pp. 44–50 (September 1983).
18. B. Stockwell, "Space Insurance: Issues and Problems," *Space Communication and Broadcasting*, **1**, pp. 261–267 (October 1983).
19. Frederick J. Tischer, *Basic Theory of Space Communications*, D. Van Nostrand Co., Princeton, NJ, 1965.
20. Walter C. Scales and Richard Swanson, "Air and Sea Rescue Via Satellite Systems," *IEEE Spectrum*, **21**, No. 3, pp. 48–52 (March 1984).
21. J. J. Spilker, Jr., *Digital Communications by Satellite*, Prentice-Hall, Englewood Cliffs, NJ, 1977.
22. James Martin, *Communications Satellite Systems*, Prentice-Hall, Englewood Cliffs, NJ, 1978.
23. K. Miya, Ed., *Satellite Communications Engineering*, Lattice Publishing Co., Tokyo, 1975.

PROBLEMS

1. An intersatellite relay allows messages to go from one satellite to another without passing through an intervening earth station. Calculate the path length and the minimum distance between the radio path and the surface of the earth for a link between two geosynchronous satellites that are exactly 110° apart. Ignoring any atmospheric effects, what is the maximum possible separation between two geosynchronous satellites with a line-of-sight link between them?

2. A new weather-research satellite is to be placed in a circular equatorial orbit so that it moves in the same direction as the earth's rotation. Using an active radar system, the satellite will store data on surface barometric pressure and play the data back to a controlling earth station after each trip around the world. The orbit is to be designed so that the station is directly above the controlling earth station, located on the equator, once every 4 h. The earth station's antenna cannot receive below 10° above local horizontal in any direction. Taking the earth's radius to be exactly 6370 km and the earth's rotational period to be exactly 24 h, find the following quantities:

   a. The spacecraft's angular velocity in radians per second.
   b. The orbital period in hours.
   c. The orbital radius in kilometers.
   d. The orbital height in kilometers.
   e. The spacecraft linear velocity in meters per second.

f. The time interval in minutes for which the controlling earth station can communicate with the spacecraft on each pass.

3. A satellite is in a 322-km high circular orbit. Determine
a. The orbital angular velocity in radians per second.
b. The orbital period in minutes.
c. The orbital linear velocity in meters per second.
This same satellite is the subject of Problem 13.

4. Here are some orbital measurements for the Italian satellite SIRIO at 00:00:00 UT on March 26, 1979.

Semimajor axis: 42,167.911 km
Eccentricity: 0.00033
Mean anomaly: 28.3866°

Determine
a. The orbital period in hours, minutes, and seconds.
b. The mean orbital angular velocity in radians per second.
c. The maximum and minimum distances of the spacecraft from the center of the earth during each orbital revolution.
d. The time (expressed as a date, hour, minute, and second) of the next perigee passage after 00:00:00 on March 26, 1979.

5. The Virginia Tech earth station is located at 80.438° W longitude and 37.229° N latitude. Calculate its look angle and range to a geosynchronous satellite whose subsatellite point is located at 121° W longitude.

6. The state of Virginia may be represented roughly as a rectangle bounded by 39.5° N latitude on the north, 36.5° N latitude on the south, 76.0° W longitude on the east, and 83.6° W longitude on the west. If a geosynchronous satellite must be visible throughout Virginia at an elevation angle no lower than 20°, what is the range of longitudes within which the subsatellite point must lie?

7. Find the look angles from the earth stations listed below to each of the geosynchronous satellites listed. If a satellite is not visible from an earth station, then so indicate.
Earth Stations:
a. Andover, Maine   44 deg 48 min 59 s N
                            70 deg 42 min 52 s W
b. Carnarvon, Australia   24 deg 52 min 13 s S
                           113 deg 42 min 13 s E

Satellites:
a. COMSTAR D-3 (U.S. domestic)   87° W
b. COMSTAR D-4 (U.S. domestic)   127.5° W
c. BSE-2 (Japan)   110° E
All subsatellite points are on the equator.

8.   Determine the maximum and minimum range in kilometers from an earth station to a geosynchronous satellite. To what round-trip propagation times do these correspond?

9.   Determine the maximum north latitude that an earth station can have and still communicate with a geosynchronous satellite.

10.   WESTAR III is a geosynchronous satellite located at 91° W longtitude.
- a.   Determine the look angles to WESTAR III from an earth station near Los Angeles at 120° W, 34° N.
- b.   Determine the latitude and longitude of (i) the northern-most and (ii) the western-most point at which WESTAR III is visible at a 10° elevation angle.

11.   Geosynchronous satellites do not provide good coverage for places in the far north and communications satellites in inclined elliptical orbits are an attractive alternative. In this problem you will work out the orbit for one such hypothetical satellite for Norway. Here are its specifications.

orbital period: 24 h
subsatellite point: 60° N, 10° E (Oslo) *at 2100 UT daily*
apogee: occurs at 2100 UT every day when the satellite is directly above 60° N, 10° E
Height of perigee: 500 km above the surface of the earth

Determine the following quantities:
- a.   semimajor axis a
- b.   time of perigee
- c.   eccentricity e
- d.   distance from the surface of the earth at apogee
- e.   orbital inclination i that would provide the longest daily period of visibility at Oslo.

12.   Prove that the smallest value that the inclination angle can have is equal to the latitude of the launch site if the launch site is in the plane of the orbit. *Hint:* Before writing equations, think about what other points must also be in the plane of the orbit.

13.   The satellite of Problem 3 (322-km circular orbit) carries a 300-MHz transmitter.
- a.   Determine the maximum frequency range over which the received signal would Doppler shift if received by a stationary observer suitably located in space. In the notation of Section 2.6 this is $2\Delta f$, since the signal frequency can shift from $+\Delta f$ to $-\Delta f$.
- b.   Assume that an earth station on the equator can receive down to 0° elevation in all directions. The station is at mean sea level 6370 km from the center of the earth. Calculate the maximum Doppler shift $(2\Delta f)$ that

this station will observe on the 300-MHz signal from the satellite. Include earth rotation and be sure you consider the *maximum possible* Doppler shift for a 322-km circular orbit.

14. Most commercial communications satellites must maintain their orbital positions to within plus or minus 0.1° of arc. If a satellite meets this condition and is in an orbit with an eccentricity of 0.001, describe the "box" in which the satellite is constrained to move and calculate the maximum variation in range to an earth station that could occur as the satellite moves about in the box.

3

SPACECRAFT

3.1 INTRODUCTION

Maintaining a microwave communication system 35,800 km out in space is not a simple problem, so communications satellites are very complex, extremely expensive to purchase, and also expensive to launch. An INTELSAT V spacecraft, for example, is estimated to cost around $50M, on station [1]. The cost of the spacecraft and launch are increased by the need to dedicate an earth station to the monitoring and control of the satellite, at a cost of several million dollars per year. The revenue to pay these costs is obtained by selling the communication capacity of the satellite to users, either by way of leasing circuits or transponders, or by charging for circuit use, as in the international telephone service.

Communications satellites are designed to have an operating lifetime of 5 to 10 years. The operator of the system hopes to recover the initial and operating costs well within the expected lifetime of the spacecraft, and the designer must provide a satellite that can survive the hostile environment of outer space for that long. In order to support the communications system, the spacecraft must provide a stable platform on which to mount the antennas, be capable of station-keeping, provide the required electrical power for the communication system, and also provide a controlled-temperature environment for the communications electronics. In this chapter we discuss the subsystems needed on a spacecraft to support its primary mission of communications. We also discuss the communications subsystem itself in some detail, and other problems such as reliability.

3.2 SPACECRAFT SUBSYSTEMS

The major subsystems required on the spacecraft are given below. Figure 3.1 shows an exploded view of a typical spacecraft with several of the subsystems indicated.

Attitude and Orbit Control System (AOCS) This subsystem consists of rocket motors that are used to move the satellite back to the correct orbit when external forces cause it to drift off station and gas jets or inertial devices that control the attitude of the spacecraft.

Telemetry, Tracking, and Command (TT&C) These systems are partly on the satellite and partly at the controlling earth station. The telemetry system sends data derived from many sensors on the spacecraft, which monitor the spacecraft's "health," via a telemetry link to the controlling earth station. The tracking system is located at this earth station and provides information on the range and the elevation and azimuth angles of the satellite. Repeated measurement of these three parameters permits computation of orbital elements, from which changes in the orbit of the satellite can be detected. Based on telemetry data received from the satellite and orbital data obtained from the tracking system, the control system is used to correct the position and attitude of the spacecraft. It is also used to control the antenna pointing and communication system configuration to suit current traffic requirements, and to operate switches on the spacecraft.

Power system All communications satellites derive their electrical power from *solar cells*. The power is used by the communication system, mainly in its transmitters, and also by all other electrical systems on the spacecraft. The latter use is termed *housekeeping*, since these subsystems serve to support the communications system.

Communications Subsystems The communications subsystem is the major component of a communications satellite, and the remainder of the spacecraft is there solely to support it. Frequently, the communications equipment is only a small part of the weight and volume of the whole spacecraft. It is usually composed of one or more antennas, which receive and transmit over wide bandwidths at microwave frequencies, and a set of receivers and transmitters that amplify and retransmit the incoming signals. The receiver–transmitter units are known as *transponders*.

Spacecraft Antennas Although these form part of the complete communication system, they can be considered separately from the transponders. On advanced satellites such as INTELSAT V, the antenna systems are very complex and produce beams with shapes carefully tailored to match the areas on the earth's surface served by the spacecraft.

The subsystems listed above are discussed in more detail in Section 3.7. The reader who is interested in spacecraft design should refer to the literature of that field, particularly the *IEEE Transactions on Aerospace and Electronic Systems* [2] and the *American Institute of Aeronautics and Astronautics Transactions* and annual *Conference Proceedings* [3, 4]. Only a brief review of the spacecraft subsystems that support the communication mission is included here.

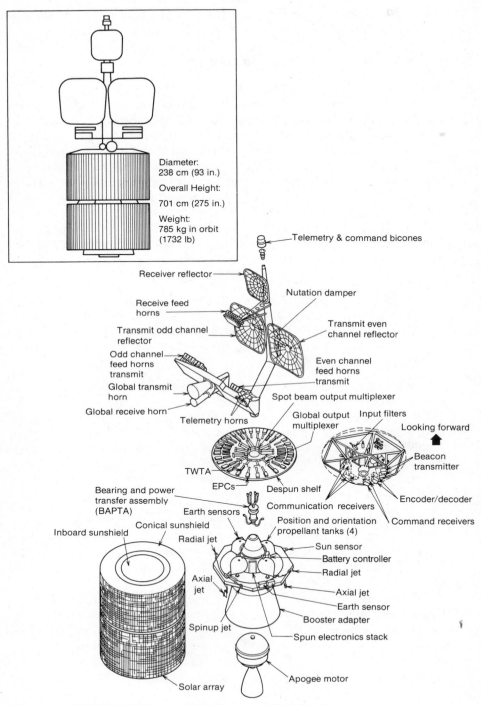

Diameter:
238 cm (93 in.)

Overall Height:
701 cm (275 in.)

Weight:
785 kg in orbit
(1732 lb)

Telemetry & command bicones

Receiver reflector

Nutation damper

Receive feed
horns

Transmit even
channel reflector

Transmit odd channel
reflector

Odd channel
feed horns
transmit

Even channel
feed horns
transmit

Global transmit
horn

Spot beam output multiplexer

Global receive horn

Global output
multiplexer

Input filters

Looking forward

Telemetry horns

Beacon
transmitter

TWTA

EPCs

Despun shelf

Encoder/decoder

Bearing and power
transfer assembly
(BAPTA)

Communication receivers

Command receivers

Earth sensors

Position and orientation
propellant tanks (4)

Inboard sunshield

Conical sunshield

Radial jet

Sun sensor

Battery controller

Radial jet

Axial
jet

Axial jet

Earth sensor

Booster adapter

Spinup jet

Spun electronics stack

Apogee motor

Solar array

Figure 3.1 INTELSAT IV-A satellite. (Courtesy of Intelsat.)

3.3 ATTITUDE AND ORBIT CONTROL SYSTEM (AOCS)

Control of the attitude of a spacecraft is necessary so that the antennas, which often have narrow beams, are pointed correctly at the earth. A number of factors tend to cause the spacecraft to rotate; some of these were discussed in Chapter 2. Gravitational forces from the sun, moon, and planets will set up rotational moments if the spacecraft is not perfectly balanced. Solar pressure acting on the antennas, spacecraft body, and solar sails may also create rotational forces. The earth's magnetic field can produce forces on the spacecraft if there is a net magnetic moment present. Because the satellite moves round the earth's center in its orbit, the forces described above all vary cyclically through a 24-hour period. This tends to set up nutation (a wobble) of the spacecraft, which must be damped out mechanically.

The variation in gravitational field that causes attitude changes will also create accelerations on the spacecraft tending to change its orbit. The major influence is the moon's gravitational field, which is about three times stronger than the sun's at geostationary altitudes. Since neither the earth's orbital plane around the sun nor the moon's orbital plane around the earth lie in the earth's equatorial plane, there is a resultant gravitation force that acts to modify the inclination of the spacecraft's orbital plane. In the 1970–1980 time frame, the rate of change of inclination was about 0.85° per year for a spacecraft initially in geostationary orbit [5]. If no North–South stationkeeping corrections were applied, the orbit inclination would increase to a maximum of 14.67° from an initial 0° in 26.7 years. The inclination would then decrease back to zero in a similar time period.

The earth is not truly spherical; there is a bulge in the equator region of about 65 m at longitudes of 15° W and 165° E, with the result that the spacecraft experiences an acceleration toward a stable point in the geostationary orbit at longitude 105° W or 75° E, as shown in Figure 3.2. For accurate station keeping, the spacecraft must be accelerated in the opposite direction by firing rocket motors, or *thrusters*, at periodic intervals. The earth is also flattened at the poles by about 20 km, but this has little influence on a geostationary satellite.

Attitude Control System

There are two ways to make a spacecraft stable when it is in orbit and weightless. The entire body of the spacecraft can be rotated at 30 to 100 rpm to provide a powerful gyroscopic action, which maintains the spin axis in the same direction; such spacecraft are called *spinners*. Alternatively, three *momentum wheels* can be mounted on three mutually orthogonal axes of the spacecraft to provide stability. The momentum wheel is usually a solid disk driven by a motor, rotating at high speed within a sealed, evacuated housing. Increasing the speed of the wheel increases its angular momentum, which causes the spacecraft to precess in the opposite direction, according to the principle of conservation of angular momentum. With three momentum wheels, rotation of the spacecraft about each axis can be commanded from earth by increasing or decreasing the appropriate momentum wheel speed. This is called *three-axis stabilization*. An example of a

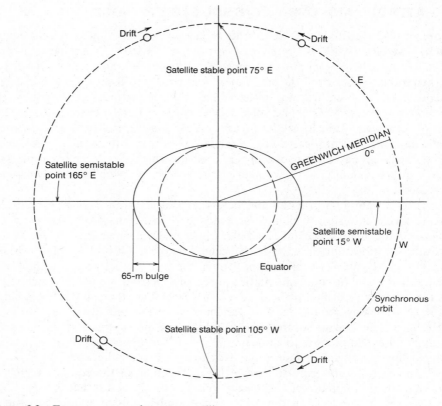

Figure 3.2   Forces on a synchronous satellite.

three-axis stabilized satellite is INTELSAT V, which is illustrated in Figure 3.3*b*. Satellites with three-axis stabilization are said to be *body stabilized*. An alternative system uses a single momentum wheel mounted on pivots so that the spin axis can be shifted to control spacecraft attitude [6, 7].

The spinner design of spacecraft is typified by INTELSAT IV-A and many similar spacecraft built by the Hughes Aircraft Corporation for domestic satellite communication systems (PALPA I, COMSTAR D1-4, HS 376 etc.). As shown in Figures 3.1 and 3.3*a*, the satellite consists of a cylindrical drum covered in solar cells that contains the power systems and the rocket motors. The communications system is mounted at the top of the drum and is driven in the opposite direction to the drum to keep the antennas pointing toward earth. An electric motor is used to drive the communication package around; such systems are called *despun*. In the early days of satellite communication *despun antennas* were not used, so antennas with a circular symmetric pattern were employed. These antennas have low gain and are now used only for basic TT&C systems that must operate regardless of spacecraft orientation. By despinning the antennas and transponders, no RF rotating joints are needed in the communication system. Electrical power

Figure 3.3a   INTELSAT IV-A satellite.

Figure 3.3b   INTELSAT V satellite. (Photographs courtesy of Intelsat, Washington, DC 20008.)

and some signals must be brought through a despin bearing, however, and the design of the bearing to guarantee friction-free operation for 10 years in the total vacuum of outer space is a challenge.

The satellite is *spun up* by operating small radial gas jets mounted on the periphery of the drum, at an appropriate point in the launch phase. The despin system is then brought into operation so that the main TT&C antennas point toward earth. The main TT&C system operates at 6/4 GHz on the Intelsat spacecraft, with a 2 GHz backup system for use during the launch phase.[1] The gas jets are usually powered by hydrazine or propane. Hydrazine (N_2H_4) is easily liquified under pressure, but readily decomposes when passed over a catalyst. Increased power can be obtained from the gas jets by electrically heating the catalyst and gas. This saves fuel and extends the life of the spacecraft. The conversion to gas produces pressure at the gas jet, and by pulsing the gas flow with a control valve at appropriate points in the spin cycle, control can be effected along two axes. Thus two nozzles and two control valves are all that is needed to provide initial spin-up and control on two axes.

In a three-axis stabilized satellite, one pair of gas jets is needed for each axis to provide for rotation in both directions of pitch, roll, and yaw. An additional set of controls, allowing only one jet on a given axis to be operated, provides for velocity increments in the X, Y, and Z directions.

Let us define a set of reference Cartesian axes (X_R, Y_R, Z_R) with the spacecraft at the origin, as shown in Figure 3.4. The Z_R axis is directed toward the center of the earth and is in the plane of the spacecraft orbit. It is aligned along the local vertical at the spacecraft subsatellite point. The X_R axis is tangent to the orbital plane and lies in the orbital plane. The Y_R axis is perpendicular to the orbital plane. For a spacecraft serving the Northern Hemisphere, the directions of the X_R and Y_R axes are nominally East and South.

Rotation about the X_R, Y_R, and Z_R axes is defined as *roll* about the X_R axis, *pitch* about the Y_R axis, and *yaw* about the Z_R axis. The spacecraft must be stabilized with respect to the reference axes to maintain accurate pointing of its antenna beams. The axes X_R, Y_R, Z_R are defined with respect to the location of the spacecraft; a second set of Cartesian axes, X, Y, Z, as shown in Figure 3.4, define the orientation of the spacecraft. Changes in spacecraft attitude cause the angles θ, ϕ, and ψ in Figure 3.4 to vary as the X, Y, Z axes move relative to the fixed reference axes X_R, Y_R, Z_R. The Z axis is usually directed toward a reference point on earth, called the *Z-axis intercept*. The location of the Z-axis intercept defines the pointing of the spacecraft antennas; the Z-axis intercept point may be moved to repoint all the antenna beams by changing the attitude of the spacecraft with the attitude control system.

---

[1] Satellite communications systems are conventionally identified by the frequency bands used. The uplink frequency band is given first, followed by the downlink band. Thus "6/4 GHz" indicates a system using the 6 GHz band for the uplink from the earth station to the satellite, and the 4 GHz band for the downlink from the satellite to the earth station. The other important frequency bands are 14/11 GHz and 30/20 GHz.

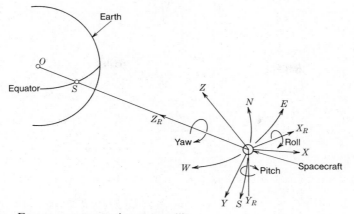

Figure 3.4a   Forces on a geostationary satellite.

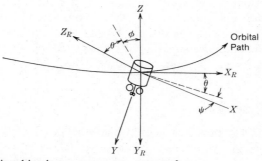

Figure 3.4b   Relationships between axes at spacecraft.

In a spinner-type spacecraft, the axis of rotation is usually the Y axis, which is maintained close to the Y_R axis, perpendicular to the orbital plane. Pitch correction is required only on the despun antenna system and can be obtained by varying the speed of the despin motor. Yaw and roll are controlled by pulsing radially mounted jets at the appropriate instant as the body of the spacecraft rotates.

If a constant torque exists about one axis of the spacecraft, a continual increase or decrease in momentum wheel speed is necessary to maintain the correct attitude. When the upper or lower speed limit of the wheel is reached, it must be *unloaded* by operating a pair of gas jets and simultaneously reducing or increasing the wheel speed.

Closed-loop control of attitude is employed on the spacecraft to maintain the correct attitude. When large, narrow beam antennas are used, the whole spacecraft may have to be stabilized within $\pm0.1°$ on each axis. The references for the attitude

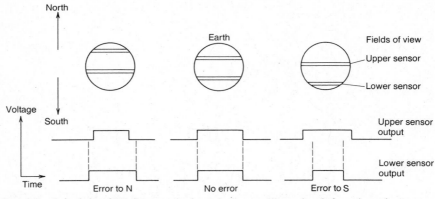

Figure 3.5   Principle of N–S control of a spinner satellite using infrared earth sensors.

control system may be the outer edge of the earth's disk, as observed with infrared sensors, the sun, or one or more stars. Figure 3.5 illustrates how an infrared sensor on the spinning body of a spacecraft can be used to control pointing toward the earth. Figure 3.6 shows a typical control system loop using the technique illustrated in Figure 3.5. The control system will be more complex for a three-axis stabilized spacecraft and may employ an on-board computer to process the sensor data and command the gas jets and momentum wheels.

Orbit Control System

As discussed in Chapter 2, a geostationary satellite is subjected to several forces that tend to accelerate it away from its required orbit. The most important, for the geostationary spacecraft, are the gravitation forces of the moon and the sun, which cause inclination of the orbital plane, and the elliptical shape of the earth around the equator, which causes drift of the subsatellite point. There are many other smaller forces that act on the spacecraft causing the orbit to change. Accurate prediction of the spacecraft position a week or two weeks ahead requires a computer program with up to 20 force parameters; we shall restrict our discussion here to the two major effects.

Figure 3.7 shows a diagram of an inclined orbital plane close to the geostationary orbit. For the orbit to be truly geostationary, it must lie in the equatorial plane, be circular, and have the correct altitude. The various forces acting on the spacecraft will steadily pull it out of the correct orbit; it is the function of the orbit control system to return it to the correct orbit. This cannot be done with momentum wheels since linear accelerations are required. Gas jets that can impart velocity changes along the three references axes of the spacecraft are required.

If the orbit is not circular, a velocity increase or decrease will have to be made along the orbit, in the X-axis direction in Figure 3.4. On a spinning spacecraft, this is achieved by pulsing the radial jets when they point along the X axis. On

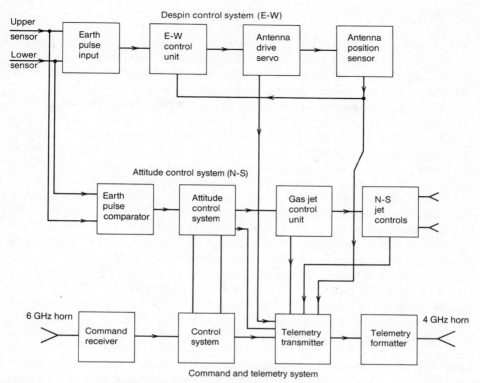

Figure 3.6 Typical on-board control system for a spinner satellite.

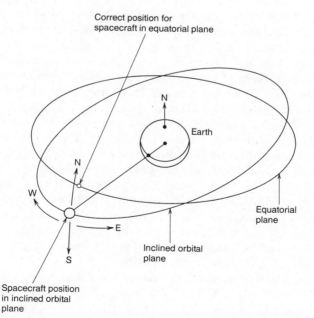

Figure 3.7 Spacecraft in inclined orbit.

a three-axis stabilized spacecraft, there will usually be two pairs of X_R axis jets acting in opposite directions, one pair of which will be operated for a predetermined length of time to provide the required velocity change. The orbit of a geostationary satellite remains approximately circular for long periods of time and does not need frequent velocity corrections to maintain circularity. Altitude corrections are made by operating the Z axis gas jets.

The inclination of the orbit of a satellite that starts out in a geostationary orbit increases at about 0.85° per year. A satellite that must be held within a 0.1° "box," so that fixed pointing antennas can be used at earth stations, will have to be corrected for inclination drift at least once every six weeks. In practice, corrections are made every two to four weeks to keep the error small. To correct for inclination drift, a velocity increase at right angles to the orbital plane is required, in the Y_R direction in Figure 3.4. This will cause the satellite to move at right angles to the orbital plane, reducing the inclination. However, when the inclination has reached zero degrees, an opposing jet must be operated to stop the satellite at that position. This procedure is known as a *North–South station keeping maneuver*.

Correcting the inclination of a spacecraft orbit requires more fuel to be expended than for any other orbital correction. This places a weight penalty on those spacecraft that must maintain very accurate stationkeeping, and reduces the communications payload they can carry. A typical spacecraft weighing 1000 kg may require 30 or 40 kg of fuel to maintain inclination within $\pm 0.1°$ over the satellite's lifetime. In satellite communication systems where all the earth stations have steerable tracking antennas, inclination control may be abandoned. The satellite is then launched into an inclined orbit of 2.5° or 3° inclination, in the opposite sense to the direction of drift. After three or four years of operation, the orbit lies in the equatorial plane and then continues to increase its inclination. Provided that East–West stationkeeping is maintained, this does not affect neighboring satellites and the saving in weight of fuel can be used to carry another communications payload. This policy may be adopted on some INTELSAT V spacecraft to allow them to carry a maritime satellite communication system in addition to the main global communication system.

East–West stationkeeping is effected by use of the X axis jets of the spacecraft. For spacecraft located away from the stable points at 75° E and 105° W, a slow drift toward these points will occur. Typically, the X axis jets are pulsed every two or three weeks to counter the drift and add a small velocity increment in the opposite direction. The spacecraft then drifts through its nominal position, stops at a point a fraction of a degree beyond it, and then drifts back again. East–West stationkeeping requires only a modest amount of fuel and is necessary on all geostationary communications satellites to maintain the spacing between adjacent satellites. With orbital locations separated by two or three degrees, East–West drifts in excess of a fraction of a degree cannot be tolerated, and most satellites are held within $\pm 0.1°$ of their allotted position.

Some communications satellites such as the Russian MOLNIYA series are not in geostationary orbit. MOLNIYA spacecraft were launched into a highly elliptical 12-hour orbit with a large (65°) inclination angle to provide communications to the northern regions of the USSR, above 70° N latitude. Other spacecraft carry earth observation equipment to study cloud cover and surface temperature, or for military surveillance purposes, and are launched into lower orbits that take the spacecraft over much of the earth's surface. All of these satellites require AOC systems to enable them to point their antennas or sensors correctly, and all need to be able to maintain the correct orbit.

3.4 TELEMETRY, TRACKING, AND COMMAND (TT&C)

The TT&C system is essential to the successful operation of a communications satellite. It is part of the spacecraft management task, which also involves an earth station, usually dedicated to that task, and a group of personnel. The main functions of spacecraft management are to control the orbit and attitude of the satellite, monitor the status of all sensors and subsystems on the spacecraft, and switch on or off sections of the communication system. On large geostationary satellites, some repointing of individual antennas is also possible, under the command of the TT&C system. Tracking is performed primarily by the earth station. Figure 3.8 illustrates the functions of a controlling earth station.

Telemetry

The telemetry system collects data from many sensors within the spacecraft and sends these data to the controlling earth station. Typically, as many as 100 sensors may be located on the spacecraft to monitor pressure in the fuel tanks, voltage and current in the power conditioning unit, current drawn by each subsystem, and critical voltages and currents in the communications electronics. The temperature of many of the subsystems is important and must be kept within predetermined limits, so many temperature sensors are fitted. The status of each subsystem and the positions of switches in the communication system are also reported back by the telemetry system. The sighting devices used to maintain spacecraft attitude are also monitored via the telemetry link: this is essential in case one should fail and cause the satellite to point in the wrong direction. The faulty unit must then be disconnected and a spare brought in, via the command system, or some other means of controlling attitude devised.

The telemetry data are usually digitized and transmitted as frequency or phase shift keying (*FSK* or *PSK*) of a low-power telemetry carrier using time division techniques. A low data rate is normally used to allow the receiver at the earth station to have a narrow bandwidth and thus maintain a high carrier-to-noise ratio. The entire TDM frame may contain thousands of bits of data and take several seconds to transmit. At the controlling earth station a computer can be used to monitor, store, and decode the telemetry data so that the status of any system or sensor on the spacecraft can be determined immediately by the controller on earth. Alarms can also be sounded if any vital parameter goes outside allowable limits.

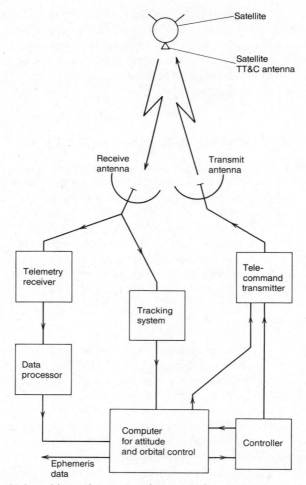

Figure 3.8 Typical tracking, telemetry, and command system.

Tracking

A number of techniques can be used to determine the current orbit of a spacecraft. Velocity and acceleration sensors on the spacecraft can be used to establish the change in orbit from the last known position, by integration of the data. The earth station controlling the satellite can observe the Doppler shift of the telemetry carrier or beacon transmitter carrier to determine the rate at which range is changing. Together with accurate angular measurements from the earth station antenna, range is used to determine the orbital elements. Active determination of range can be achieved by transmitting a pulse, or sequence of pulses, to the satellite and observing the time delay before the pulse is received again. The propagation delay in the spacecraft transponder must be accurately known, and more than one earth station may make range measurements. If a sufficient number of earth stations with an adequate separation are observing the satellite, its position can be established by triangulation from the earth station look angles or by simultaneous range measurements. With precision equipment at the earth stations, the position of the spacecraft can be determined within 100 m.

Command

A secure and effective command structure is vital to the successful launch and operation of any communications satellite. The command system is used to make changes in attitude and corrections to the orbit and to control the communication system. During launch, it is used to control the firing of the apogee boost motor and to spin up a spinner or extend the *solar sails* of a three-axis stabilized spacecraft.

The command structure must possess safeguards against inadvertent operation of a control due to error in a received command. A typical system of the type shown in Figure 3.8 will originate commands at the control terminal of the computer. The *control code* is converted into a *command word*, which is sent in a TDM frame to the satellite. After checking for validity in the spacecraft, the word is sent back to the control station via the telemetry link where it is checked again in the computer. If it is found to have been received correctly, an *execute* instruction will be sent to the satellite so that the command is executed. The entire process may take 5 or 10 s, but minimizes the risk of erroneous commands causing a satellite malfunction.

The command and telemetry links are usually separate from the communication system, although they may operate in the same frequency band (6 and 4 GHz). Two levels of command system are used in the Intelsat spacecraft: the main system operates in the 6-GHz band, in a gap between the communication channel frequencies; the main telemetry system uses a similar gap in the 4-GHz band. The TT&C antennas for the 6/4 GHz system can be seen in Figure 3.1 on an INTELSAT IV-A satellite. These are earth-coverage horns, so the main system can be used only after correct attitude of the spacecraft is achieved.

During the launch phase and injection into geostationary orbit, the main TT&C system may be inoperable because the spacecraft does not have the correct attitude or has not extended its solar sails. A backup system is used at this time, which controls only the most important sections of the spacecraft. A great deal of redundancy is built into this system, since its failure will jeopardize the entire mission. Near omnidirectional antennas are used at either UHF or S band (2–4 GHz), and sufficient margin is allowed in the signal-to-noise ratio (S/N) at the satellite receiver to guarantee control under the most adverse conditions. The backup system provides control of the apogee boost motor, the attitude control system and orbit control thrusters, the solar sail deployment mechanism (if fitted), and the power conditioning unit. With these controls, the spacecraft can be injected into geostationary orbit, turned to face earth, and switched to full electrical power so that hand-over to the main TT&C system is possible. In the event of failure of the main TT&C system, the backup system can be used to keep the satellite on station. It is also used to eject the spacecraft from geostationary orbit and to switch off all transmitters when it eventually reaches the end of its useful life.

3.5 POWER SYSTEMS

All communications satellites obtain their electrical power from solar cells, which convert incident sunlight into electrical energy. Some military surveillance satellites built by the USSR have used thermonuclear generators to supply electrical power, but because of the danger to people on the earth if the launch should fail and the nuclear fuel be spread over an inhabited area, communications satellites have not used nuclear generators.

The sun is a powerful source of energy. In the total vacuum of outer space, at geostationary altitude, the radiation falling on a spacecraft has an intensity of 1.39 kW/m^2. Solar cells do not convert all this incident energy into electrical power; their efficiency is typically 10 to 15 percent but falls with time because of aging of the cells and etching of the surface by micrometeor impacts. Since sufficient power must be available at the end of the lifetime of the satellite to supply all the systems on board the spacecraft, about 15 percent extra area of solar cells is usually provided as an allowance for aging.

A spin-stabilized spacecraft usually has a cylindrical body covered in solar cells. Early spacecraft were of small dimensions and had relatively small areas of solar cells. More recently, a large communications satellite such as INTELSAT IV-A has 20 m^2 of solar cells generating 900 W of electrical power at the beginning of the satellite's lifetime. Figure 3.9 shows a comparison of the electrical power available on some INTELSAT communications satellites.

Future generations of satellites now planned have even higher power requirements. For example, RCA Americom's *Ku-band* satellites K1 through K4, due for launch in 1985–1987, will generate 2450 to 2900 W at the end of life from solar arrays with total area of 27 to 30 m^2. These are very large satellites weighing up to 2270 kg, with total RF output power of 960 W (K3, K4) [8].

A three-axis stabilized spacecraft can make better use of its solar cell area, since the cells can be arranged on flat panels that can be rotated to maintain normal incidence of the sunlight. Only one-third of the total area of solar cells is needed relative to a spinner, with some saving in weight. A primary advantage, however, is that by unfurling a folded solar array when the satellite reaches geostationary orbit, power in excess of 1 kW can be generated. To obtain 1000 W from a spinner requires a very large body on which to place the solar cells, which may then exceed the maximum payload dimensions of the launch vehicle. Thus INTELSAT V uses a three-axis stabilized configuration with unfurling solar sails that produce 1220 W, each of which is 15 ft long. However, INTELSAT VI uses

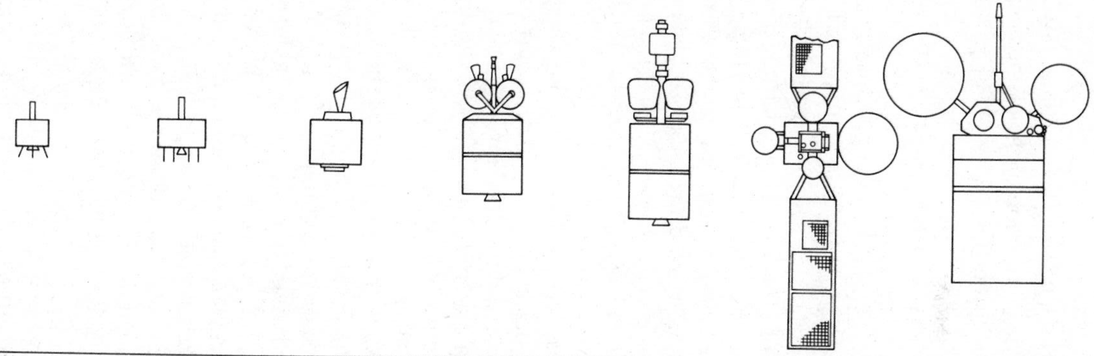

Spacecraft	INTELSAT I	INTELSAT II	INTELSAT III	INTELSAT IV	INTELSAT IV-A	INTELSAT V	INTELSAT VI
Year of first launch	1965	1967	1968	1971	1975	1980	1986 (planned)
Dimensions	0.71 m dia × 0.59 m high	1.42 m dia × 0.67 m high	1.42 m dia × 1.98 m high	2.38 m dia × 7.01 m high	2.38 m dia × 7.01 m high	15.27 m across solar sails × 6.71 m high	3.6 m dia × 11.7 m high
On orbit weight	34 kg	76 kg	152 kg	595 kg	786 kg	1020 kg	1800 kg
End of life primary power	46 W	85 W	125 W	569 W	708 W	1220 W	2100 W
Total bandwidth	50 MHz	130 MHz	360 MHz	450 MHz	720 MHz	2250 MHz	3360 MHz
Notional capacity two-way telephone circuits	240	240	1500	5000	11,000 plus 2 TV channels	24,000 plus 2 TV channels	33,000 plus 2 TV channels
Design lifetime	1.5 years	3 years	5 years	7 years	7 years	10 years	10 years
Spacecraft cost	$3.6 M	$3.5 M	$4.5 M	$14 M	$18 M	$25 M	$140 M (first five satellites)
Launch cost	$4.6 M	$4.6 M	$6 M	$20 M	$20 M	$23 M	?
Cost per telephone circuit year	$23,000	$11,000	$1,600	$810	$494	$200	?
Contractor	Hughes	Hughes	TRW	Hughes	Hughes	Ford Aerospace	Hughes

Figure 3.9 Illustration of the growth of Intelsat satellites over two decades from the launch of Intelsat I (Early Bird) in 1965.

a cylindrical body much larger than that of INTELSAT IV, as larger dimensions are possible with launch by the Space Shuttle. The lower section of the cylindrical body slides up around the upper section for launch and is then extended when the spacecraft is in orbit.

Solar sails must be rotated by an electric motor once per 24 h to keep the cells in full sunlight. This causes the cells to heat up, typically to 50 to 80°C, which causes a drop in output voltage at the rate of 2 mV/°C temperature rise. In the spinner design, the cells cool down when in shadow and run at 20 to 30°C, with somewhat higher efficiency. The bombardment of the sails by protons and electrons is also more severe, and a thicker layer of glass may be needed to slow down deterioration of the cells, with a consequent weight penalty. The Hughes Aircraft Company has favored the spin-stabilized design and has constructed many satellites of this type. Most other companies now build three-axis stabilized spacecraft.

The spacecraft must carry batteries to power the subsystems during launch and during eclipses. Eclipses occur twice per year, around the spring and fall equinoxes, when the earth's shadow passes across the spacecraft, as illustrated in Figures 2.22 and 2.23. The longest duration of eclipse is 70 min, occurring around March 21 and September 21 each year. To avoid the need for large, heavy batteries, part or all of the communications system load may be shut down during eclipse, but this technique is rarely used when telephony or data traffic is carried. TV broadcast satellites will not carry sufficient battery capacity to supply their high-power transmitters during eclipse, and must shut down. By locating the satellite 20° W of the longitude of the service area, the eclipse will occur after 1 A.M. local time for the service area, when shutdown is more acceptable.

Batteries are usually of the sealed nickel cadmium type, which do not gas when charging and have good reliability and long life. A power-conditioning unit controls the charging current and dumps excess current from the solar cells into heaters or load resistors on the cold side of the spacecraft. Sensors on the batteries, power regulator, and solar cells monitor temperature, voltage, and current and supply these data to both the on-board control system and the controlling earth station via the telemetry downlink. Typical battery voltages are 20 to 50 V with capacities of 20 to 50 ampere-hours.

The power-generating and control systems account for a substantial part of the weight of a communications satellite, typically 10 to 15 percent. In an effort to reduce battery weight, RCA has developed a nickel hydrogen battery for use on ADVANCED SATCOM spacecraft launched from 1984 onward. The nickel hydrogen battery can sustain a greater depth of discharge than the nickel cadmium battery, and this allows a lower total battery capacity to be used on the spacecraft. If a battery is discharged too deeply, it may not recharge properly, jeopardizing the entire spacecraft. Nickel cadmium batteries are usually not allowed to discharge beyond 50 percent of capacity, whereas nickel hydrogen can safely be discharged to 60 percent of capacity. RCA claims that the reduced battery weight in the ADVANCED SATCOM spacecraft allows sufficient extra stationkeeping fuel to be carried to provide an additional year in orbit [8].

3.6 COMMUNICATIONS SUBSYSTEMS

Description of the Communication System

A communications satellite exists to provide a platform in geostationary orbit for the relaying of voice, video, and data communications. All other subsystems on the spacecraft exist solely to support the communications system, although this may represent only a small part of the volume, weight, and cost of the spacecraft in orbit. Since it is the communication system that earns the revenue for the system operator, communications satellites are designed to provide the largest traffic capacity possible. The growth in capacity is well illustrated in Figure 3.9 for the Intelsat system. Successive satellites have become larger, heavier, and more costly, but the rate at which traffic capacity has increased has been much greater, resulting in a lower cost per telephone circuit with each succeeding generation of satellite. The introduction of switched-beam technology and onboard processing in high-capacity satellites will offer a further increase in capacity in the late 1980s and 1990s. These topics are discussed in Section 3.7.

In the design of a satellite communication system, the downlink from the satellite to the earth station is usually the most critical part. The satellite transmitter has limited output power and the earth station is at least 36,000 km away, so the received power level, even with large aperture earth station antennas, is very small and rarely exceeds 10^{-10} W. For the system to perform satisfactorily, the signal power must exceed the power of the noise generated in the receiver by between 5 and 25 dB, depending on the bandwidth of the transmitted signal and the modulation scheme used. With low-power transmitters, narrow receiver bandwidths have to be used to maintain the required signal-to-noise ratios. Higher power transmitters and satellites with directional antennas enable wider bandwidths to be utilized, increasing the capacity of the satellite.

Early communications satellites were fitted with transponders of 250 or 500 MHz bandwidth, but had low-gain antennas and transmitters of 1 or 2 W output power. The earth station receiver could not achieve an adequate signal-to-noise ratio when the full bandwidth was used with the result that the system was *power limited*.

Later generations of communications satellites have also been power limited because they cannot use the RF bandwidth as efficiently as a terrestrial microwave communication system, but have steadily improved in bandwidth utilization efficiency, as seen in Figure 3.9. The total channel capacity of a satellite that uses a 500-MHz band at 6/4 GHz can be increased only if the bandwidth can be increased or reused. The trend in high-capacity satellites has been to reuse the available bands by employing several directional beams at the same frequency (*spatial frequency reuse*) and orthogonal polarizations at the same frequency (*polarization frequency reuse*). INTELSAT V and some domestic satellites also use both the 6/4 GHz and 14/11 GHz bands to obtain more bandwidth; for example, INTELSAT V achieves an effective bandwidth of 2250 MHz in its communication system within a 500 MHz band at 6/4 GHz and 250 MHz band at 14/11 GHz by a combination of spatial and polarization frequency reuse.

The designer of a satellite communication system is not free to select any frequency and bandwidth he or she chooses. International agreements restrict the frequencies that may be used for particular services, and the regulations are administered by the appropriate agency in each country—the Federal Communication Commission (FCC) in the United States, for example. Frequencies allocated to satellite services are listed in Tables 4.1 and 4.2 in Chapter 4. The bands currently used for the majority of services are 6/4 GHz and 14/11 GHz. The 6/4 GHz band was expanded from 500 MHz bandwidth in each direction to 1000 MHz by a World Administrative Radio Conference in 1979, but other services share the new part of the band and may cause interference to satellite communication links during the 1980s, so most systems designed before 1985 have used a 500-MHz total bandwidth. A similar bandwidth is available at 14/11 GHz, but because of propagation problems in heavy rain, this band is less popular than 6/4 GHz. A different frequency is required for the transmit path (the higher fre-

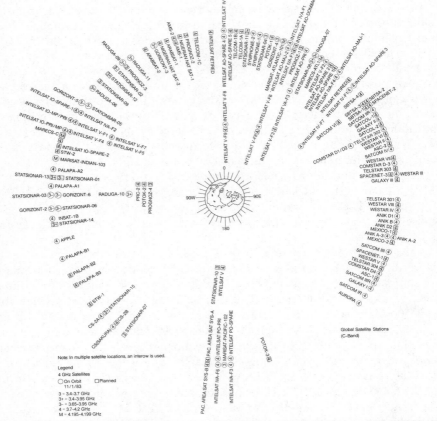

Figure 3.10 The crowded geosynchronous orbit. (*Source:* W. L. Morgan, "Global Satellite Stations," *Satellite Communications*, **7**, 22–32 (December 1983). Reprinted with permission of *Satellite Communications*. 6530 So. Yosemite St., Englewood, Colorado, 80111.)

quency) and the receive path in all satellite links because the satellite transmitter generates a signal that would jam its own receiver if both shared the same frequency.

The 500 MHz band originally allocated for 6/4 GHz satellite communications has become very congested and is now completely filled for some segments of the geostationary orbit such as that serving North America. Extension of the bands to 1000 MHz at 6/4 GHz will provide greater capacity as the new frequencies come into use, but many systems are now being designed that will use 14/11 GHz. Figure 3.10 illustrates the crowded nature of the geostationary orbit in 1984. The

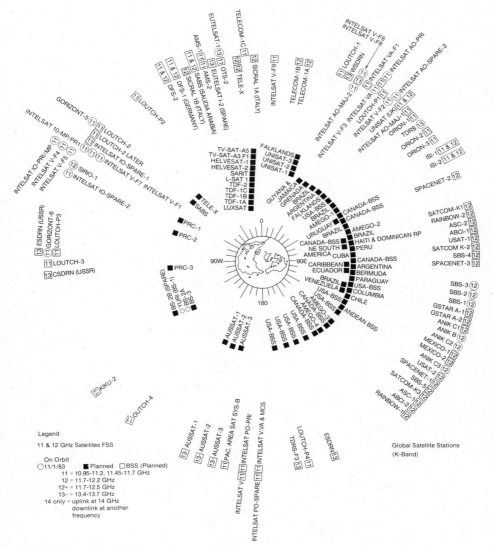

Figure 3.10   *(continued)*

standard spacing between satellites was originally set at 3°, but under new regulations covering North America, the spacing will be reduced to 2° for domestic satellites by 1985. This will require some earth stations to modify their antennas to ensure narrower transmit beams so that interference with adjacent satellites is avoided. The move to 2° spacing opens up extra slots for new satellites in the 6/4 GHz band.

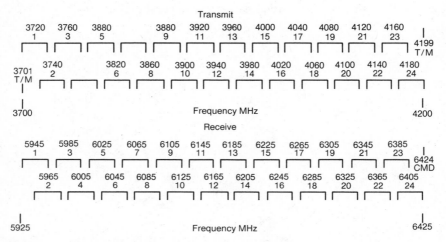

Frequency plan. Note: Number below channel center frequency refers to transponder identity.

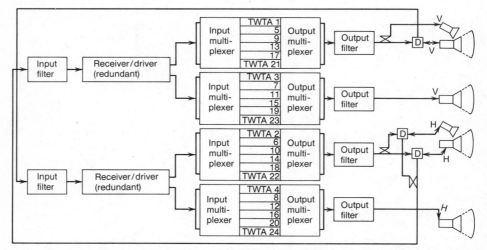

Figure 3.11 Transponder arrangement of RCA's SATCOM satellites and frequency plan. The translation frequency is 2225 MHz. (Reproduced with permission from W. H. Braun and J. E. Keigler, "RCA Satellite Networks: High Technology and Low User Cost," *Proceedings of the IEEE*, **72**, 1483–1505 (November 1984). Copyright © 1984 IEEE.)

Many new satellite systems for the late 1980s and 1990s are being designed for Ku band (14/11 GHz), where many more orbital locations are available. The higher frequencies make possible narrower antenna beams, and better control of coverage patterns can be achieved at Ku band than at C band (4–8 GHz).

Figure 3.11 shows a simplified block diagram of a satellite communication subsystem for the 6/4 GHz band. The 500 MHz bandwidth is divided up into channels, often 36 or 40 MHz wide, which are each handled by a separate transponder. A transponder consists of a band pass filter to select the particular channel's band of frequencies, a downconverter to change the frequency from 6 GHz at the input to 4 GHz at the output, and an output amplifier. The communication system has many transponders, some of which may be spares; typically 12 to 24 active transponders are carried by a high-capacity satellite. The transponders are supplied with signals from one or more receive antennas and send their outputs to a switch matrix that directs each transponder band of frequencies to the appropriate antenna or antenna beam. In a large satellite such as INTELSAT V there may be four or five beams to which any transponder can be connected. The switch setting can be controlled from the ground to allow reallocation of the transponders between the downlink beams as traffic patterns change.

The following sections discuss the transponder and antenna design in more detail.

Transponders

Signals (known as *carriers*) transmitted by an earth station are received at the satellite by either a *zone beam* or a *spot beam* antenna. Zone beams can receive from transmitters anywhere within the coverage zone, whereas spot beams have limited coverage. The received signal is often taken to two low-noise amplifiers and is recombined at their output to provide *redundancy*. If either amplifier fails, the other one can still carry all the traffic. Since all carriers from one antenna must pass through a low-noise amplifier, a failure at that point is catastrophic. Redundancy is provided wherever failure of one component will cause the loss of a significant part of the satellite's communication capacity.

In the early satellites such as INTELSAT I and II, one or two 250-MHz bandwidth transponders were employed. This proved unsatisfactory because of the nonlinearity of the traveling wave tube transmitter used at the output of the transponder, and later satellites have used 12, 20, or 24 transponders each with 36, 40, or 72 MHz bandwidth. The reason for using narrower bandwidth transponders is to avoid excessive intermodulation problems when transmitting several carriers simultaneously with a nonlinear transmitter, as discussed in Chapter 6.

Intermodulation distortion is likely to occur whenever we use traveling wave tube amplifiers as wideband amplifiers for more than one signal and drive the amplifier into saturation. Since we generally want to have more than one earth station transmitter sending signals via a satellite, one solution would be to provide one transponder for each earth station's signal. In the case of the Intelsat global system, this could result in a requirement for as many as 100 transponders per satellite. As a compromise, 36 MHz has been widely used for transponder bandwidth, with 72 MHz now adopted for some transponders on INTELSAT V satellites.

Many domestic satellites operating in the 6/4 GHz band carry 24 active transponders. The center frequencies of the transponders are spaced 40 MHz apart, to allow guard bands for the 36-MHz-wide filter skirts. With a total of 500 MHz available, a single polarization satellite can accommodate 12 transponders across the band. When frequency reuse by orthogonal polarizations is adopted, 24 transponders can be accommodated in the same 500-MHz bandwidth.

Figure 3.12 shows a simplified diagram of the communication system carried by INTELSAT V satellites. The bulk of the traffic is carried by the 6/4 GHz section, with a total bandwidth of 2000 MHz available by frequency reuse. The switch matrix allows a very large number of variations in connecting the 6-GHz receivers to the 4-GHz transmitters, and also interconnects the 6/4 and 14/11 GHz sections. This provides Intelsat with a great deal of flexibility in setting up links through the satellite.

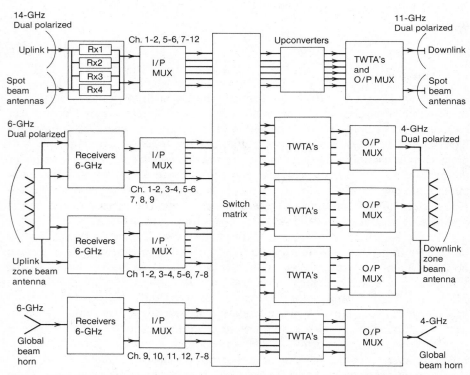

Figure 3.12 Simplified block diagram of an INTELSAT V communication system. Note that the switch matrix allows many possible interconnections between uplink beams and downlink transmitters. (Courtesy C.F. Hoeber, Ford Aerospace and Communications Corporation.)

The choice of transponder bandwidth also depends on the nature of the signals to be carried by the satellite and the multiple-access technique used. These aspects are discussed in some detail in Chapter 6. Digital modulation using time division multiple access (TDMA) allows the transponder to be allocated to only one signal at any instant of time, so nonlinearity in the transponder output amplifier is not important and the output power amplifier can be driven into its nonlinear region, with greatly improved efficiency in converting DC power into RF transmitted power. When analog modulation (FM) is used and more than one signal shares a transponder (frequency division multiple access, $FDMA$), the power amplifier must be run well below its maximum output power to maintain linearity and reduce intermodulation products. The degree to which the transmitter output power is reduced below its peak output is known as "output backoff": in FDMA systems 3 to 7 dB of output *backoff* is used, typically, depending on the number of accesses to the transponder. Backoff results in a lower downlink carrier-to-noise ratio at the earth station, with the result that more data can be transmitted using TDMA than with FDMA when multiple accesses to each transponder are required.

At 4 GHz, transistor amplifiers are now available with power outputs in excess of 20 W, and these will steadily replace $TWTs$ as the output amplifiers for 4-GHz transponders. At 11 GHz, solid-state amplifiers are now being developed with output power up to 45 W [9]. Most 14/11 GHz systems are based on digital modulation and TDMA, so the use of a nonlinear power amplifier presents less of a problem.

Figure 3.13 shows a typical single-conversion transponder of the type used on Intelsat spacecraft for the 6/4 GHz band. The output power amplifier may be a $TWTA$ or a $GaAsFET$ amplifier (often called a solid-state power amplifier, $SSPA$). The local oscillator is at 2225 MHz to provide the appropriate shift in frequency from the 6-GHz uplink frequency to the 4-GHz downlink frequency, and the bandpass filter after the mixer removes unwanted frequencies resulting from the down conversion operation. The attenuator can be controlled via the uplink command system to set the gain of the transponder and provides the necessary control

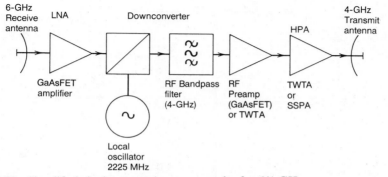

Figure 3.13   Simplified single conversion transponder for 6/4 GHz.

to back off the output TWT. Redundancy is provided for the high-power amplifiers (*HPA*) in each transponder by including a spare TWT or GaAsFET amplifier that can be switched into circuit if the primary power amplifier fails. The lifetime of TWTs is limited, and they represent the least reliable component in most transponders. Providing a spare TWT in each transponder greatly increases the probability that the satellite will reach the end of its working life with all its transponders still operational.

Transponders for use in the 14/11 GHz bands normally employ a double-frequency conversion scheme as illustrated in Figure 3.14. It is easier to make filters, amplifiers, and equalizers at an intermediate frequency (*IF*) such as 1100 MHz than at 14 or 11 GHz, so the incoming 14 GHz carrier is translated to an IF around 1 GHz. The amplification and filtering are performed at 1 GHz and a relatively high-level carrier is translated back to 11 GHz for amplification by the HPA.

Stringent requirements are placed on the filters used in transponders, since they must provide good rejection of unwanted frequencies, such as intermodulation products, and also have very low amplitude and phase ripple in their pass bands. Frequently a filter will be followed by an equalizer that smooths out amplitude and phase variations in the pass band. Phase variation across the pass band produces *group delay distortion*, which is particularly troublesome with wideband FM signals and high-speed phase shift keyed data transmission. The group delay

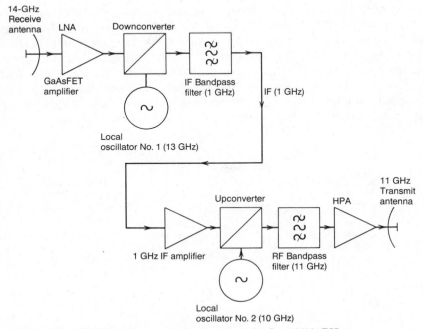

Figure 3.14 Simplified double conversion transponder for 14/11 GHz.

allowable in the transponder depends to some extent on the nature of the signal being carried. In the INTELSAT IV satellites, for example, two transponders were reserved for FM TV transmission and had group delay equalizers specifically designed for TV signals [10].

A considerable increase in the communications capacity of a satellite can be achieved by combining on-board processing with switched-beam technology. A switched-beam satellite generates a narrow transmit beam for each earth station with which it communicates, and then transmits sequentially to each one using time division multiplexing of the signals. The narrow beam has to cover only one earth station allowing the satellite transmit antenna to have a very high gain compared to a zone-coverage antenna. A narrow scanning beam can also be used, or a combination of fixed and scanning beams, as proposed for NASA's ACTS satellite illustrated in Figure 3.15. Unless the spacecraft has a zone-coverage receiver antenna, data storage is required at the satellite since it communicates with only one earth station at a time. The high-gain antennas used in switched-beam systems raise the *EIRP* (effective isotropically radiated power) of the satellite transmitter and thus increase the capacity of the downlink.

The downlink usually limits the capacity of a transponder because of the low transmit power available from the satellite. The uplink is not limited by transmitter power in the same way, because much larger and heavier transmitters can be used, without concern for the power required to operate them. It is possible to conserve uplink bandwidth by using different modulation techniques on the uplink and downlink and by providing a baseband processor on the spacecraft. The processor can provide the data storage needed for a switched-beam system and also can perform error correction on uplink data. An example of an advanced design of transponder for NASA's ACTS satellite incorporating these ideas is shown in Figure 3.16.

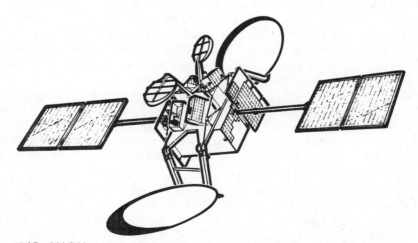

Figure 3.15   NASA's proposed 20/30 GHz ACTS satellite. (Courtesy NASA.)

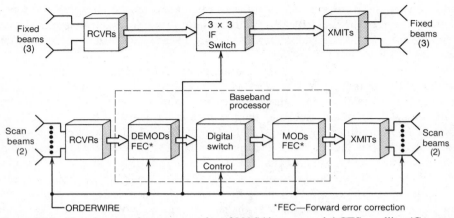

Figure 3.16 Signal processing transponder of NASA's proposed ACTS satellite. (Courtesy of NASA.)

3.7 SPACECRAFT ANTENNAS

Basic Antenna Types and Relationships

Four main types of antennas are used on spacecraft. These are

1. *Wire* antennas: monopoles and dipoles.
2. *Horn* antennas.
3. *Reflector* antennas.
4. *Array* antennas.

Wire antennas are used primarily at VHF and UHF to provide communications for the TT&C systems. They are positioned with great care on the body of the spacecraft in an attempt to provide *omnidirectional* coverage. Most spacecraft measure only a few wavelengths at VHF frequencies, which makes it difficult to get the required antenna patterns, and there tend to be some orientations of the spacecraft in which the sensitivity of the TT&C system is reduced by *nulls* in the antenna pattern.

An *antenna pattern* is a plot of the field strength in the far field of the antenna when the antenna is driven by a transmitter. It is usually measured in *decibels* (dB) below the maximum field strength. The *gain* of an antenna is a measure of the antenna's capability to direct energy in one direction, rather than all around. *Antenna gain* is defined in Chapter 4, Section 4.1. At this point, it will be used with the simple definition given above.

Figure 3.17 shows typical satellite antenna coverage zones. The pattern is frequently specified by its *3-dB beamwidth*, the angle between the directions in which the radiated (or received) field falls to half the power in the direction of maximum field strength. However, a satellite antenna is used to provide coverage of a certain area, or *zone* on the earth's surface, and it is more useful to have contours of

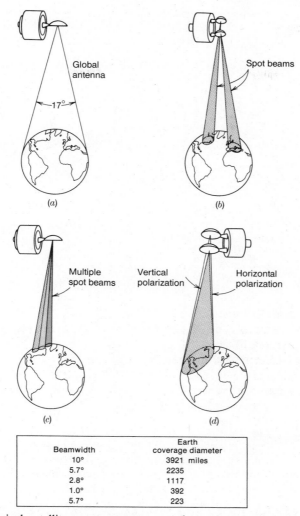

Figure 3.17 Typical satellite antenna patterns and coverage zones (*Source:* Robert M. Gagliardi, *Satellite Communications*, Lifetime Learning Publications, Belmont, CA, 1983, p. 127. Copyright © 1984 Wadsworth, Inc. Reprinted by permission of Van Nostrand Reinhold Co., Inc.)

antenna gain as shown in Figure 3.18. In this particular figure it is the EIRP of the satellite antenna and transmitter that is shown: EIRP is simply antenna gain multiplied by transmitter output power, and is usually expressed in dBW, decibels greater than 1 W.

When computing the signal power received by an earth station from the satellite, it is important to know where the station lies relative to the satellite transmit antenna contour pattern, so that the exact EIRP can be calculated. If the pattern

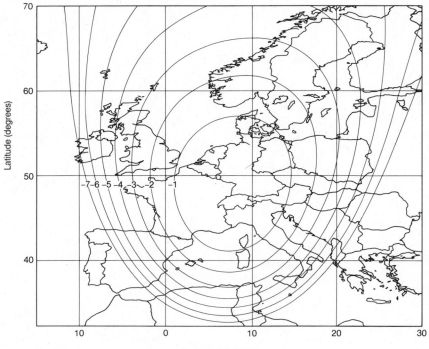

Figure 3.18   Contour plot of the spotbeam of ESA's OTS satellite projected onto the earth. The contours are in 1 dB steps, normalized to 0 dB at the center of the beam. (Courtesy of ESA.)

is not known, it may be possible to estimate the antenna gain in a given direction if the antenna *boresight* or *beam axis* direction and its beamwidth are known.

Horn antennas are used at microwave frequencies when relatively wide beams are required, as for global coverage. A horn is a flared section of waveguide that provides an aperture several wavelengths wide and a good match between the waveguide impedance and free space. Horns are also used as feeds for reflectors, either singly or in clusters. Horns and reflectors are examples of *aperture antennas* that launch a wave into free space from a waveguide. It is difficult to obtain gains much greater than 23 dB or beamwidths narrower than about 10° with horn antennas. For higher gains or narrow beamwidths a reflector antenna or array must be used.

Reflector antennas are usually illuminated by one or more horns and provide a larger aperture than can be achieved with a horn alone. For maximum gain, it is necessary to generate a plane wave in the aperture of the reflector. This is achieved by choosing a reflector profile that has equal path lengths from the feed to the aperture, so that all the energy radiated by the feed and reflected by the reflector reaches the aperture with the same phase angle and creates a uniform phase front. One reflector shape that achieves this with a point source of radiation is the paraboloid, with a feed placed at its focus. The paraboloid is the basic shape for most reflector antennas, although many spacecraft antennas use modified paraboloidal reflector profiles to tailor the beam pattern to a particular coverage zone.

Some basic relationships in aperture antennas can be used to determine the approximate size of a spacecraft antenna for a particular application, as well as the antenna gain. More accurate calculations are needed to determine the exact gain, efficiency, and pattern of a spacecraft antenna, and the interested reader should refer to one of the many excellent texts in this field for details [11, 12, 13].

The following approximate relationships will be used here to illustrate the selection of antennas for a communications satellite.

An aperture antenna has a gain G given by

$$G = \eta \frac{4\pi A}{\lambda^2} \tag{3.1}$$

where A is the area of the antenna aperture, λ is the operating wavelength, and η is the *aperture efficiency* of the antenna. The efficiency η is not easily determined but is typically in the range 50 to 65 percent for reflector antennas with single feeds, lower for antennas with shaped beams. Horn antennas tend to have higher efficiencies than reflector antennas, typically in the range 65 to 80 percent.

The beamwidth of an antenna is related to the aperture dimension in the plane in which the pattern is measured. A useful rule of thumb is that the 3-dB beamwidth in a given plane for antennas with dimension D in that plane is

$$\theta_{3dB} \simeq \frac{75\lambda}{D} \text{ degrees} \tag{3.2}$$

where θ_{3dB} is the beamwidth between half power points on the antenna pattern and D/λ is the aperture dimension in wavelengths. The beamwidth of a horn antenna may depart from Eq. (3.2) quite radically. A small rectangular horn will produce a narrower beam than suggested by Eq. (3.2) in its E plane and a wider beamwidth in the H plane.

Since both Eqs. (3.1) and (3.2) contain antenna dimension parameters, the gain and beamwidth of an aperture antenna are related. For antennas with $\eta = 55$ percent, the gain is approximately

$$G \simeq \frac{30,000}{(\theta_{3dB})^2} \tag{3.3}$$

where θ_{3dB} is in degrees. If the beam has different beamwidths in orthogonal planes, $(\theta_{3dB})^2$ should be replaced by the product of the two 3-dB beamwidths.

Example 3.7.1

The earth subtends an angle of 17° when viewed from geostationary orbit. What are the dimensions and gain of a horn antenna that will provide global coverage at 4 GHz?

If we design our horn to give a circularly symmetric beam with a 3-dB beamwidth of 17°, using Eq. (3.2)

$$D/\lambda = 75/\theta_{3dB} = 4.4$$

At 4 GHz, $\lambda = 0.075$ m, so $D = 0.33$ m (just over 1 ft). If we use a circular horn excited in the TE_{11} mode, the beamwidths in the E and H planes will not be equal and we may be forced to make the aperture slightly smaller to guarantee coverage in the E plane. A *corrugated horn* designed to support the HE hybrid mode has a circularly symmetric beam and could be used in this application. Waveguide horns are generally used for global beam coverage. Reflector antennas are not efficient when the aperture diameter is less than 8λ.

Using Eq. (3.3), the gain of the horn is approximately 100, or 20 dB, at the center of the beam. However, in designing our communication system we will have to use the edge-of-beam gain figure of 17 dB, since those earth stations close to the earth's horizon, as viewed from the satellite, are close to the 3-dB contour of the transmitted beam.

Example 3.7.2

The continental United States (48 contiguous states) subtends an angle of approximately $6° \times 3°$ when viewed from geostationary orbit. What dimension must a reflector antenna have to illuminate half this area with a circular beam 3° in diameter at 11 GHz? Can a reflector be used to produce a $6° \times 3°$ beam? What gain would the antenna have?

Using Eq. (3.2), we have for a 3° circular beam

$$D/\lambda = 75/3 = 25$$

and with $\lambda = 0.0272$ m, $D = 0.68$ m (just over 2 ft). The gain of this antenna, from Eq. (3.3) is approximately 35 dB.

To generate a beam with different beamwidths in orthogonal planes we need an aperture with different dimensions in the two planes. In this case, a rectangular aperture $25\lambda \times 12.5\lambda$ would generate a beam $3° \times 6°$, and would have a gain of 32 dB, approximately. In order to illuminate such a reflector, a horn with unequal beamwidths is required, since the reflector must intercept most of the radiation from the feed if it is to have an acceptable efficiency. Rectangular, or more commonly elliptical, outline reflectors are used to generate unequal beamwidths, but have poorer polarization characteristics. When orthogonal polarizations are to be transmitted or received, it is better to use a circular reflector with a distorted profile to broaden the beam in one plane, or a feed cluster to provide the appropriate amplitude and phase distribution across the reflector.

Spacecraft Antennas in Practice

The antennas of a communications satellite are often a limiting element in the complete system. In an ideal spacecraft, there would be one antenna beam for each earth station, completely isolated from all other beams, for transmit and receive. However, if two earth stations are 300 km apart on the earth's surface and the satellite is in geostationary orbit, their angular separation at the spacecraft is 0.5°.

For θ_{3dB} to be 0.5°, D/λ must be 150, which requires an aperture diameter of 11.3 m at 4 GHz. Although antennas this large have been flown on spacecraft (ATS-6 deployed a 2.5 GHz, 10-m diameter antenna), they are not yet in regular use.

To provide a separate beam for each earth station would also require one antenna feed per earth station if a multiple-feed antenna with a single reflector were used. A compromise between one beam per station and one beam for all stations has been used in many satellites by using zone-coverage beams and orthogonal polarizations within the same beam to provide more channels per satellite. Figure 3.3*b* shows an INTELSAT V spacecraft that has four reflector antennas. Each reflector is illuminated by a complex feed that provides the required beam shape to permit communication between earth stations within a given coverage zone. Figure 3.19 shows the coverage zones provided by INTELSAT V. The largest reflector on the spacecraft transmits at 4 GHz and produces the "peanut" shaped patterns for the zone beams, which are designed to concentrate the transmitted energy onto densely populated areas such as North America and Western Europe where much telecommunications traffic is generated. The smaller antennas are used to provide hemisphere transmit and receive beams, and the 14/11 GHz spot beams. In addition, there are horn antennas providing global beam coverage.

The antennas on INTELSAT V are highly complex and provide seven independent beams, with four-fold frequency reuse in the 6/4 GHz band. This gives

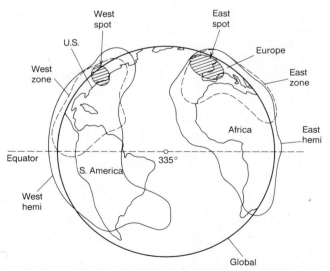

Figure 3.19 Beam coverage patterns for INTELSAT V at 335° (over the Atlantic Ocean). (Courtesy C.F. Hoeber, Ford Aerospace and Communications Corporation.)

an available bandwidth of 2000 MHz, with a further 250 MHz provided by the 14/11 GHz system. Most domestic satellites do not have such complex antenna systems, but nevertheless use orthogonal polarization frequency reuse to double the effective bandwidth at 6/4 GHz. Figure 3.20 shows an example of a domestic satellite with beams tailored to provide communication within the 48 contiguous states of the United States, with subsidy coverage to Hawaii and Alaska. This spacecraft is the satellite of GTE Spacenet, which provides video, voice, and data communication services to customers within the United States.

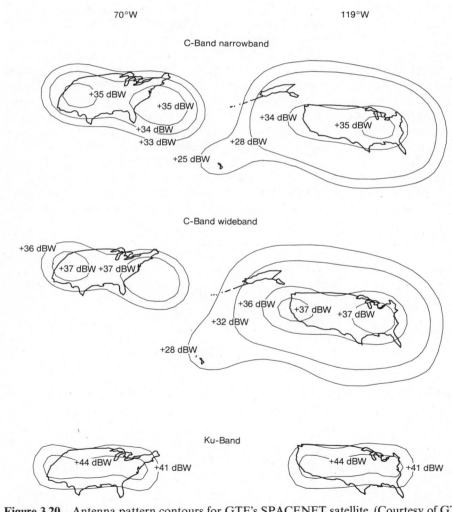

Figure 3.20   Antenna pattern contours for GTE's SPACENET satellite. (Courtesy of GTE Spacenet Corp. All rights reserved.)

Countries such as the United States create an enormous demand for communication services, and a number of domestic satellite communication systems have been established to meet that demand. In 1984, the geostationary orbit had domestic satellites spaced every 3°, operating at 6/4 GHz, from longitude 60° W to longitude 140° W. This encompasses all orbital locations that can be simultaneously viewed by earth stations in the United States and Canada, and each operator has been given a limited number of orbital "slots" in which to place a spacecraft. As a result, there is a great deal of pressure on the operating companies to obtain the maximum number of channels per spacecraft in order to give the operator the greatest possible revenue-earning capacity. This has encouraged the development of *frequency reuse* antennas by means of orthogonal polarizations, the combination of 6/4 GHz and 14/11 GHz communication systems on one spacecraft, and the use of digital modulation and TDMA to increase capacity. However, video channels, which are used to distribute program material to cable TV companies throughout the United States, use FM and typically need a full 36-MHz transponder to carry one video channel so that the cable TV companies can employ relatively small, low-cost earth station antennas.

Frequency Reuse Antennas

There are two main types of frequency reuse: *spatial beam separation* and *orthogonal polarization.*

In spatial beam frequency reuse, the same frequency bands are used for transmission and reception to geographically separated regions of the earth's surface. The positioning of each satellite in the Intelsat global system has been selected to provide an east and a west zone by locating each spacecraft over an ocean. The angular separation of the east and west zones is about 10°, and the coverage zones are approximately 5° wide. By generating very carefully shaped beams using large clustered feeds, it is possible to reduce the sidelobes from one beam to an acceptable level in the other zone. For FM transmission, it is necessary to keep interference between beams down to about −25 dB. With digital transmission systems, higher interference levels can be tolerated, perhaps up to −17 dB.

An example of the degree to which spacecraft antenna beams can be shaped to provide specific coverage areas is shown in Figure 3.21 for a satellite antenna developed by RCA Astro-Electronics [8]. The antenna is designed to be reconfigured in orbit so that the beams can be tailored to a specific mission. The antenna uses a reflector and 13 horns, arranged in groups and fed by 15 variable-power dividers. The transmitter power is split between the feed horns according to the variable power-divider settings, which can be controlled from earth by telecommand. This allows a single design of antenna to be fitted to spacecraft that will operate at different longitudes and serve different geographical areas.

A logical extension of the shaped zone beam, which aims to direct as much transmit power as possible into its coverage zone, is the switched-beam concept discussed in Section 3.6. By combining a switched-beam antenna with TDMA of

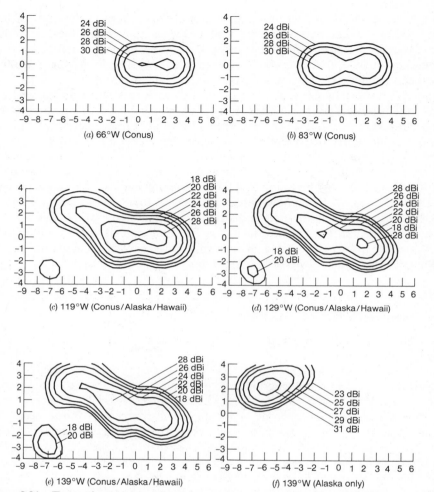

Figure 3.21 Transmit-antenna gain pattern for the reconfigurable antenna system. This sample of possible antenna patterns shows how the shaped-beam coverages can be changed as a function of orbital longitude and specified service area. (Reproduced with permission from J. F. Balcewicz, "In-orbit Reconfigurable Communications Satellite Antennas," in *RCA Engineer*, **28**, p. 40 (1983).)

the communications channel, much higher transmit EIRP can be achieved. Figure 3.22 shows one conceptual design of an offset *Cassegrain antenna* with multiple fixed beams using a waveguide beam-forming matrix.

In orthogonal polarization frequency reuse, each beam is generated in two polarizations. The INTELSAT 6/4 GHz system uses circular polarization and INTELSAT V uses left-hand circular polarization (*LHCP*) for each of the "hemi" beams shown in Figure 3.19. The two zone beams each use right-hand circular polarization (*RHCP*) and overlay the hemi beams. This places a very stringent requirement on the antenna, which must maintain an isolation between the two polarizations of at least 25 dB throughout the coverage zones of both beams. A single antenna is used for the transmit beams at 4 GHz, and a second antenna for reception at 6 GHz.

Reflector antennas tend to generate cross-polarized beams due to cross-polarized radiation from the feed system and curvature of the reflector. Careful matching of the feed radiation characteristics to the curvature of the reflector, which is usually offset to avoid blockage by the large feed cluster, can achieve cross-polar isolations between beams of the required 25 dB. The development of these multiple-beam dual polarization antennas was the result of a great deal of work by microwave antenna specialists throughout the 1970s. Reference 12 is a good survey of the problems and solutions in designing such antennas for use on spacecraft and at earth stations.

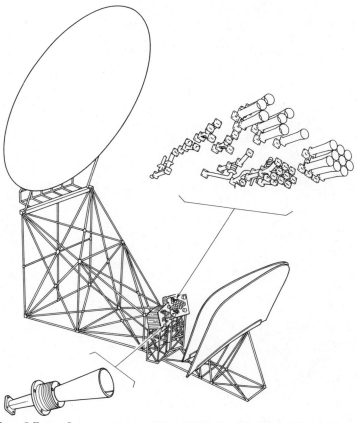

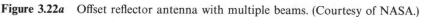

Figure 3.22a   Offset reflector antenna with multiple beams. (Courtesy of NASA.)

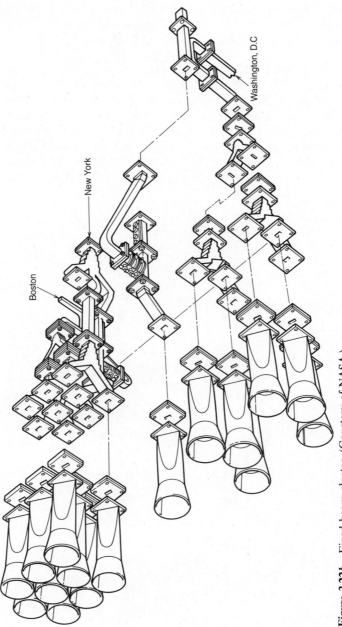

Figure 3.22b   Fixed beam cluster. (Courtesy of NASA.)

The requirements of narrow antenna beams with high gain over a small coverage zone leads to large antenna structures on the spacecraft. Frequently, the antennas are too large to fit within the shroud dimensions of the launch vehicle, and must be folded down during the launch phase. Once in orbit, the antennas can be deployed. Figure 3.23 shows the deployment sequence used for the 30-ft

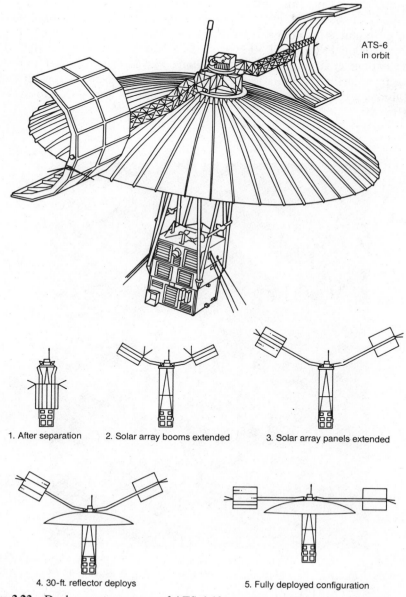

ATS-6
in orbit

1. After separation   2. Solar array booms extended   3. Solar array panels extended

4. 30-ft. reflector deploys   5. Fully deployed configuration

Figure 3.23   Deployment sequence of ATS-6 10-m antenna. (Courtesy of NASA.)

antenna carried by ATS-6: the antenna was built as a series of petals that folded over each other to make a compact unit during launch. The solar sails folded down over the antenna, and were deployed first. Springs or pyrotechnic devices can be used to provide the energy for deployment of antennas or solar sails, with a locking device to assist correct positioning after deployment.

In many larger satellites, the antennas use offset paraboloidal reflectors with clusters of feeds to provide carefully controlled beam shapes. The feeds mount on the body of the spacecraft, close to the communications subsystem, and the reflector is mounted on a hinged arm. Figure 3.24 shows an example of this design of antenna for the INTELSAT VI satellite. For launch, the reflector folds down to provide a compact structure; in orbit, the hinged arms are swung out and locked in place to hold the reflectors in the correct position. When the spacecraft is in geostationary orbit it is weightless, so very little energy is required to move the large reflector.

Figure 3.24 INTELSAT VI on station. (Courtesy of Hughes Aircraft Company, Space and Communications Group, Public Relations Department, P. O. Box 92919, Los Angeles, CA. 90009.)

3.8 EQUIPMENT RELIABILITY AND SPACE QUALIFICATION

Communications satellites built in the 1970s and 1980s have provided operational lifetimes of up to 10 years. Once a satellite is in geostationary orbit, there is little possibility of repairing components that fail or adding more fuel for station-keeping. The components that make up the spacecraft must therefore have very high reliability in the hostile environment of outer space, and a strategy must be devised that allows some components to fail without causing the entire communication capacity of the satellite to be lost. Two separate approaches are used: *space qualification* of every part of the spacecraft to ensure that it has a long life expectancy in orbit and *redundancy* of the most critical components to provide continued operation when one component fails.

Space Qualification

Outer space, at geostationary orbit distances, is a harsh environment. There is a total vacuum and the sun irradiates the spacecraft with 1.4 kW of heat and light on each square meter of exposed surface. Where surfaces are in shadow, heat is lost to the infinite sink of space, and surface temperature will fall toward absolute zero. Electronic equipment cannot operate at such extremes of temperature and must be housed within the spacecraft and heated or cooled so that its temperature stays within the range 0°C to 75°C. This requires a thermal control system that manages heat flow throughout the spacecraft as the sun moves around once every 24 hours.

The first stage in ensuring high reliability in a spacecraft is by selection and screening of every component used. Past operational and test experience of components indicates which components can be expected to have good reliability. Only components that have been shown to have high reliability under outer space conditions will be selected. Each component is then tested individually (or as a subsystem) to ensure that it meets its specification. This process is known as *quality control* or quality assurance and is vital in building any equipment that is to be reliable. Once individual components and subsystems have been space qualified, the complete spacecraft must be tested as a system to ensure that its many systems are reliable.

When a spacecraft is designed, three prototype models are often built and tested. The *mechanical model* contains all the structural and mechanical parts that will be included in the spacecraft and is tested to ensure that all moving parts operate correctly in a vacuum, over a wide temperature range. It is also subjected to vibration and shock testing to simulate vibration levels and G forces likely to be encountered on launch. The *thermal model* contains all the electronics packages and other components that must be maintained at the correct temperature. It is illuminated by powerful heaters to simulate the sun and is tested for proper temperature control. The antennas are usually included on the thermal model to check for distortion of reflectors and displacement or bending of support structures. In orbit, an antenna may cycle in temperature from above 100°C to below −100°C as the sun moves around the spacecraft. The *electrical model* contains all the electronic parts of the spacecraft and is tested for correct electrical performance under total vacuum and a wide range of temperatures. The antennas of the electrical model must provide the correct beamwidth, gain, and polarization properties.

Testing carried out on the prototype models is designed to overstress the system and induce failure in any weak components: temperature cycling will be carried out to 10° beyond expected extremes; structural loads and G forces 50 percent above those expected in flight may be applied. Electrical equipment will be subjected to excess voltage and current drain to test for good electronic and thermal reliability. The prototype models used in these tests will not usually be flown. A separate flight model (or several models) will be built and subjected to the same tests as the prototype, but without the extremes of temperature, stress, or voltage. Preflight testing of flight models, while exhaustive, is designed more to cause failure of parts, rather than to check that they will operate under worst-case conditions.

Many of the electronic and mechanical components are known to have limited lifetimes, or a finite probability of failure. If failure of one of these components will jeopardize the mission or reduce the communication capacity of the satellite, a backup, or redundant, unit will be provided. The design of the system must be such that when one unit fails, the backup can automatically take over or be switched into operation by command from the ground. For example, redundancy is always provided for traveling wave tube amplifiers in the transponders of a communications satellite, as these are known to have a limited lifetime.

The success of the testing and space qualification procedures used by NASA has been well illustrated by the lifetime achieved by many of its scientific satellites. Spacecraft designed for a specific mission lasting one or two years have frequently operated successfully for two or three times that period. Sufficient reliability was designed into the spacecraft to guarantee the mission lifetime such that the actual lifetime has been much greater. In the next section we will look at how reliability can be quantified.

Reliability

We need to be able to calculate the reliability of a spacecraft subsystem for two reasons: we want to know what the probability is that the subsystem will still be working after a given time period, and we need to provide redundant components or subsystems where the probability of a failure is too great to be accepted. The owner of a satellite used for communications expects to be able to use a predetermined percentage of its communications capacity for a given length of time. Amortization of purchase and launch costs will be calculated on the basis of an expected *lifetime*. The manufacturers of satellites must provide their customers with predictions (or guarantees) of the reliability of the satellite and subsystems: to do this requires the use of *reliability theory*. Reliability theory is a mathematical attempt to predict the future and is therefore less certain than other mathematical techniques that operate in absolute terms. However, the application of reliability theory has enabled spacecraft engineers to build satellites that perform as expected, at acceptable construction costs. It should be noted, however, that the cost of a spacecraft is very high compared to other equipment with a comparable number of components: an INTELSAT V spacecraft costs around \$45M, close to the cost of a Boeing 747 jet airliner. The cost is acceptable because of the high revenue-earning capability of the satellite.

The reliability of a component can be expressed in terms of the probability of failure after time t, $P_F(t)$. For most electronic equipment, probability of failure is higher at the beginning of life—the *burn-in* period—than at some later time. As the component ages, failure becomes more likely, leading to the "bathtub" curve shown in Figure 3.25. This is a familiar trend with automobiles. A new car may have defects when it is delivered, and errors in manufacturing may lead to components failing soon after purchase. Once these defects have been overcome, by repair or replacement, reliability improves for a number of years until mechanical parts start to wear and failures occur. In an automobile, preventive maintenance can be carried out to replace parts that are known to wear most quickly. For example, spark plugs may be replaced every 10,000 miles. Preventive maintenance is not possible with a geostationary satellite, at present, so other strategies must be adopted.

Components for spacecraft are selected only after extensive testing. The aim of the testing is to determine reliability, causes of failure, and expected lifetime. The result is a plot similar to Figure 3.25. Testing is carried out under rigorous conditions, representing the worst operating conditions likely to be encountered in space, and may be designed to accelerate failure in order to shorten the testing duration needed to determine reliability. Units that are exposed to the vacuum of space are tested in a vacuum chamber, and components subjected to sunlight are tested under equivalent radiant heat conditions.

The initial period of reduced reliability can be eliminated by a burn-in period before a component is installed in the spacecraft. Semiconductors and integrated circuits that are required to have high reliability are subjected to burn-in periods from 100 to 1000 hours, often at a high temperature and excess voltages to induce failures in any suspect devices and to get beyond the initial low reliability part of the bathtub curve.

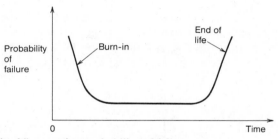

Figure 3.25   "Bathtub" curve for probability of failure.

The reliability of a device or subsystem is defined as

$$R(t) = \frac{N_s(t)}{N_0} = \frac{\text{Number of surviving components at time } t}{\text{Number of components at start of test period}} \qquad (3.4)$$

The numbers of components that failed in time t is $N_f(t)$ where

$$N_f(t) = N_0 - N_s(t) \qquad (3.5)$$

From the engineering viewpoint, what we need to know is the probability of any one of the N_0 components failing: this is related to the mean time before failure ($MTBF$). Suppose we continue testing devices until all of them fail. The ith device fails after time t_i where

$$\text{MTBF} = m = \frac{1}{N_0} \sum_{t=1}^{N_0} t_i \qquad (3.6)$$

The *average* failure rate λ, is the reciprocal of the MTBF, m. If we assume that λ is a constant, then

$$\lambda = \frac{\text{Number of failures in a given time}}{\text{Number of surviving components}}$$

$$\frac{1}{N_s} \frac{\Delta N_f}{\Delta t} = \frac{1}{N_s} \frac{dN_f}{dt} = \frac{1}{\text{MTBF}} \qquad (3.7)$$

Failure rate λ is often given as the average failure rate per 10^9 hours. The rate of failure, dN_f/dt, is the negative of the rate of survival dN_s/dt, so we can redefine λ as

$$\lambda = \frac{-1}{N_s} \frac{dN_s}{dt} \qquad (3.8)$$

By definition from Eq. (3.4), the reliability R is N_s/N_0, so

$$\lambda = \frac{-1}{N_0 R} \frac{d}{dt} (N_0 R) = \frac{-1}{R} \frac{dR}{dt} \qquad (3.9)$$

A solution of Eq. (3.9) for R is

$$R = e^{-\lambda t} \qquad (3.10)$$

Thus the reliability of a device decreases exponentially with time, with zero reliability after infinite time, that is, certain failure. However, end of useful life is usually taken to be the time t_l at which R falls to 0.37 ($1/e$), which is when

$$t_l = \frac{1}{\lambda} = m \qquad (3.11)$$

The probability of a device failing, therefore, has an exponential relationship to the MTBF and is represented by the right-hand end of the bathtub curve.

Redundancy

The equations in the preceding section allow us to calculate the reliability of a given device when we know its MTBF. In a spacecraft, many devices, each with a different MTBF, are used, and failure of one device may cause catastrophic failure of a complete subsystem. If we incorporate redundant devices, the subsystem can continue to function correctly. We can define three different situations for which we want to compute subsystem reliability: series connection, parallel connection, and switched connection. These are illustrated in Figure 3.26; also shown is a hybrid arrangement, a series/parallel connection, widely used in electronic equipment.

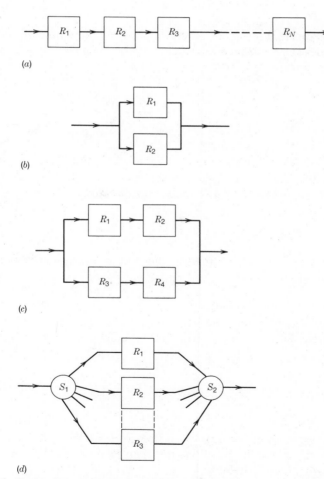

Figure 3.26 Redundancy connections. (*a*) Series connection. (*b*) Parallel connection. (*c*) Series/parallel connection. (*d*) Switched connection.

Series reliability is defined as the reliability of a subsystem made up of a number of units in series, any one of which will cause failure of the entire subsystem. In Figure 3.26a, the reliability of the individual devices is given as $R_1, R_2, \ldots, R_N$. Then the system reliability R is

$$R = \prod_{i=1}^{N} R_i = R_1 R_2 R_3 \ldots R_N \tag{3.12}$$

Clearly, such an arrangement is less reliable than the least reliable of its component devices.

If all the devices have the same reliability R_c, the system reliability R for N devices is

$$R = R_c^N \tag{3.13}$$

The situation is different if the devices have multiple failure modes. Suppose that two failure modes can be defined for the devices, an open circuit, with probability Q_{oi}, and a short circuit, with probability Q_{si}. Failure of the whole system occurs if one device goes open circuit or all devices go short circuit, giving a system reliability

$$R_s = \prod_{i=1}^{N} (1 - Q_{oi}) - \prod_{i=1}^{N} Q_{si} \tag{3.14}$$

If we connect several devices in parallel, as in Figure 3.26b, any one device can provide correct operation. Failure of the system occurs when all devices are open circuit, or one device is short circuit. Thus *parallel* reliability is given by

$$R_p = \prod_{i=1}^{N} (1 - Q_{si}) - \prod_{i=1}^{N} Q_{oi} \tag{3.15}$$

If we switch devices out of circuit as they fail, as illustrated in Figure 3.26d, the system reliability is

$$R_{SW} = R_{SD}(1 - P_{fN}) \tag{3.16}$$

where R_{SD} is the reliability of the switching device and P_{fN} is the probability that all N devices have failed, since the switch and the devices are in series.

$$P_{fN} = (P_i)^N = (1 - R_i)^N \tag{3.17}$$

assuming each device has the same reliability R_i; hence

$$R = R_{SD}(1 - (P_i)^N) \tag{3.18}$$

Note that the system is no more reliable than the switching device. The switch must clearly have a reliability much greater than R_i if switched redundancy is to provide increased system reliability.

The series/parallel connection shown in Figure 3.26c is widely used in communications satellite transponders to provide redundancy of the high-power amplifier.

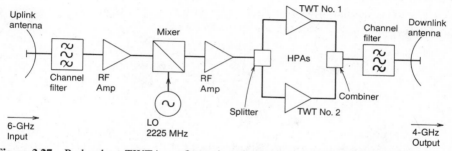

Figure 3.27 Redundant TWTA configurations in HPA of a 6/4 GHz transponder.

High-power amplifiers are much less reliable than low-power devices and are often traveling wave tube amplifiers (TWTA) in transponders. The TWTA is a thermionic device, with a heated cathode and high-voltage power supplies. In common with other thermionic devices such as cathode ray tubes and magnetrons, they have a relatively short MTBF. Although the MTBF may be 50,000 hours, this is the period after which 50 percent of such devices will have failed, on average. The parallel connection of two TWTs, as shown in Figure 3.27, raises the reliability of the amplifier stage to 0.60 at the MTBF period, assuming zero probability of a short circuit. A lifetime of 50,000 hours is approximately six years of continuous operation, which is close to the typical design lifetime of a satellite. To further improve the reliability of the transponders, a second redundant transponder may be provided with switching between the two systems. Note that a combination of parallel and switched redundancy is used to combat failures that are catastrophic to one transponder channel and to the complete communication system.

Example 3.8.1
The solar cells on a spacecraft are connected in four arrays with the series/parallel arrangement shown in Figure 3.26c.

Each of the four arrays has 100 solar cells, with the following failure rates for a group of 100 cells:

10 open circuit failures in 10^9 hours

8 short circuit failures in 10^9 hours

(i) Calculate the probability that one solar cell goes (a) open circuit, and (b) short circuit in a one-year period.

(ii) Calculate the reliability of an array of 100 solar cells in parallel, over a one-year period. Assume that the whole array fails if one cell goes short circuit or if all cells go open circuit.

(iii) Calculate the reliability of the entire system of four arrays when connected in the series/parallel arrangement of Figure 3.26c, for a one-year period.

(i) The reliability of one cell is given by Eq. (3.10), and λ can be found from Eq. (3.7). For case (a), an open circuit cell,

$$\lambda_{o/c} = \frac{1}{N_s}\frac{dN_f}{dt} = \frac{1}{90}\frac{10}{10^9}$$

$$R_{o/c} = \exp(-\lambda_{o/c}t) = \exp\left(\frac{-8760}{(90 \times 10^8)}\right) = 0.99999903$$

Hence probability of an open circuit cell within a year is $1 - R_{o/c}$, or 9.73×10^{-7}. For case (b), the short circuit cell,

$$R_{s/c} = \exp(-\lambda_{s/c}t) = \exp\left(\frac{-8760 \times 8}{(92 \times 10^9)}\right) = 0.99999924$$

Hence probability of a short circuit cell within a year is 7.6×10^{-7}.

(ii) The likelihood of all 100 cells going open circuit is very small compared to the probability that one cell goes short circuit. Thus the probability of failure of the whole array is approximately

$$P_{\text{array}} = 100 \times P_{s/c} = 100 \times 7.6 \times 10^{-7} = 7.6 \times 10^{-5}$$

Hence the reliability of the whole array is approximately $1 - P_{\text{array}}$ or 0.999924 over one year.

(iii) In the series/parallel arrangement of Figure 3.26c, we will suffer a total system failure if one array in each arm goes open circuit at the same time, or if both arrays in either arm go short circuit at the same time. The likelihood of two short circuits is much greater than that of 100 open circuits, so the probability of a total failure is approximately

$$P_{\text{system}} = 2 \times (P_{\text{array}})^2 = 5.78 \times 10^{-9} \times 2$$

and thus the reliability of the system over a one-year period is

$$R_{\text{system}} = 1 - P_{\text{system}} = (1 - 2 \times 5.78 \times 10^{-9}) = 0.9999999884.$$

3.9 SUMMARY

Spacecraft that carry communication relays must provide a stable platform in geostationary orbit for a period of 5 to 10 years.

The spacecraft must carry a number of subsystems to support its communications mission. The attitude and orbital control system keeps the satellite in the correct orbit and on station, and pointing in the correct direction. The telemetry, tracking, and command system allows an earth station to control the subsystems on the spacecraft and to monitor their health. The power system provides the electrical energy needed to run the spacecraft (housekeeping) and the communications system. Solar cells generate the electrical power, a power condition unit controls its distribution, and batteries provide power during launch and eclipses.

The communication system has two major elements, the antennas and the transponders. The spacecraft antennas generate beams with carefully designed patterns to cover specific parts of the earth's surface. The transponders, typically 12 or 24 on one spacecraft, receive the signals from the uplink, amplify them, and after frequency conversion transmit the signals to the downlink. Satellites using the 6/4 GHz band often employ frequency reuse, either by spatial beams or orthogonal polarizations or both. Frequency reuse allows the same RF frequencies to be used for a second time to send different data from separate earth stations.

Reliability is an important consideration in the design and construction of communications satellites. The spacecraft must be built from components that have been thoroughly tested and space qualified. Where components such as TWTAs are known to have a limited life, redundancy may be used to improve the reliability of the transponder. Redundancy is also used for any component whose failure would lead to failure of the complete spacecraft.

REFERENCES

1. J. Martin, *Communication Satellite Systems*, Prentice Hall, Englewood Cliffs, NJ, 1978.
2. IEEE Transactions on Aerospace and Electronic Systems, AS-1 through AS-22, 1963–1984, Institute of Electronic and Electrical Engineers, New York.
3. "Astronautics and Aeronautics," *Journal of the American Institute of Aeronautics and Astronautics*, vol. 1 through 22, 1963–1984, AIAA, 1633 Broadway, NY.
4. J. Alper and J. N. Pelton, Eds., "The Intelsat Global Satellite System," *Progress in Astronautics and Aeronautics*, vol. 93, The American Institute of Aeronautics and Astronautics, NY, 1984.
5. J. J. Spilker, *Digital Communications by Satellite*, Prentice-Hall, Engelwood Cliffs, NJ, 1977.
6. J. Keigler, W. Lindorter and L. Muhlfelder, "Stabile Attitude Control for Synchronous Communications Satellites," in *Progress in Astronautics and Aeronautics*, **33**, MIT Press, Cambridge, MA, 1974.
7. G. E. Schmidt, Jr., "Magnetic Attitude Control for Geosynchronous Spacecraft," *Proceedings of the AIAA Communications Satellite Systems Conference*, San Diego, CA, April 1978, pp. 110–112.
8. Q. H. Braun and J. E. Keigler, "RCA Satellite Networks: High Technology and Low User Cost," *Proceedings of the IEEE*, **72**, 1483–1505 (November 1984).
9. H. J. Wolkenstein and J. N. LaPadre, "Solid State Power Amplifiers Replacing TWTs in C-Band Satellites," *RCA Engineer*, **27**, 7 (October/November 1982).
10. J. Dicks and M. Brown, "INTELSAT IVA Transmission System Design," *COMSAT Technical Review*, **5**, 73–103 (Spring 1975).
11. W. L. Stutzman and G. A. Thiele, *Antenna Theory and Design*, John Wiley & Sons, NY, 1981.

12. A. W. Rudge, K. Milne, A. D. Olver and P. Knight, *Handbook of Antenna Design, Vol. 1*, IEE Electromagnetic Wave Series No. 15, Peter Perigrinus Ltd., Stevenage, Herts, UK, 1983.

13. S. Silver, Ed., *Microwave Antenna Theory and Design*, originally published as Vol. 12 of the MIT Radiation Laboratory Series, 1949. (Republished by Peter Perigrinus Ltd., Stevenage, Herts, UK, 1984.)

PROBLEMS

1. The telemetry system of a communications satellite samples 100 sensors on the spacecraft in sequence. Each sample is transmitted to earth as an eight-bit word in a TDM frame. An additional 200 bits are added to the frame for synchronization and status information. The data are then transmitted at a rate of 1 kilobit per second using FSK modulation of a low-power carrier.

    a. How long does it take to send a complete set of samples to earth from the satellite?

    b. Including the propagation delay, what is the longest time the earth station operator must wait between a change in a parameter occurring at the spacecraft and the new value of that parameter being received via the telemetry link? (Assume a path length of 40,000 km.)

2. The INTELSAT IV-A generation of spacecraft used solar cells wrapped round a cylindrical drum 2.38 m in diameter, with a height of approximately 2.92 m. The drum was rotated at about 60 rpm to spin-stabilize the satellite. At the end of life, the solar cells were designed to deliver 708 W of electrical power.

    a. Calculate the efficiency of the solar cells at end of life. Assume an incident solar power of 1.39 kW/m^2.

    b. If the cells degraded by 15 percent over the lifetime of the satellite, what was the beginning-of-life output power?

    c. How much electrical energy was dumped into load resistors at beginning of life if a margin of 80 W was provided at end of life?

    d. If the spinner design had been replaced by solar sails that rotated to face the sun at all times, what area of solar sails would have been needed? Assume that cells on solar sails generate 10 percent less power than those on a spinner due to their higher operating temperature.

3. Batteries make up a significant part of the in-orbit weight of a communications satellite but are needed to keep the communications system operating during eclipses. This question looks at some trade-offs in battery capacity. A particular satellite carries batteries that weigh 65 kg of a total in-orbit weight of 1130 kg at start of life. The communication system weighs 167 kg, and the station-keeping fuel and propellant 189 kg. The solar array and power system weigh 77 kg.

The satellite requires 140 W of housekeeping power at all times, and the communication system draws 875 W. During an eclipse of 70 min, the batteries discharge to 50 percent of their capacity.

a. If the power system operates at 50 V, what is the capacity of the batteries in ampere-hours? (One ampere-hour means the battery supplies a current of one ampere for one hour.)

b. If half of the communications load is shut down during eclipse, what reduction in battery weight can be achieved? If we shut down all the communication system in an eclipse, how much battery weight is needed?

c. If the battery weight saved in part (b) is replaced by stationkeeping fuel, what extra lifetime on-station can be provided if the original 189 kg was designed to last 5 years?

d. Nickel hydrogen batteries can be discharged to 70 percent of their capacity without damage. What extension of on-station lifetime can be achieved by using these batteries, if the full communications load is kept on during eclipse?

4. A satellite is designed to be launched into geostationary orbit with a total weight of 1500 kg on station. Fuel weighing 150 kg is provided to permit $\pm 0.1°$ N–S and E–W stationkeeping for a period of seven years. The satellite costs $35M to build, $15M to launch, and $5M per year to operate. Expected revenue is $25M per year.

a. If the weight of the spacecraft hardware can be reduced by 10 kg and extra stationkeeping fuel of 10 kg substituted, how much longer can the satellite stay on station, and how much extra profit can it earn? (Assume that equal weight of fuel is expended each year for stationkeeping.)

b. The satellite system operator pays 15 percent interest each year on the capital borrowed to buy and launch the satellite. Approximately how many years does it take to break even on the venture?

c. The weight of fuel on the spacecraft can be increased to 200 kg if a larger launch vehicle is used at a cost of $20M. After seven years in orbit, assume that the revenue-earning capacity of the satellite falls to $10M per year due to transponder failures. Is it economic to add the extra fuel?

d. If the satellite is launched with 1500-kg weight (on station) and 150 kg of fuel, and 20 kg of fuel is left at the end of seven years, estimate the total profit to the satellite operator over its eventual lifetime. (Assume revenue earning drops to $10M after seven years in orbit.)

5. Some proposals for direct broadcast satellites anticipate spacecraft prime powers as high as 4000 W (end of life). Estimate the area required for a solar array to generate this power if the satellite is (a) a spinner and (b) body-stabilized with solar sails. Would the spinner fit into any of the launch vehicles described in Chapter 2?

6. An INTELSAT V satellite communications subsystem is configured according to Table P.6, for the downlink. Fourfold frequency reuse is employed for part of the 4 GHz downlink by having four wideband transponders connected to four antenna beams, all sharing the same frequencies. Twofold frequency reuse is employed at 11 GHz by two spot beams.

a. Calculate the total bandwidth used by the downlink.
b. Assuming that each transponder could be fully loaded with the number of channels shown in Table P.6, calculate the total number of channels carried by the downlink.
c. How many simultaneous telephone conversations can be in progress through the satellite when it is (i) fully loaded, and (ii) 15 percent loaded? (Because of differences in time zones and the use of FDMA, 15 percent loading is typical of normal operating conditions.)

Table P.6
Communication Subsystem Configuration of INTELSAT V Satellite for Problem 6 (downlink only).

Frequency Band	Transponder Number	Bandwidth (MHz)	Downlink Antenna Beams Connected to Transponder	Polarization	Number of 3-kHz Telephone Channels
	1–2	77	W spot	Linear	2000
11 GHz	5–6	72			2000
	7–12	241	E spot	Linear	4000
	1–2	77	W hemi	RHC	1200
4 GHz	3–4	72	E hemi	RHC	1200
	5–6	72	W zone	LHC	2000
	7–8	72	E zone	LHC	2000
	9	36			500
4 GHz	10	36	Global	RHC	500
	11	36			500
	12	41			2 TV

7. The State of Virginia can be represented approximately on a map as a rectangle bounded by 39.5° N latitude, 36.5° N latitude, 76.0° W longitude, and 83.0° W longitude. A geosynchronous satellite to be located at 79.5° W longitude is to be designed with a spot beam that covers all of Virginia at a downlink center frequency of 11,155 MHz. In this problem you will estimate the antenna dimensions subject to two different assumptions. In both cases use an aperture efficiency of 55 percent.
a. The antenna is a circular parabolic reflector whose pattern circumscribes the Virginia rectangle. Determine its diameter in meters and its gain in decibels.
b. The antenna is an elliptical parabolic reflector whose pattern approximates the exact dimensions of the Virginia rectangle. Determine the antenna dimensions in meters and the antenna gain in decibels.

8. A satellite in geostationary orbit serves an elliptical zone on the earth's surface. The major and minor axes of the ellipse subtend angles of 3.6° and 2.4° at

the satellite. The satellite has separate antennas for the uplink and the downlink, operating at frequencies of 14.1 and 11.5 GHz.

    a. Assume that the antennas have rectangular apertures. Estimate the dimensions of the antennas when the 3-dB points of the antenna beams in the principal planes lie at the edges of the coverage zone.

    b. Estimate the gain of each antenna.

    c. If the satellite has an attitude stabilization accuracy of $\pm 0.2°$, and allowance for this pointing error is made in the antenna beamwidths, what reduction in antenna gain occurs relative to the value found in Part (b)?

9.   Traveling wave tube amplifiers (TWTAs) for space applications typically exhibit mean times before failure of 40,000 h. This figure includes both the vacuum tube and its associated power supplies.

    a. Calculate the TWTA reliability for 1-, 2-, 3-, 4-, and 5-year periods.

    b. Assuming that by themselves the tubes have a mean time before failure of 100,000 h, calculate the MTBF of the power supplies.

10.   A satellite transponder has the configuration shown in Figure 3.27. The HPA consists of two TWTAs in parallel, with the redundant connection of Figure 3.26b. The MTBF of each TWTA is 80,000 h, and the failure mode can be considered open circuit. The MTBF of the other parts of the transponder is as follows.

Antennas	500,000 h	RF amplifiers	400,000 h
Filters	1,500,000 h	LO and mixer	250,000 h

    a. Calculate the probability of the HPA failing after 2 years and after 5 years.

    b. Calculate the probability of the entire transponder failing after 2 years and after 5 years.

    c. Calculate the probability that communications can still be maintained after 5 years if a second, identical transponder can be switched in when the first fails, using a switch with a MTBF of 250,000 h.

4

SATELLITE LINK DESIGN

The design of a satellite communication system is a complex process, involving compromises between many factors in order to obtain the maximum performance at an acceptable cost. Several factors dominate the design, however, of any system using geostationary satellites. These are

1. The weight of the satellite.
2. The DC power that can be generated on board.
3. The frequency bands allocated for satellite communication.
4. The maximum dimensions of satellite and ground station antennas.
5. The multiple access technique used to share communications capacity between many earth stations.

The weight of the satellite is limited by the high cost of launching a spacecraft into geostationary orbit, typically $50,000 per kilogram per launch with final positioning (1982 figures). From a satellite of limited weight and size, a limited number of solar cells can be deployed, which places an upper limit on the DC power available for transponders. There is a maximum diameter of spacecraft that can be housed within the launching shroud, although deployable structures are often used to extend the size of the spacecraft in orbit. Larger spacecraft can be carried into low earth orbit by the Space Shuttle and then injected into geostationary orbits; assembly of large space platforms in orbit is also possible. The size limitations combine to produce a situation in which the spacecraft has a limited RF output power, which cannot be concentrated onto very small areas of the earth because of the limited size of the satellite antennas. The resulting flux density at the earth's surface for a typical communications satellite is about -127 dBW/m^2 [1]. A direct broadcast satellite using a 200-W transmitter and directional beam can achieve -100 dBW/m^2. This is a very weak signal compared to that found in terrestrial microwave links, and all satellite systems are constrained by this problem. In this chapter, we shall examine the way in which the small signal can be used to carry many hundreds of telephone channels.

In general, it is easiest and cheapest to use the lowest available frequency. However, bandwidth is limited in the lower frequency bands, and interference is more likely, as these frequencies (especially 6/4 GHz) are already heavily used by terrestrial systems. The higher frequencies offer the advantages of wider bandwidths—3.5 GHz at 30/20 GHz against the 1000 MHz available in the 6/4 GHz band, for example—but incur propagation difficulties above 10 GHz.

In addition, antennas of a given diameter can produce narrower beams at higher frequencies, so regional satellites can be made more directive at the upper end of the frequency range. The frequency does not appear in the link equation, however, if the antenna on the satellite is required to provide coverage of a given area on the earth's surface.

Tables 4.1 and 4.2 show the frequencies allocated for satellite communication systems.

The 6 and 4 GHz frequency bands have been the most heavily used for the first 15 years of satellite communication systems. Until the World Administrative Radio Conference (WARC) in 1979 [2], the main allocations were 5.925 to 6.425 GHz for uplinks and 3.700 to 4.200 GHz for downlinks for commercial systems. The SBS-I satellite launched in 1979 was the first commercial satellite to use the 14/11 GHz bands [3].

The 6/4 GHz bands have been the most popular because they offer the fewest propagation problems and, historically, RF components for these bands have been readily available. Attenuation of radio waves by rain is a major factor in the selection of an RF frequency: attenuation is not a significant factor on a satellite path with an elevation angle of 5° or higher for frequencies below about 10 GHz, because of the short path length through the earth's atmosphere and the low attenuation per kilometer for frequencies below 10 GHz. (Chapter 8 discusses the problems of propagation on slant paths.) Sky noise is also low at 4 GHz, so it is possible to build receiving systems with lower noise temperatures at 4 GHz than at 11 GHz. For these reasons, commercial systems have used 6/4 GHz extensively, and the 1979 WARC extended the original 500 MHz up and down allocations to a total of 1000 MHz. However, interference may occur to systems using the newly allocated frequencies from other users already transmitting in these bands. Full use of the new frequencies may not be possible until the early 1990s.

In addition to the frequencies listed in Tables 4.1 and 4.2, meteorological satellites use 137 MHz (APT) and 1700 MHz (NOAA), and there are other allocations for experimental and amateur use. Many regional variations in use also exist.

There are often a large number of earth stations sharing one satellite and providing many interconnecting paths. As explained in Chapter 3, several of these paths may be through one transponder, with the result that the available communication capacity of the transponder must be shared between several earth stations. The sharing technique is called "multiple access" and may be achieved by sharing the transponder bandwidth in separate frequency slots (FDMA) or the transponder's availability in discrete time slots (TDMA). A third technique, code

Table 4.1

Frequency Allocations for Fixed Satellite Service and Broadcasting Satellites

Frequency	Fixed Satellite Service		Broadcasting Satellites	
2500–2535 MHz	Down	Region II and III	Down	
2535–2655	Down	Region II	Down	
2655–2690	Up	Region II and III	Down	
3400–3700	Down			
3700–4200	Down			
4500–4800	Down			
5725–5850	Up	Region I		
5850–5925	Up			
5925–7075	Up			
7250–7450	Down			
7450–7550	Down			
7550–7750	Down			
7900–8025	Up			
8025–8400	Up			
10.7 –11.7 GHz	Down		Up	Region I
11.7 –12.1	Down	Region II	Down	Region I and III
12.1 –12.2	Down	Region II	Down	
12.2 –12.3	Down	Region II	Down	Region I and II
12.3 –12.5			Down	Region I and II
12.5 –12.75	Up/Down		Down	Region II
12.75–13.25	Up			
14.0 –14.5	Up			
14.5 –14.8			Up	
17.3 –17.7			Up	
17.7 –18.1			Up	
18.1 –18.6	Down			
18.6 –18.8	Down			
18.8 –20.2	Down			
20.2 –21.2	Down			
22.5 –23.0			Down	Region II and III
27.0 –27.5	Up	Region II and III		
27.5 –29.5	Up			
29.5 –31.0	Up			
37.5 –39.5	Down			
39.5 –40.5	Down			
40.5 –42.5			Down	
42.5 –43.5	Up			

Regions I, II, and III are regions of the earth's surface defined in the International Telecommunication Union's Radio Regulations. Region I covers Europe, Africa, and northern Asia. Region II covers North and South America, and Region III covers the remainder of Asia.

Source: Reprinted from the *Manual of Regulations and Procedures for Radio Frequency Management*, Office of the Superintendent of Documents, Washington, D.C. 20402, 1984

Table 4.1   (*continued*)

Frequency	Fixed Satellite Service	Broadcasting Satellites
47.2 −49.2		Up
49.2 −50.2	Up	
50.4 −51.4	Up	
71.0 −75.5	Up	
81–84	Down	
84–86		Down
92–95	Up	
102–105	Down	
149–164	Down	
202–217	Up	
231–241	Down	
265–275	Up	

Table 4.2
Frequency Allocations for Mobile Satellite Services

Frequency	Aeronautical Mobile	Maritime Mobile	General Mobile
806—890 MHz			Region II and III (Limited use)
1530—1535		Down, shared	
1535—1544		Down, exclusive	
1544—1545			Down
1545—1559	Down, exclusive		
1626.5—1645.5		Up, exclusive	
1645.5—1646.5			Up
1646.5—1660	Up, exclusive		
1660—1660.5	Up, shared		
19.7—21.2 GHz			Down
29.5—31.0			Up
39.5—40.5			Down
43.5—47			Up
66—71			Up/Down
71—74			Up/Down
81—84			Up/Down
95—100			Up/Down
134—142			Up/Down
190—200			Up/Down
252—265			Up/Down

Regions I, II, and III are regions of the earth's surface defined by the International Telecommunications Union. (See Table 4.1 for an explanation of their geographic locations.)

Source: Reprinted from the *Manual of Regulations and Procedures for Radio Frequency Management*, Office of the Superintendent of Documents, Washington, D.C., 20402, 1984

division multiple access, or spread spectrum, shares the transponder by allowing coded signals to overlap in time and frequency. The earth stations then separate the signals by recognizing which of the codes is destined for each station.

Multiple access is described in much greater detail in Chapter 6. The technique used can radically affect the total number of channels carried by a single transponder: when the number of accesses per transponder is large, TDMA has a distinct advantage.

A communication system must be designed to meet certain minimum performance standards, within limitations of transmitter power and RF bandwidth. The most important performance criterion is the *signal-to-noise ratio (S/N)* in the *information channel*, which carries the signal in the form in which it is delivered to the user. (Usually, the information channel is at *baseband* frequency and carries an audio or television signal, or digital data.) In designing a satellite communication system, we must try to guarantee a minimum signal-to-noise ratio in the receiver's baseband channels and also meet constraints on satellite transmitter power and RF bandwidth.

Signal-to-noise ratio in a baseband channel depends on a number of factors; the carrier-to-noise ratio (C/N) of the RF or IF signal in the receiver, the type of modulation used to impress the baseband signal onto the carrier, and the IF and baseband channel bandwidths in the receiver are the most important. In this chapter we are concerned mainly with the design and analysis of satellite communication links in terms of the carrier-to-noise ratio. Thus we need to be able to calculate carrier (received) power in an earth station receiver, and also the noise power in the receiver, to establish the (C/N). Chapter 5 is concerned with the modulation techniques used in satellite communication systems. Invariably, a method is used that makes the baseband (S/N) larger than the IF (C/N).

4.1 BASIC TRANSMISSION THEORY

The calculation of the power received by an earth station from a satellite transmitter is fundamental to the understanding of satellite communications. In this section, we discuss two approaches to this calculation: the use of flux density and the Friis transmission equation.

Consider a transmitting source, in free space, radiating a total power P_t W uniformly in all directions as shown in Figure 4.1. Such a source is called *isotropic*; it is an idealization that cannot be realized physically because it could not create transverse polarized electromagnetic waves. At a distance R from the hypothetical isotropic source, the flux density crossing the surface of a sphere, radius R, is given by

$$F = \frac{P_t}{4\pi R^2} \ W/m^2 \tag{4.1}$$

In practice, we use directive antennas to constrain our transmitted power to be radiated primarily in one direction. The antenna has a gain $G(\theta)$ in a direction

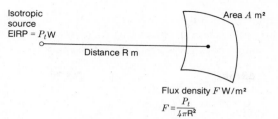

Figure 4.1 Flux density produced by an isotropic source.

θ, defined as the ratio of power per unit solid angle radiated in a given direction to the average power radiated per unit solid angle:

$$G(\theta) = \frac{P(\theta)}{P_0/4\pi} \qquad (4.2)$$

where

    $P(\theta)$ is the power radiated per unit solid angle by the test antenna

    P_0 is the total power radiated by the test antenna

    $G(\theta)$ is the gain of the antenna at an angle θ [4]

The reference for the angle θ is usually taken to be the direction in which maximum power is radiated, often called the *boresight* of the antenna. The *gain* of the antenna is then the value of $G(\theta)$ at angle $\theta = 0°$, and is a measure of the increase in power radiated by the antenna over that from an isotropic radiator emitting the same total power.

    For a transmitter with output P_t W driving a lossless antenna with gain G_t, the flux density in the direction of the antenna boresight at distance R m is

$$F = \frac{P_t G_t}{4\pi R^2} \text{ W/m}^2 \qquad (4.3)$$

The product $P_t G_t$ is often called the effective isotropically radiated power or EIRP, and it describes the combination of transmitter and antenna in terms of an equivalent isotropic source with power $P_t G_t$ W, radiating uniformly in all directions.

    If we had an ideal receiving antenna with an aperture area of A m^2, as shown in Figure 4.2, we would collect power P_r W given by

$$P_r = FA \qquad (4.4)$$

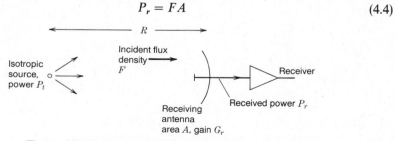

Figure 4.2 Power received by antenna with area A.

A practical antenna with a physical aperture area of A_r m^2 will not deliver the power given in Eq. (4.4). Some of the energy incident on the aperture is reflected away from the antenna, and some is absorbed by lossy components. This reduction in efficiency is described by using an effective aperture A_e, where

$$A_e = \eta A_r \qquad (4.5)$$

and η is the aperture efficiency of the antenna [5]. The aperture efficiency η accounts for all the losses between the incident wavefront and the antenna output port: these include *illumination efficiency* or *aperture taper efficiency* of the antenna, which is related to the energy distribution produced by the feed across the aperture, and also other losses due to spillover, blockage, phase errors, diffraction effects, polarization, and mismatch losses. For parabolodial reflector antennas, η is typically in the range 50 to 75 percent, lower for small antennas and higher for large Cassegrain antennas. Horn antennas can have efficiencies approaching 90 percent. Thus the power received by a real antenna with a physical receiving area A_r and effective aperture area A_e m^2 is

$$P_r = \frac{P_t G_t A_e}{4\pi R^2} \qquad (4.6)$$

Note that this equation is essentially independent of frequency within a given band; the power received at an earth station depends only on the EIRP of the satellite, the effective area of the earth station antenna, and the distance R.

A fundamental relationship in antenna theory [5] is that the gain and area of an antenna are related by

$$G_r = \frac{4\pi A_e}{\lambda^2} \qquad (4.7)$$

Substituting for A_e in Eq. (4.6) gives

$$P_r = P_t G_t G_r \left[\frac{\lambda}{4\pi R} \right]^2 \qquad (4.8)$$

This expression is known as the Friis transmission equation, and it is essential in the calculation of power received in any radio link. The frequency (as wavelength, λ) appears in this equation for received power because we have used the receiving antenna gain, instead of effective area.

The term $[4\pi R/\lambda]^2$ is known as the *path loss*, L_p. It is not a loss in the sense of power being absorbed; it accounts for the way energy spreads out as an electromagnetic wave travels away from a transmitting source.

Collecting the various factors together, we can write

$$\text{Power received} = \frac{\text{EIRP} \times \text{Receiving antenna gain}}{\text{Path loss}} \qquad (4.9)$$

In communication systems, the decibel is used very widely to simplify expressions such as that in Eq. (4.9). In decibel terms, we have

$$P_r = (\text{EIRP} + G_r - L_p) \text{ dBW} \qquad (4.10)$$

where

$$\text{EIRP} = 10 \log_{10}(P_t G_t) \text{ dBW}$$

$$G_r = 10 \log_{10}(4\pi A_e/\lambda^2) \text{ dB}$$

$$L_p = \text{path loss} = 20 \log_{10}(4\pi R/\lambda) \text{ dB}$$

Equation (4.10) represents an idealized case, in which there are no additional losses in the link. In practice, we will need to take account of a more complex situation in which we have losses in the atmosphere due to attenuation by rain, losses in the antennas at each end of the link, and possible loss of gain due to antenna mispointing. All of these factors are taken into account by the *system margin* but need to be calculated to ensure that the margin allowed is adequate. More generally, Eq. (4.10) can be written

$$P_r = \text{EIRP} + G_r - L_p - L_a - L_{ta} - L_{ra} \text{ dBW} \tag{4.11}$$

where

L_a = attenuation in atmosphere

L_{ta} = losses associated with transmitting antenna

L_{ra} = losses associated with receiving antenna

The conditions in Eq. (4.11) are illustrated in Figure 4.3. The expression dBW means decibels greater or less than 1 W (0 dBW). The units dBW and dBm (dB

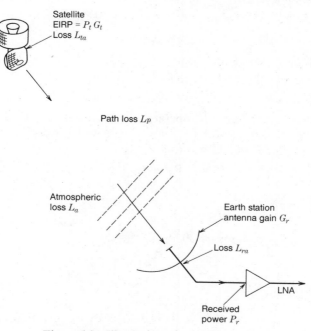

Figure 4.3 Illustration of a satellite link.

greater or less than 1 mW) are widely used in communications engineering. EIRP, being the product of transmitter power and antenna gain is often quoted in dBW.

(Note that once a value has been calculated in dB, it can readily be scaled if one parameter is changed. For example, if we calculated G_r for an antenna to be 48 db, at a frequency of 4 GHz, and wanted to know the gain at 6 GHz, we could multiply G_r by $(6/4)^2$. Using dB, we simply add 20 log $(6/4)$ or 20 log $(3) - 20$ log $(2) = 9.5 - 6 = 3.5$ dB. Thus the gain of our antenna at 6 GHz is 51.3 dB.)

Appendix I gives more information on the use of decibels in communications engineering.

Example 4.1.1

A satellite at a distance of 40,000 km from a point on the earth's surface radiates a power of 2 W from an antenna with a gain of 17 dB in the direction of the observer. Find the flux density at the receiving point, and the power received by an antenna with an effective area of 10 m^2.

Using Eq. (4.3)

$$F = \frac{P_t G_t}{4\pi R^2} = \frac{2 \times 50}{4\pi (4 \times 10^7)^2} = 4.97 \times 10^{-15} \text{ W/m}^2$$

The power received with an effective collecting area of 10 m^2 is therefore

$$P_r = 4.97 \times 10^{-14} \text{ W}$$

The calculation is more easily handled using decibels:

$$F \text{ in dB} = 10 \log_{10}(P_t G_t) - 20 \log_{10}(R) - 11.0 \text{ dB}$$
$$= 20 - 152 - 11 = -143 \text{ dBW/m}^2$$

Then $P_r = -143 + 10 = -133$ dBW.

Here we have put the antenna effective area into decibels greater than 1 m^2 (10 m^2 = 10 dB greater than 1 m^2).

Example 4.1.2

The satellite in Example 4.1.1 operates at a frequency of 11 GHz. The receiving antenna has a gain of 52.3 dB. Find the received power.

Using Eq. (4.10) and working in decibels

$$P_r = \text{EIRP} + G_r - \text{path loss (dB)}$$

$$\text{EIRP} = 20 \text{ dBW}$$

$$G_r = 52.3 \text{ dB}$$

$$\text{path loss} = 20 \log_{10}\left(\frac{4\pi R}{\lambda}\right)$$

$$= 20 \log\left(\frac{4\pi \times 4 \times 10^7}{2.727 \times 10^{-2}}\right) \text{dB}$$

$$= 205.3 \text{ dB}$$

$$P_r = 20 + 52.3 - 205.3 = -133 \text{ dBW}$$

We have the same answer as in Example 4.1.1 because the figure of 52.3 dB is the gain of a 10 m^2 aperture at a frequency of 11 GHz.

Equation (4.10), with other parameters for antenna and propagation losses, is commonly used for calculation of received power in a microwave link and is set out as a *link power budget* in tabular form using decibels. This allows the system designer to adjust parameters such as transmitter power or antenna gain and quickly recalculate the received power.

The received power, P_r, calculated by Eqs. (4.6) and (4.8) is commonly referred to as carrier power, C. This is because most satellite links use either frequency modulation for analog transmission or phase modulation for digital systems. In both of these modulation systems, the amplitude of the carrier is not changed when the data are modulated onto the carrier, so carrier power C is always equal to received power P_r.

4.2 SYSTEM NOISE TEMPERATURE AND *G/T* RATIO

Noise Temperature

Noise temperature is a useful concept in communications receivers, since it provides a way of determining how much thermal noise is generated by active and passive devices in the receiving system. At microwave frequencies, all objects with physical temperature, T_p, greater than 0°K generate electrical noise at the receiver frequency. The noise power is given by [6]

$$P_n = kT_nB \qquad (4.12)$$

where

$$k = \text{Boltzmann's constant} = 1.38 \times 10^{-23} \text{ J/K}$$
$$= -228.6 \text{ dBW/K/Hz}$$

T_n = noise temperature of source in Kelvins

B = bandwidth of power measurement device in hertz

P_n is the *available noise power* and will be delivered only to a device that is imped-ance matched to the source.

The term kT_n is a noise spectral density, in watts per hertz. The density is constant for all radio frequencies up to 300 GHz.

In satellite communication systems we are always working with weak signals, because of the large distances involved. Consequently, we must reduce the noise in our receivers as far as possible to ensure that we maintain the best possible carrier-to-noise ratio and hence the highest quality of communication. This is done by setting the bandwidth in the receiver, usually in the IF amplifier stages, to be just large enough to allow the signal (carrier and sidebands) to pass unrestricted, while keeping the noise power to the lowest value possible. The bandwidth used in Eq. (4.12) should be the equivalent noise bandwidth. Frequently we do not know the equivalent noise bandwidth and use the 3-dB bandwidth of our receiving system instead. The error introduced by using the 3-dB bandwidth is small when the filter characteristic of the receiver has steep sides.

A second objective in designing receiving systems must be to keep the noise temperature as low as possible. In large earth stations this may be done by immersing the front end amplifier of the receiver in liquid helium to hold its physical temperature around 4°K. Such an approach is effective, but expensive to install and maintain. Noise temperatures from 70 K to 200 K can be achieved without physical cooling if GaAsFET amplifiers or uncooled parametric amplifiers (*paramps*) are employed. In large earth stations thermoelectric cooling may be used. GaAsFET amplifiers can be built with noise temperatures of 70 K at 4 GHz and 180 K at 11 GHz, without cooling.

To determine the performance of a receiving system we need to be able to find the total thermal noise against which the signal must be demodulated. We do this by determining the *system noise temperature*, T_s. T_s is the noise temperature of a noise source, located at the input of a noiseless receiver, which gives the same noise power as the original receiver, measured at the output of the receiver. The equivalent noise source T_s is usually located at the input to the receiver, replacing the antenna. If the overall RF and IF gain of the receiver is G (G is a ratio, not in decibels) and its narrowest bandwidth is B, the noise power at the demodulator input is

$$P_n = kT_sBG \tag{4.13}$$

where G is the gain of the receiver from RF input to demodulator input.

Let the antenna deliver a signal power P_r to the receiver at the input to the RF section. The signal power at the demodulator input is P_rG, representing the power contained in the carrier and sidebands after amplification and frequency conversion within the receiver. Hence, the carrier-to-noise ratio at the demodulator is given by

$$\frac{C}{N} = \frac{P_rG}{kT_sBG} = \frac{P_r}{kT_sB} \tag{4.14}$$

This demonstrates how useful it is to replace several sources of noise in the receiver by a single system noise temperature, T_s.

Calculation of System Noise Temperature

Figure 4.4 shows a typical communications receiver with an RF amplifier and single frequency conversion, from its RF input to the IF output. The equivalent circuits in Figure 4.5 can be used to represent this receiver. The noisy devices are replaced by a single noise source with temperature T_s, plus noiseless amplifiers and frequency converter. The total noise power at the output of the IF amplifier is given by

$$P_n = G_{IF}kT_{IF}B + G_{IF}G_mkT_mB + G_{IF}G_mG_{RF}kB(T_{RF} + T_{in}) \tag{4.15}$$

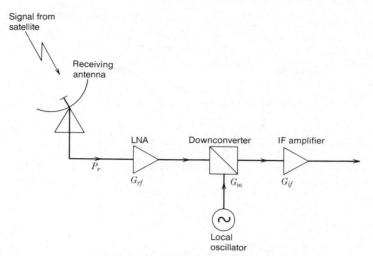

Figure 4.4 Earth station receiver.

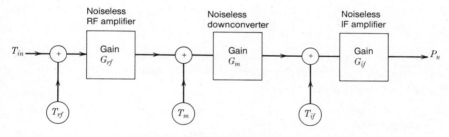

Figure 4.5a Equivalent circuit of receiver. The noisy amplifiers and downconverter have been replaced by noiseless units, with equivalent noise generators at their inputs.

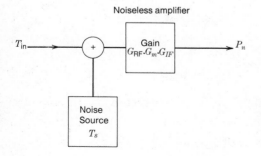

Figure 4.5b Equivalent circuit of receiver. All noisy units have been replaced by one noiseless amplifier, with a single noise source T_s as its input.

where G_{IF}, G_m, G_{RF} are the gains of the IF amplifier, mixer, and RF amplifier, and T_{IF}, T_m, and T_{RF} are their equivalent noise temperatures. T_{in} is the noise temperature of the antenna, measured at the receiver input.

Equation (4.15) can be rewritten as

$$P_n = G_{IF}G_mG_{RF}\left[\frac{kT_{IF}B}{G_mG_{IF}} + \frac{kT_mB}{G_{RF}} + kB(T_{RF} + T_{in})\right]$$

$$= G_{IF}G_mG_{RF}kB\left[T_{RF} + T_{in} + \frac{T_m}{G_{RF}} + \frac{T_{IF}}{G_mG_{RF}}\right] \qquad (4.16)$$

A single source of noise, with noise temperature T_s, would generate the same noise power P_n at the output of the IF amplifier, if

$$G_{IF}G_mG_{RF}kT_sB = P_n \qquad (4.17)$$

Then

$$kT_sB = kB\left[T_{RF} + T_{in} + \frac{T_m}{G_{RF}} + \frac{T_{IF}}{G_mG_{RF}}\right]$$

or

$$T_s = \left[T_{RF} + T_{in} + \frac{T_m}{G_{RF}} + \frac{T_{IF}}{G_mG_{RF}}\right] \qquad (4.18)$$

Succeeding stages of the receiver contribute less and less noise as the gain from each stage is added in. Frequently, the noise contributed by the IF amplifier and later stages can be ignored.

Example 4.2.1

Suppose we have a 4-GHz receiver with the following gains and noise temperatures:

$$T_{in} = 50 \text{ K} \qquad G_{RF} = 23 \text{ dB}$$

$$T_{RF} = 50 \text{ K} \qquad G_m = 0 \text{ dB}$$

$$T_m = 500 \text{ K} \qquad G_{IF} = 30 \text{ dB}$$

$$T_{IF} = 1000 \text{ K}$$

Calculate the system noise temperature.

The system noise temperature is given by

$$T_s = [50 + 50 + (500/200) + (1000/200)] = 107.5 \text{ K}$$

If the mixer had a loss, as is usually the case, the effect of the IF amplifier would be greater: suppose $G_m = -10$ dB, i.e., $G_m = 0.1$ as a ratio. Then

$$T_s = [50 + 50 + (500/200) + (1000/20)] = 152.5 \text{ K}$$

Example 4.2.2

In the system illustrated in Example 4.2.1, a section of lossy waveguide is inserted between the antenna and the RF amplifier.
Find the new system noise temperature.
Let the waveguide loss be 2 dB ($G_t = 1/1.58 = 0.63$). The lossy waveguide attenuates the noise sent through it and adds noise generated by its ohmic loss. The equivalent noise generator that represents the waveguide has a noise temperature T_l where

$$T_l = T_p(1 - G_l)$$

where G_l is a gain less than unity, and T_p is the physical temperature of the lossy device.
Thus for $G_l = 0.63$, $T_a = 50$ K, $T_p = 290°$K (assumed)

$$T_l = G_l T_a + T_p(1 - G_l)$$

$$T_{in} = 31.5 + 107.3 = 138.8 \text{ K}$$

This is the noise temperature at the input to the *LNA* due to the antenna and waveguide. The insertion of the lossy waveguide increases the overall system noise from 107.5 K to 196.3 K, measured at the LNA input. Since the lossy waveguide will reduce the signal by 2 dB, the carrier-to-noise ratio will fall by 4.6 dB, a substantially larger reduction than that caused by attenuation of the signal alone.

The noise temperature can be translated to the antenna output port by dividing T_s by G_l. This transfers the equivalent noise generator from the LNA input to the antenna output terminals, where the antenna gain is usually given. In the example above, the system noise temperature referred to the antenna terminals is $196.3/0.63 = 306.3$ K.

Note that when the system noise temperature is low, each 0.1 dB of attenuation ahead of the RF amplifier will add approximately 7 K to the system noise temperature. [Using the formula in Example 4.2.2 with $T_p = 290°$K, $G_l = -0.1$ dB $= 0.977$, $T_l = 290 \times 0.023 = 6.6$ K.] In low-noise systems, it is important to keep losses ahead of the RF amplifier to an absolute minimum.

Noise Figure and Noise Temperature

Noise figure is frequently used to specify the noise generated within a device. The operational noise figure is defined by the following formula [6]:

$$NF = \frac{(S/N)_{in}}{(S/N)_{out}} \tag{4.18}$$

Because noise temperature is more useful in satellite communication systems, it is best to convert noise figure to noise temperature, T_d. The relationship is

$$T_d = T_0(NF - 1) \tag{4.19}$$

Table 4.3

Comparison of Noise Temperature and Noise Figure

Noise temperature (K)	0	20	40	60	80	100	120	150	200	290
Noise figure (dB)	0	0.29	0.56	0.82	1.06	1.29	1.50	1.81	2.28	3.0
Noise temperature (K)	400	600	800	1000	1500	2000	3000	5000	10,000	
Noise figure (dB)	3.8	4.9	5.8	6.5	7.9	9.0	10.5	12.6	15.5	

where T_0 is the reference temperature used to calculate the standard noise figure—usually 290 K. [Japan uses 293 K.]

NF is frequently given in decibels and must be converted to a ratio before being used in Eq. (4.19). Table 4.3 gives a comparison between noise figure and noise temperature over the range encountered in typical systems.

Example 4.2.3

An amplifier has a quoted noise figure of 2.5 dB. What is its equivalent noise temperature?

Using Eq. (4.19):

$$T_d = 290(1.78 - 1) = 226 \text{ K}$$

This value of noise temperature could then be used in Eq. (4.17), with other appropriate data, to calculate system noise temperature.

G/T Ratio for Earth Stations

The link equation can be rewritten in terms of (C/N) at the earth station:

$$(C/N) = \frac{P_t G_t G_r}{k T_s B} \left[\frac{\lambda}{4\pi R} \right]^2$$

$$= \frac{P_t G_t}{k B} \left[\frac{\lambda}{4\pi R} \right]^2 \frac{G_r}{T_s} \tag{4.20}$$

Thus $(C/N) \propto G_r/T_s$ and the terms in the square brackets are all constants for a given satellite system. G_r/T_s is a ratio that can be used to specify the "quality" of an earth station, since increasing G_r/T_s increases the (C/N).

G_r/T_s is usually shortened to G/T ratio, sometimes called *figure of merit*. A Standard A earth station used in the Intelsat network is required to have a G/T ratio of 40.7 dBK^{-1}, at 4.0 GHz and 5° elevation angle [7]. It is necessary to specify the frequency and elevation angle in 4-GHz earth stations because G_r varies as f^2 across the frequency band and T_s depends on sky noise temperature, which increases as the elevation angle is reduced below 10°. At higher frequencies, for example, 11 GHz, the variations in G_r and T_s are smaller because of the higher low-noise amplifier (LNA) contribution and narrower fractional bandwidths employed.

Satellite antennas may be quoted as having a negative G/T, which is below 0 dBK^{-1}. This simply means that the numerical value of G_r is smaller than the numerical value of T_s.

Example 4.2.4

An earth station antenna has a diameter of 30 m, has an overall efficiency of 68 percent, and is used to receive a signal at 4150 MHz. At this frequency, the system noise temperature is 79 K when the antenna points at the satellite at an elevation angle of 28°. What is the earth station G/T under these conditions? If heavy rain causes the sky temperature to increase so that the system noise temperature rises to 88 K, what is the new G/T value?

First calculate the antenna gain. For a circular aperture:

$$G_r = \eta \frac{4\pi A}{\lambda^2} = \eta \frac{\pi^2 D^2}{\lambda^2}$$

at 4150 MHz, $\lambda = 0.0723$ m. Then

$$G_r = 0.68 \times \left(\frac{\pi \times 30}{0.0723}\right)^2 = 1.16 \times 10^6$$

$$= 60.6 \text{ dB}$$

Converting T_s into dBK,

$$T_s = 10 \log 79 = 19.0 \text{ dBK}$$

$$G/T = 60.6 - 19.0 = 41.6 \text{ dBK}^{-1}$$

If $T_s = 88$ K in heavy rain,

$$G/T = 60.6 - 19.4 = 41.2 \text{ dBK}^{-1}$$

4.3 DESIGN OF DOWNLINKS

The downlink of any satellite communication system must be designed with the following objectives:

1. To guarantee continuity of the link for a specified percentage of the time (typically 99.9 percent), with a given (S/N).
2. To carry the maximum number of channels at a minimum capital and maintenance cost.

The first objective requires a minimum (C/N) at the receiver input for 99.9 percent of the time, and will almost certainly require a modulation scheme or signal-processing technique that gives a (S/N) improvement over the receiver (C/N). Modulation is usually FM in analog systems and PSK in digital systems, but companded SSB is also used.

The second objective brings in a series of compromises between antenna cost, receiver cost, tracking accuracy, station manning, modulation, and multiple-access techniques. In the Intelsat system, standard earth station specifications have been produced, which go some way to optimizing the earth segment of that system. In a maritime or a military system, a totally different earth station would be needed. An Intelsat standard A earth station costs several million dollars; a home TV system using a broadcast satellite might cost less than $500.

INTELSAT IV-A Downlink

In order to obtain an idea of signal levels and antenna dimensions, let us first consider the design of a link for use with an INTELSAT IV-A satellite with the configuration shown in Figure 4.6. The major parameters of an INTELSAT IV-A are listed in Table 4.4. This satellite is typical of many international and domestic satellites built by the Hughes Aircraft Company and is used here as a convenient model to illustrate how satellite communication links are designed.

For a transponder operating into a global beam antenna, the beam edge antenna gain is 16.0 dB; the transponder output power is 8 dBW, giving a satellite EIRP of 24.0 dBW (nominal). The bandwidth of the transponder is 36 MHz, and the downlink frequency is between 3.7 aand 4.2 GHz. We can calculate the flux density at the earth's surface from this data, assuming a path length of 40,000 km, using Eq. (4.3)

$$F = \frac{P_t G_t}{4\pi R^2}$$
$$= \text{EIRP in dBW} - 20\log_{10}(4 \times 10^7) - 11 \text{ dBW/m}^2$$
$$= -139 \text{ dBW/m}^2 \tag{4.21}$$

CCIR has placed a limit on the maximum permissible flux density at the earth's surface due to a satellite transmitting in the 3.7 to 4.2 GHz band, to prevent interference with terrestrial links [8]. In shared frequency channels in

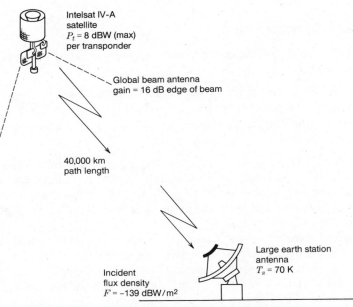

Figure 4.6 Intelsat IV-A link via global beam.

Table 4.4
INTELSAT IV-A Details

Design life	7 years
Body size	2.38 m diameter × 2.82 m high
Overall height	6.98 m
Weight (on station)	818 kg
Antenna configuration:	
Global	1 Tx and 1 Rx Horn
Spot	4:NW, NE, SW, SE, Tx only
Hemispheric	2:W, E, Tx; W, E, Rx
TT&C	Dual mode torroidal beam bicone
Solar cells:	
Area	20.5 m²
DC power	596 W (end of life)
Spin rate	45–75 rpm
Pointing accuracy	±0.35° global beams
	±0.10° spot beams
Beacons	3947 and 3952 MHz
Assigned frequencies:	
Up link	5925–6425 MHz
Down link	3700–4200 MHz
Polarizations:	
Receive global	LHCP
Transmit global	RHCP
Receive hemi/spot	LHCP
Transponders:	
Number	20 on line, 12 redundant
Bandwidth (per transponder)	36 MHz
Saturated output power	6.3 W (8 dBW)
(per transponder)	
Antenna gain:	
Global	18.5 dB
Hemi	22.7 dB } at beam center
Spot	25.7 dB
EIRP:	
Global	22.0 dBW
Hemi	26.0 dBW } at beam edge
Spot	29.0 dBW
Receiver G/T:	
Global	−17.6 dB/K
Hemi	−11.6 dB/K
Saturation flux density:	
Beam edge, uplink	−75.0 to −67.5 dBW/m²

Source: Reprinted from J. Dicks and M. Brown, Jr., "INTELSAT IV-A Transmission System Design," *Comsat Technical Review,* **5,** 1975, with permission.

Table 4.4   *(continued)*

Capacity per repeater:	
FDMA/FM	450 channels ⎫
SPADE/TDMA	800 ⎪
One TV (30 MHz)	S/N = 54 dB ⎬ global beam
Two TV (17.5 MHz each)	S/N = 49 dB ⎭
FDM/FM/FDMA	600 hemi beam
FDM/FM/FDMA	700 spot beam
Total satellite capacity:	12,500 channels in 18 FDM/FM/FDMA repeaters
	800 SPADE channels
	One or two TV channels
Bandwidth used:	800 MHz with spatial frequency reuse
	in hemi and spot beams

the 1 to 10 GHz band, the maximum permitted flux density, F_{max}, in any 4 kHz slot, for a wave arriving at an angle Θ degrees above the horizon, is given by

$$F_{max} = -152 + \frac{\Theta}{15} \ \text{dBW/m}^2 \qquad (4.22)$$

At 4 GHz, 0° elevation, this limits the flux density to -152 dBW/m² in a 4-kHz band, so energy dispersal of 23 dB is required to prevent all the transponder power being radiated at one frequency if no modulation were applied to the carrier. (See Chapter 5 for details of energy dispersal techniques.)

The noise at the receiver input is given by Eq. (4.12). The lowest noise temperatures are obtained with helium-cooled paramps, typically 20 K over a 500-MHz band. However, we must add the antenna noise temperature, T_a, to get the system noise temperature. With a large antenna at 4 GHz and 5° elevation, $T_a = 50$ K, typically, and $T_s = 70$ K. Assuming that only one carrier is transmitted, with 36-MHz bandwidth, we can calculate the system noise using decibels:

Botzmann's constant, k: -228.6 dBW/K/Hz

System noise temp. T_s: 18.4 dBK

Bandwidth, 36 MHz: 75.6 dB Hz

$$N = -134.6 \ \text{dBW} \qquad (4.23)$$

In order to maintain satisfactory communications, the (C/N) must remain above threshold under all conditions; for an FM system, the threshold is in the range 4 dB to 15 dB depending on the type of demodulator used. For PSK systems it is typically 8 dB to 15 dB. In the Intelsat system, using FM and a 36 MHz bandwidth, a threshold of 11 dB is usually taken. An allowance, called the *system margin*, of 7 dB is usually made in Intelsat systems for propagation and equipment degradation. (At 4 GHz, fading of 3 dB and sky noise increasing to 90 K can be experienced for 0.01 percent of time; this degrades C by 3 dB, increases N by 3 dB, giving 6 dB total degradation due to propagation factors. The remaining 1 dB is allowed for equipment degradation.)

When using frequency modulation (FM) and frequency division multiple access (FDMA), the transponder cannot be run at its maximum output power, as the characteristic becomes nonlinear at high output levels; backoff is employed, usually 3 to 7 dB at the output of the TWTA, to keep intermodulation products down to an acceptable level. In a multichannel system, more backoff is needed as the number of multiple accesses to the transponder increases.

Intermodulation arises from multiplication of signals because of the nonlinearity of the transponder output amplifier. When a large number of signals are trying to share a nonlinear amplifier, a great many unwanted intermodulation products are generated, some of which fall within the transponder passband. Usually, the intermod products are not intelligible signals, but they do increase the overall noise level and thus reduce the (C/N) of the wanted signals at the receiving earth stations. Chapter 6 discusses intermodulation in detail.

For an INTELSAT IV-A transponder using the global beam, a typical output backoff level would be 3 dB with one access to that transponder, so the radiated power from the transponder is 5 dBW, giving a satellite EIRP of 21 dBW for an observer at the edge of the transmit antenna beam. Using Eq. (4.3) we can calculate the flux density at the earth's surface for an observer at the edge of the coverage zone. Assuming a slant path length of 40,000 km

$$F = \frac{P_t G_t}{4\pi R^2} = -142 \text{ dBW/m}^2 \tag{4.24}$$

Using Eqs. (4.4) and (4.5), we can write

$$\frac{C}{N} = \frac{P_r}{N} = \frac{F \times \eta A_r}{N} \tag{4.25}$$

The signal into the receiver must remain above the 11-dB (C/N) threshold for 99.99 percent of the time, so allowing for the 7-dB system margin, a (C/N) of 18 dB is needed with clear sky conditions. Hence a carrier level of -116.6 dBW at the input to the earth station low-noise amplifier is required.

Putting (C/N) = 18 dB and $C = -116.6$ dBW in Eqs. (4.24) and (4.25) gives

$$\eta A_r = A_e = -116.6 \text{ dBW} + 142 \text{ dBW/m}^2$$
$$= 25.4 \text{ dBm}^2 \text{ or } 346 \text{ m}^2 \tag{4.26}$$

Assuming an antenna aperture efficiency of 65 percent, the actual collecting area of a dish with an effective area of 346 m² is 533 m², giving a diameter of 26.1 m, (or 86 ft). Many Standard A earth stations use antennas with diameters of 30 m. The diameter of 26.1 m represents the smallest antenna with 65 percent efficiency that can be used in an INTELSAT IV-A global beam link, to provide an 18-dB (C/N) with a system noise temperature T_s of 70 K.

We cannot increase the (C/N) by decreasing the bandwidth in order to decrease the noise power unless we are prepared to reduce the total number of channels transmitted through the transponder. If we reduce the bandwidth of our carrier, we must use only part of the transponder output power so that the remaining bandwidth can be used by another carrier. The carrier power C must be shared in proportion to the bandwidth used by each carrier, and (C/N) remains constant for all carriers. The number of channels per carrier falls in proportion to the bandwidth used, but even the smallest capacity carrier must be received by a 28-m antenna to guarantee a (C/N) of 18 dB. This is a disadvantage of FDMA systems that makes the cost of low-capacity earth stations in the Intelsat system using FDMA almost as high as the cost of high-capacity stations.

An alternative multiple access technique, time division multiple access (TDMA) allows large and small earth stations to share the same transponder with increased efficiency, but at the cost of greater earth station complexity. Mixing earth stations with widely different G/T ratios is not possible, though, since each station must receive timing information with the same (C/N).

The weak signal received by the earth station may be used to carry many hundreds of telephone or data channels by multiplexing either in frequency (*FDM*) or in time (*TDM*), as discussed in Chapters 5 and 6. Each telephone (or data) channel requires a signal-to-noise ratio (S/N) much greater than 18 dB for satisfactory operation. As we shall see in the next chapter, frequency modulation, or phase shift keying, can be used as techniques to increase the signal-to-noise ratio in a single channel above the received (C/N). Typically, an improvement of 35 dB is obtained, giving a clear air (S/N) of 53 dB in a telephone channel when the received (C/N) is 18 dB.

In systems using frequency modulation, the improvement in (S/N) is obtained by using wide-deviation FM. The RF bandwidth required to transmit 132 FDM telephone channels via a global beam is typically 7.5 MHz, an average of 57 kHz per channel [9]. If we used single sideband amplitude modulation, we could send 132 FDM telephone channels in a bandwidth of 528 kHz, using 4 kHz per channel. Clearly, the (S/N) improvement obtained by using FM increases the RF bandwidth per channel a great deal and therefore reduces the communications capacity of the satellite correspondingly. One global transponder on an INTELSAT IV-A has a 36-MHz bandwidth, but can carry only 450 channels when a large number of multiple accesses are made to that transponder. We have traded bandwidth (and capacity) for (S/N) improvement because we cannot achieve a 50-dB (C/N) at the earth station.

Digital systems using time division multiplexing and PSK modulation can achieve improvements in transponder capacity. For example, the *SPADE* system can transmit 800 telephone channels through a global transponder of INTELSAT IV-A that carries only 450 channels in FDMA [9]. The operators of satellite communication systems are always anxious to transmit the largest possible number of channels through each transponder since the revenue earned by the system increases in direct proportion to the number of channels carried.

The Intelsat system is designed with generous system margins and uses conservative design techniques to ensure excellent quality of the transmitted signals. Considerable increases in capacity can be obtained by trading system margin and signal quality for a larger number of channels. The result may be barely noticeable to the user but can double or quadruple the capacity of a transponder with considerable impact on the economics and pricing of the system. For example, syllabic compandors and time assigned or digital speech interpolation make use of the fact that a telephone channel carries voice sounds (one way) only 40 percent of the time, on average. If the channel is allocated to a given speaker only when a signal is present (i.e., if the speaker is talking), it can be used by another speaker for part of the 60 percent of time it is not needed by the first speaker. Brief overloads occur in these systems but are not particularly noticeable to the user.

Single sideband systems that use *companding* of speech have achieved throughputs up to 6000 speech channels in one 36 MHz transponder in a domestic U.S. satellite communication system [10]. Compared to the 450 channels carried by a global beam transponder of the same bandwidth, this is a dramatic increase in capacity, which can be reflected in the price charged to the user for each telephone channel.

4.4 DOMESTIC SATELLITE SYSTEMS USING SMALL EARTH STATIONS

In the previous section we found that we had to use a 28-m diameter antenna and a 20 K LNA to receive signals from INTELSAT IV-A's global beam, in order to guarantee an 18-dB (C/N) in clear air. Large antennas of this type are extremely expensive, costing several million dollars each, and restrict satellite communications of this type to trunk routes carrying many hundreds of telephone or data channels, where revenues are high. There are many applications in which we wish to use satellites to carry only one or two telephone or data channels, or a direct-broadcast TV signal and want to use small, low-cost earth stations. In these cases, earth stations costing a few hundred or a few thousand dollars are needed. There are only two parameters in the equation for received power that we can adjust at the satellite to allow us to use a smaller receiving antenna: satellite transmitted power and satellite antenna beamwidth.

In domestic satellite systems, narrow beams can be used for transmitting from the spacecraft to provide coverage over only the region that the system is designed to serve. Because the dimensions of the antennas that can be mounted on most spacecraft are limited, the coverage zone cannot be made arbitrarily small. The earth's disk subtends an angle of 17° when viewed from geostationary orbit, and can be illuminated with a microwave horn having an aperture a few wavelengths in diameter. At 4 GHz, to obtain a 4° spot beam, a dish 1.4 m in diameter is needed. As the frequency is increased, the diameter of the spacecraft antenna in wavelengths is increased for a given dish diameter, making it feasible to use more directive beams. However, unless a switched or multiple beam system is used, the single transmit (or receive) beam must cover the whole region that the domestic satellite serves. The topic of spacecraft antennas is explored in more detail in Section 3.7.

As an example consider the problem of providing service to the 48 contiguous states of the United States, as illustrated in Figure 4.7; the constraints then become

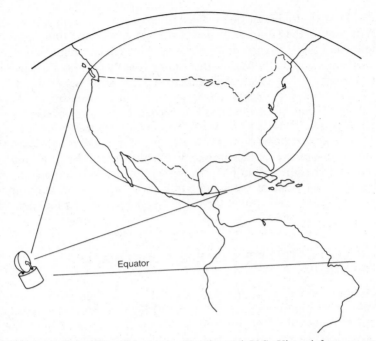

Figure 4.7 Domestic satellite link to the Continental U.S. Viewed from geostationary orbit, a beam approximately 6° × 3° is required for coverage of the 48 contiguous states. Additional beams may be provided for Alaska, Hawaii, and possessions.

apparent. Viewed from a geostationary orbit, at a longitude of around 100° W, the continental United States subtends an angle of about 3° in the latitude plane and 6° in the longitude plane. Regardless of the frequency used, an aperture antenna to produce a single beam with 3-dB beamwidths of 6° by 3° has dimensions approximately 13λ by 26λ, and a gain of 32 dB. For an earth station at the edge of the coverage zone, the gain is typically 3 dB lower, or 29 dB.

For the system shown in Figure 4.7 the received carrier level and (C/N) can readily be calculated for an earth station with a 6-m antenna, using a transponder with an output power of 5 W at 4 GHz, transmitting a single carrier. Ignoring all losses, and using a receiving antenna gain of 45.6 dB (60 percent efficiency) and a system noise temperature of 120 K

$$P_r = 7 + 29 - 196 + 45.6 = -114.4 \text{ dBW} \qquad (4.27)$$

The noise power at the input to a low noise receiver with bandwidth 36 MHz and system noise temperature 120 K is

$$N = kTB = -228.6 + 20.8 + 75.6 \text{ dBW}$$
$$= -132.2 \text{ dBW} \qquad (4.28)$$

Thus for this system the (C/N) is 17.8 dB. This is some 6.8 dB above an FM threshold of 11 dB and provides an adequate margin for an operational system.

These figures are typical of those used in U.S. domestic satellite systems designed for distribution of television programs to cable TV networks and broadcast TV stations. (See Chapter 11 for details.)

The diameter of the receiving antenna can be reduced if an FM demodulator with a lower threshold is used. A typical threshold figure for a high-quality demodulator producing very few noise impulses at threshold is 11 dB. Threshold extension demodulators have been developed for 30-MHz bandwidths, which can operate down to a threshold of 7 dB. (See Figure 9.38 for an example.) The extra 4 dB of margin above threshold with a (C/N) of 17.8 dB can be traded for antenna gain, allowing the 6-m antenna to be replaced by one with a 3.8-m diameter. The cost of the smaller antenna is much lower than that of the larger one (antenna cost goes up somewhere between the square and the cube of dish diameter), and because the smaller antenna has a wider beam, it is easier to align and does not need as rigid a mount as the larger antenna.

Threshold is recognized by an increase in noise at the FM demodulator output caused by noise spikes in the IF input to the demodulator. When noise spikes approach the carrier magnitude, sudden rapid changes in phase of the FM signal can occur. The FM demodulator produces an output voltage proportional to the rate of change of phase (i.e., the instantaneous frequency) and interprets the rapid phase changes caused by noise as large voltage spikes. The voltage spikes at the FM demodulator output occur at random time intervals, with increasing numbers of spikes as the (C/N) falls toward the threshold. In a telephone circuit, the noise pulses have been described as clicks or crackles; they can easily be seen on an oscilloscope or counted by a pulse counter.

In FM TV systems, noise spikes that appear close to threshold cause black and white dots across an otherwise good quality TV picture. (In satellite TV jargon, these noise spikes are called *sparklies*.) The interested reader should refer to a text on communication system theory for an analysis of threshold effects in FM demodulators [11, 12].

Figure 4.8 shows the characteristics of a threshold extension demodulator of a satellite FM TV receiver [13]. The threshold is at a (C/N) of 8.5 dB for a nominal 30 MHz RF bandwidth, where threshold is defined as an extra loss of 1 dB in (S/N) as (C/N) is reduced.

By allowing a much smaller operating margin, it is possible to receive domestic TV transmissions with a 3-m diameter dish. The gain of a 3-m dish at 4 GHz, for 60 percent efficiency, is 39.6 dB. With a system noise temperature of 120 K, a 36-MHz bandwidth signal can be received with a (C/N) of 11.8 dB, in a system with no losses. This is 3.3 dB above the margin for an 8.5-dB threshold in an FMTV system. Many earth terminals sold for satellite TV reception in the United States use 3-m dishes and operate close to the threshold of the FM demodulator.

A 3-m dish was the smallest earth station that could be used for reception of TV transmission from the first generation of U.S. domestic satellites. With satellite antenna dimensions fixed by the beamwidth required for regional coverage, the earth station antenna size could only be reduced by increasing the power radiated by the satellite. This is a technique employed in direct-broadcast satellite TV systems where very small earth station antennas are used.

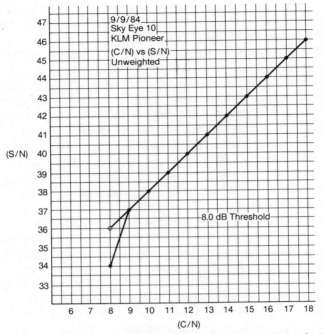

Figure 4.8 Characteristics of an FM threshold extension demodulator for TV reception. (Reproduced with permission from H. L. Gustafson, "The KLM Sky Eye 10 Receiver," *Satellite Television*, **2**, pp. 66–70 (November 1984). Reprinted with permission from *STV* (TM) *Satellite Television Magazine*.)

Direct Broadcast Television

Direct broadcast satellite (DBS) TV is intended to supplement existing terrestrial TV broadcasting by using a high-power satellite and regional coverage antennas. Because of differences in time zones across the United States, most proposed U.S. DBSTV systems have more than one beam so that separate programs can be transmitted in each beam. If we assume a simple system with two beams only, each $3° \times 3°$ and a downlink frequency of 12 GHz, we can calculate the power required to provide an 11-dB (C/N) in a 27-MHz bandwidth at the receiver with a noise temperature of 700 K, with a receiving antenna diameter of 0.7 m. The downlink power budget and noise calculation are shown in Table 4.5. Note that the margin over an FM threshold of 9 dB is only 2.1 dB, so heavy rain in the slant path from the satellite will cause the (C/N) to go below threshold. The transmit power of 155 W per TV channel is lower than proposed for many system, which require 200 or 300 W. The notes in Table 4.5 show some of the ways in which the user can upgrade the receiving system to obtain a greater margin.

A DBSTV satellite carrying 12 TV channels, designed on the above basis and using 200-W transponders, would provide a good quality service to a large region of the United States. The satellite would transmit a total of 2.4 kW of RF power and would need to generate about 10 kW of DC power. This is an order of magnitude greater than an INTELSAT V satellite, which generates about 1220 W; 2.4 kW

Table 4.5
Downlink Power and Noise Budgets for Typical DBS TV System

DBS Satellite

200-W transmitter output power per channel
$3° \times 2°$ zone coverage from geostationary orbit
(2000 km × 1400 km area)
12.2 GHz channel frequency

Satellite transmit power per channel	200 W
Transmit antenna gain (peak)	37 dB
Satellite EIRP/channel	60 dBW
Path length to receiving station (typical)	38,000 km
Theoretical flux density in center of coverage zone	-102.5 dBW/m^2
Clear air atmospheric loss	0.5 dB
Actual flux density (F)	-103.0 dBW/m^2

Receiving Station

Antenna diameter	0.7 m
Efficiency	60 %
Effective area of receiving aperture (A_e)	0.24 m^2
Theoretical received power (FA_e)	-109.2 dBW

Losses

Station at edge of coverage zone	-3 dB
Polarization loss in receive antenna	-0.5 dB
Pointing error in receive antenna	-1 dB
Losses in receiver before LNA	-1 dB
Actual received power (C)	-114.7 dBW

Noise Power Budget

Boltzmann's constant	-228.6 dBW/K/Hz
Receiving system noise temperature (700 K)	28.5 dBK
Channel bandwidth (IF) (27 MHz)	74.3 dB Hz
Noise power (N)	-125.8 dBW
Hence worst case design (C/N)	11.1 dB
Margin over 9 dB (C/N) threshold	2.1 dB

1. A lot of people live within the 2-dB contour of the satellite's transmit antenna pattern. If they point their receive antennas accurately so that the loss is only 0.5 dB instead of 1 dB, the system margin becomes 3.6 dB.

2. The 0.7-m diameter receiving antenna has a beamwidth of approximately 2.5°. To point this antenna with a 1 dB pointing loss requires a positional accuracy of $\pm 0.7°$. For a 0.5 dB loss, the antenna must be pointed within $\pm 0.5°$.

3. For a receiving station in the East of the United States, a 2.1-dB margin would result in the signal falling below 9-dB (C/N) for 0.5 percent of the year, typically, due to rain attenuation. A 3.6-dB margin would give a yearly outage of about 0.28 percent, or 24 hours.

4. The G/T ratio of the receiving station is 9.5 dBK^{-1}. A system noise temperature of 700 K is achievable with an image rejection mixer and no 12.2 GHz low-noise amplifier. With a 12.2 GHz LNA, a system noise temperature of 500 K could easily be achieved, giving an increase in (C/N) of 1.5 dB.

5. The user of the DBSTV earth station can elect to pay more for the receiving system in order to provide a larger margin. A combination of the options listed in 1–4 above gives a margin 5.1 dB for a user on the 2-dB contour of the coverage zone. Larger antennas will be available for people who live at the edge of a coverage zone, just as higher gain Yagi antennas are available for fringe areas in the terrestrial TV broadcast service.

of RF power could theoretically be used to transmit about 40 TV channels to 30-m earth terminals. The complexity, weight, and cost of the satellite have all been increased enormously to make possible the use of millions of small, low-cost earth stations. This is an example of the trade-offs available to the satellite communications system designer; however, the decisions are not all technical since economic, marketing, and political factors are also involved, as discussed in Chapter 11.

Design of Low-Capacity Satellite Links

In the previous two sections we saw how the satellite transmitter power could be increased and the transmitting antenna beamwidth reduced to increase the flux density at the earth's surface and allow the use of small earth station antennas. The cost of antennas is approximately constant for dish diameters less than 1 m, so this represents the lower limit for economical systems. As an alternative to increasing the incident flux density by raising the EIRP of the satellite, the noise at the receiver can be reduced by narrowing the receiver bandwidth. This is an effective technique for improving the (C/N), and the only method available when the satellite EIRP cannot be increased. However, reducing the bandwidth of the receiver results in a reduction in communications capacity. It cannot be employed in television reception, for example, without a corresponding reduction in picture quality. In FDM or digital telephony, reducing the receiver bandwidth means that fewer channels can be carried.

There are many communication systems in which a single two-way (duplex) telephone channel, or data channel, is all that is required. Ships and aircraft have used the HF band for voice communications for many years, using only a single channel of frequently very poor quality. Since the early 1970s, voice and data transmission has become available to ships via the Marisat system, operated by Comsat, and, more recently, through the Inmarsat system. Such systems illustrate very well how a single voice channel can be established through a satellite with small, low-cost earth stations and low-power, global-coverage satellites.

Inmarsat was established in 1979 by a consortium of maritime nations and is the marine equivalent of Intelsat, with headquarters in London, England. The Inmarsat system uses three geostationary satellites over the three main ocean regions with global beam coverage of each zone. A number of shore-based large-antenna earth stations provide interconnection with the international telephone system. Because up to 10,000 ships may eventually be fitted with terminals, it is important that the cost of the terminal should be low and its operation and maintenance simple [14].

Figure 4.9 shows the general concept of the Inmarsat system.

The frequencies used by the maritime systems are specifically allocated for that service, primarily 1530 to 1544 MHz and 1626.5 to 1646 MHz. The satellite–ship link is the most critical, so these two bands are used for the down and up path between the satellite and the ship. The link between the shore stations and the satellite operates in the 6/4 GHz bands, shared with other systems with large earth stations. The lower frequencies used by the ship–satellite links help to reduce the cost of the antenna, receiver, and transmitter. Recalling Eq. (4.3), the flux

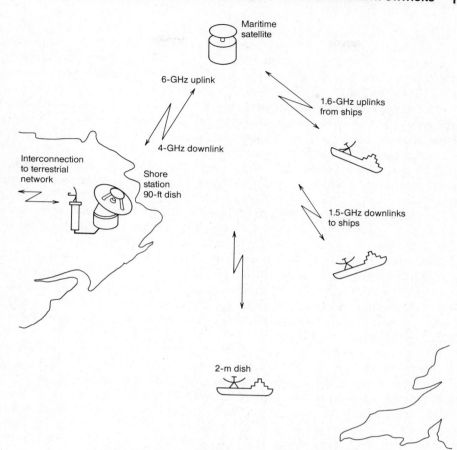

Figure 4.9 Illustration of a maritime satellite system.

density of the carrier transmitted by the satellite is

$$F = \frac{P_t G_t}{4\pi R^2} \text{ W/m}^2 \qquad (4.29)$$

A minimum carrier level is required in the receiver to maintain (C/N) above threshold, and, from Eq. (4.6), the received power, C, is

$$C = \frac{P_t G_t A_e}{4\pi R^2} \qquad (4.30)$$

The antenna diameter (for a circular dish) will be given by

$$A_e = \eta A_r = \eta \cdot \frac{\pi D^2}{4} \qquad (4.31)$$

which is independent of frequency if η is constant.

Thus we are free to choose our frequency for reasons other than receiver power level, since none of the parameters in Eqs. (4.29) and (4.30) are frequency dependent. (G_t is fixed because we require global coverage from the transmit antenna on the satellite, and gain is related to beamwidth—a useful rule of thumb is $G = 30,000/(\Theta_{3dB})^2$.)

The beamwidth of our receiving antenna increases as the frequency is reduced, making the pointing of the antenna less critical. On a ship in heavy seas, maintaining accurate pointing of the antenna is expensive. A 2-m dish operating at 1500 MHz has a beamwidth of about 8°, so a pointing error of 4° would result in a 3-dB reduction of carrier. The longer wavelength is also less affected by rain and seawater spray than short wavelengths, ensuring little loss from propagation factors.

The receivers used in the Inmarsat system employ FET amplifiers to give low noise figures, and high-power transistor amplifiers in the transmitter generate 10 or 20 W of transmit power. These components are in quantity production, unlike many microwave devices, and can be obtained at much lower cost, allowing low-cost earth stations to be built for shipboard use.

A single voice channel in the Inmarsat system can use FM analog or PSK digital modulation. An Inmarsat Standard A ship earth station (SES-A) has a G/T ratio of -4 dBK^{-1} and is designed to carry one voice channel [15]. An alternative higher capacity earth station (SES-D) with a G/T ratio of $+5$ dBK^{-1} can carry up to 10 voice channels in the same RF bandwidth by allocating less bandwidth per channel by virtue of having a higher G/T on receive and EIRP on transmit.

Table 4.6 gives up and downlink power budgets for an example of a maritime system using a receiving terminal with a 2-m dish having $G/T = -4$ dBK^{-1}. The satellite has 40 voice channels occupying an RF bandwidth of about 2.0 MHz. Note that the 10 W of RF power must be shared between the 40 channels giving 0.25 W per channel. Service is also available via a maritime communications package carried aboard several INTELSAT V spacecraft.

The Inmarsat Standard A earth station can accommodate one two-way telephone circuit and uses single channel per carrier (SCPC) demand assignment of frequency channels in the spacecraft transponder, and frequency modulation for telephony. This multiple-access scheme, which is discussed in detail in Chapter 6 (Multiple Access), makes efficient use of the small bandwidth available in the transponder. A threshold (C/N) as low as 4.5 dB is possible by using threshold extension techniques in the FM demodulator.

The low (C/N) for the satellite–ship link results from the use of a small earth station antenna. As a consequence, a large RF bandwidth per channel is needed to permit the use of wide deviation FM to give a large (S/N) improvement, as discussed in Chapter 5. Companding is also used to improve the average (S/N) in the voice channel. Calculation of the complete link (C/N) requires the addition of the noise added by the transponder to that at the earth station. However, by using a large earth station for the satellite–shore link, a high (C/N) ratio can be guaranteed for that half of the link and the overall (C/N) for the complete link will be dominated by the ship–satellite portion.

Table 4.6

Power Budget for Single Voice Channel in a Maritime System

1. Satellite–ship downlink	
Frequency band	1535–1543.5 MHz
RF bandwidth occupied	2.0 MHz
Number of channels	40
Channel RF bandwidth	30 kHz
Satellite transponder output power	10 W
Power per RF channel	0.25 W = −6 dBW
Satellite transmit antenna gain	17 dB
at edge of coverage zone	
Path loss at 1535 MHz (38,000 km range)	187.5 dB
Earth station G/T ratio	−4 dBK^{-1}
∴ C/T for received signal	−180.5 dBWK^{-1}
Boltzmann's constant, k	−228.6 dBW/Hz/K
Receiver bandwidth, 20 kHz	43.0 dB Hz
kB for receiver	−185.6 dBWK^{-1}
Hence (C/N) = C/kTB	5.1 dB
2. Ship–satellite uplink	
Frequency band	1636.5–1645 MHz
RF bandwidth occupied	2.0 MHz
Number of channels	40
Channel RF bandwidth	30 kHz
EIRP of ship terminal (10-W, 2-m dish)	37 dBW
Path loss to satellite at 1640 MHz	188.3 dB
Satellite receiving antenna gain	16 dB
edge of coverage zone	
∴ Received power at satellite, C =	−135.3 dBW
Satellite transponder system noise temperature (500 K)	27 dBK
Channel RF bandwidth, 30 kHz	44.8 dBHz
Boltzmann's constant, k	−228.6 dBW/Hz/K
∴ Noise power at transponder input, N =	−156.8 dBW
Hence (C/N) (per channel) in transponder =	21.5 dB

With an Inmarsat Standard D earth station with a receive G/T of $+5\ dB$ and higher EIRP, the extra 9 dB can be used to send 10 channels in place of one channel in the Standard A system.

4.5 UPLINK DESIGN

The uplink design is rather easier than the downlink in most cases, since an accurately specified carrier power density must be presented at the satellite transponder and it is feasible to use much higher power transmitters at earth stations than can be used on a satellite in most cases.

The cost of transmitters tends to be high compared to the cost of receiving equipment in satellite communication systems. Generation of high-stability, high-power microwave carriers is invariably expensive, and considerable care is needed throughout the transmitter to control spurious emissions and group-delay effects. As a result, the major growth in satellite communications has been in point-to-multipoint transmission, as in cable TV distribution. One high-cost transmit earth station provides service to many low-cost receive-only stations. The situation worsens as antenna diameter is reduced. Smaller antenna gain requires greater transmitter power for a given EIRP, and the use of TDMA requires still more power if the satellite transponder is to be driven into saturation. For example, a TV receive-only earth station for the 4 GHz band costs as little as $1500 with a 10-ft dish (1984 prices). To add a 6 GHz transmitter to this station would increase the cost by a factor of 10 or 15, even for a low output power.

At *L-band*, (1000–2000 MHz), where the maritime systems operate, transistor amplifiers can be used to generate 50 W of output power, and costs are not so high as at 14 or 30 GHz, where the traveling wave tube is the main HPA in use. The development of a low-cost, high-power, high-stability microwave source that could operate at millimeter wavelengths would open up a vast range of new satellite communication services from nationwide mobile communications to wristwatch personal two-way radios.

The satellite transponder is a quasilinear amplifier and the received carrier level determines the output level. Where a traveling wave tube is used as the output high-power amplifier (HPA) in the transponder, as is often the case, and FDMA is employed, the HPA must be run with a predetermined backoff to avoid inter-modulation products appearing at the output. The output backoff is typically 3 to 7 dB and is determined by the uplink carrier power level received at the space-craft. Accurate control of the power transmitted by the earth station is therefore essential, and Intelsat specifies ± 0.5 dB for Standard A stations when operated in the FDMA mode.

As an example, for an INTELSAT IV-A satellite transponder operated in the FDMA mode, an input carrier density of -73.7 dBW/m^2 to -67.5 dBW/m^2, depending on the transponder gain setting, which can be controlled via the TT&C system, is required to saturate the output stage [1]. If we take the lowest setting, -73.7 dBW/m^2, and a slant range of 40,000 km (5° elevation at the ground station) and use Eq. (4.11), we can establish an uplink power budget:

Flux density at satellite	-73.7 dBW/m^2
Path length factor	152 dB
Constant (4π)	11 dB
Required EIRP	89.3 dBW

The Intelsat Standard A earth station has a gain of 60 dB at 6 GHz, so a transmitter power level of 29.3 dBW, or 900 W, is needed. In practice, with several accesses to one transponder and 6 or 7 dB output backoff, an individual station

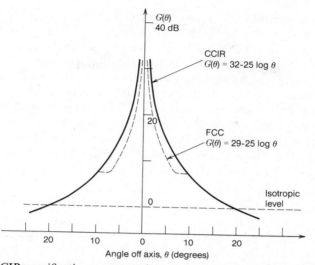

$G(\theta)$
40 dB

CCIR
$G(\theta) = 32\text{-}25 \log \theta$

20

FCC
$G(\theta) = 29\text{-}25 \log \theta$

Isotropic
level

0

20   10   0   10   20

Angle off axis, θ (degrees)

Figure 4.10 CCIR specification on transmit antenna patterns, and FCC specification for 2° satellite spacing.

would transmit a maximum power of 180 W in one carrier. However, just as in the satellite transponder, the earth station HPA must be run well below saturation to avoid the generation of intermodulation products when multiple carriers are transmitted.

With small-diameter earth stations, a higher power transmitter is required to achieve a similar EIRP. This has the disadvantage that the interference level at adjacent satellites rises, since the smaller earth station antenna inevitably has a wider beam. There is a CCIR specification for transmit station antenna patterns, designed to minimize interference from adjacent uplinks [16]. It is the uplink interference problem that determines satellite spacing and limits the capacity of the geostationary orbit in any frequency band. Figure 4.10 shows the CCIR specification, $G(\theta) = (32 - 25 \log \theta)$ dB, where θ is in degrees, for $\theta > 1$ degree off axis, for satellites spaced by 3°.

To increase the capacity of the crowded geostationary orbit arc south of the United States, the FCC introduced new regulations in 1983 requiring better control of 6-GHz earth station antenna transmit patterns so that intersatellite spacing could be reduced to 2° [17]. The new FCC requirement is for the transmit antenna pattern to lie below $G(\theta) = 29 - 25 \log \theta$ dB in the range $1° < \theta < 7°$ from the antenna boresight and $G(\theta) = 32 - 25 \log \theta$ dB beyond 7°. The FCC pattern is shown as a dotted line in Figure 4.10.

The uplink (C/N) for the quoted flux density for an INTELSAT IV-A satellite is in the range 20.9 to 29.0 dB. The transponder input stage is often a GaAsFET amplifier, and the front end G/T of the satellite is -17.6 dBK^{-1} on global beam and -11.6 dBK^{-1} on spot beam. The noise contribution from the uplink is generally constant, since the input noise temperature does not change significantly during propagation disturbances, being dominated by transponder receiver noise and noise radiated by the "hot" earth, which fills the receiving antenna beam. The uplink contributes rather less to the link noise budget than the downlink.

At frequencies above 10 GHz, for example, 14.6 GHz and 30 GHz, propagation disturbances in the form of fading in rain cause variation in received power level at the satellite. It may be necessary for each station, or a central station, to monitor uplink attenuation by measurement of the carrier output power from the satellite, so that an increase in the transmitter power can be made to compensate for the fade. This is called *uplink power control*. Automatic monitoring and control of transmitted power is used in 6-GHz earth stations, but not in an adaptive sense at the present time.

Example 4.5.1

The B transponder of the OTS-II satellite has a linear gain of 127 dB and a nominal output power at saturation of 5 W. The 14-GHz receiving antenna has a gain of 26 dB on axis, and the beam covers Western Europe.

Calculate the power output of an uplink transmitter that gives an output power of 1 W from the satellite transponder at a frequency of 14.45 GHz when the earth station antenna has a gain of 50 dB and there is a 1.5-dB loss in the waveguide run between the transmitter and antenna. Assume that the atmosphere introduces a loss of 0.5 dB under "clear-sky" conditions and that the earth station is located on the -2 dB contour of the satellite's receiving antenna.

If rain in the path causes attenuation of 7 dB for 0.01 percent of the year, what output power rating is required for the transmitter to guarantee that a 1-W output can be obtained from the satellite transponder for 99.99 percent of the year if uplink power control is used?

The input power required by the transponder is simply the output power divided by the transponder gain, so

$$P_{in} = 0 \text{ dBW} - 127 \text{ dB} = -127 \text{ dBW}$$

The uplink power budget is given by Eq. (4.11)

$$P_r = P_t + G_t + G_r - L_p - L_a - L_{ta} - L_{ra} \text{ dBW}$$

Rearranging and putting in the appropriate losses

$$P_t = P_{in} + L_p - G_t - G_r + L_{wg} + L_{at} + L_{pt} \text{ dBW}$$

where L_{wg} is the waveguide loss, and L_{at} is the atmospheric loss, and L_{pt} is the pointing loss (antenna pattern loss). Then assuming a path length of 38,500 km

$$P_t = -127 \text{ dBW} + 207.2 - 50 - 26 + 1.5 + 0.5 + 2.0$$

That is

$$P_t = 7.2 \text{ dBW or } 5.2 \text{ W.}$$

If we provide an extra 7 dB of output power to compensate for fading on the path due to rain, the output power will be

$$P_{t_{\text{rain}}} = 7.2 + 7 = 14.2 \text{ dBW} \quad \text{or} \quad 26.3 \text{ W.}$$

4.6 DESIGN OF SATELLITE LINKS FOR SPECIFIED (C/N)

We can now consider the design of a complete satellite link between two earth stations to achieve a specified performance. This requires consideration of both up and down links, and also the effect of time-dependent factors such as atmospheric propagation losses on the (C/N). The carrier level at a receiving earth station depends on the EIRP at the satellite, the earth station antenna gain, and the path loss but is reduced by any losses that occur along the way. For frequencies above about 8 GHz, rain in the slant path can cause attenuation of the carrier, and the attenuation rises rapidly with frequency. Methods for predicting the attenuation on the path are discussed in Chapter 8, but it is the statistical nature of the attenuation that is important in calculating (C/N).

The noise generated in the receiving system depends on the system noise temperature and the bandwidth used for communication. However, there are external sources of noise that contribute to the noise power in the (C/N). The satellite itself radiates noise; when the uplink transmitter has a low EIRP, the (C/N) in the transponder may be low and a significant reduction in (C/N) at the receiving station can result. Where quasilinear operation of the transponder is used in an FDMA system, the high-power amplifier within the transponder may generate intermodulation products that are considered to be noiselike. Interference from other earth stations and terrestrial microwave links can also occur, contributing further to the noise entering the earth station receiver. In most systems, however, the thermal noise generated at the receiving earth station is the major contributor to the noise power budget.

Specification of (C/N)

The limiting factor for (C/N) in almost all satellite systems is the threshold effect in the receiver demodulator. In both analog and digital radio systems there is a linear relationship between the signal-to-noise ratio (S/N) or the bit error rate (*BER*) of the demodulator output and the (C/N) at the demodulator input, provided that the (C/N) is above a specified threshold. Figure 4.11a illustrates the behavior of a typical FM demodulator. The output (S/N) after the demodulator is given by

$$(\text{S/N})_{\text{out}} = (\text{C/N})_{\text{in}} + \text{FM improvement} \tag{4.32}$$

The FM improvement term is the main reason for using FM: we can obtain adequate (S/N) with low (C/N) at the receiving system. Chapter 5 discusses the design of analog systems to achieve a required improvement factor.

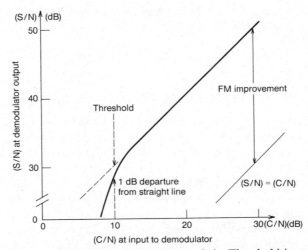

Figure 4.11a Typical FM demodulator characteristic. Threshold is at 10 dB (C/N).

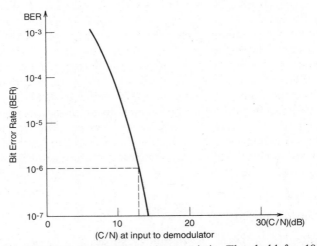

Figure 4.11b Typical PSK demodulator characteristic. Threshold for 10^{-6} BER is at 12.6 dB (C/N).

Threshold in an FM demodulator is usually defined as the (C/N) level at which the (S/N) at the demodulator output departs from the linear relationship of Eq. (4.32) by 1 dB, as illustrated in Figure 4.11a.

The exact level of the FM threshold depends on the type of demodulator and the deviation and bandwidth of the FM signal. Typically, wide-bandwidth high-capacity carriers such as FDM telephony with hundreds of channels and wide-deviation TV require demodulators with higher thresholds in the 8 to 12 dB range. Single channel per carrier systems, or FDM carriers with narrower bandwidths, can use threshold extension demodulators that have thresholds down to 5 dB.

In digital satellite links, phase shift keying (PSK) is the modulation method most often used. The relationship between demodulator output BER and the (C/N) at the demodulator input depends on the type of PSK used. PSK demodulators have different characteristics from FM demodulators and do not have the same threshold effect. The calculation of BER from (C/N) is discussed in detail in Chapter 5, which shows how to determine the BER for a particular system. The threshold (C/N) is determined by specifying a minimum allowable BER, typically between 10^{-8} and 10^{-3}, and calculating the corresponding (C/N). An example of the characteristics of a PSK demodulator is shown in Figure 4.11b.

The performance objectives of the satellite link must be specified either in terms of minimum allowable (S/N) or BER for a given signal, or as a minimum allowable (C/N). In the former cases, the minimum acceptable (C/N) must be calculated from the corresponding (S/N) or BER using the techniques described in Chapter 5. Usually, the minimum (C/N) is specified for a percentage of time. In 6/4 GHz systems where rain attenuation is small, the outage time (i.e., time for which (C/N) falls below the minimum allowed and the link is deemed inoperable) may be specified at 0.01 percent of a year or 0.03 percent of a month. At higher frequencies, where rain attenuation may be high, much larger outage times may be specified to avoid the necessity for very large system margins. At 14/11 GHz, outage times of 0.1 percent of the year may have to be tolerated. At 30/20 GHz, rain attenuation of 10 dB for 0.2 percent of the time is common for many locations in the eastern United States, requiring the system designer to consider the use of alternate paths (site diversity) for the link or to design a communication system that can tolerate occasional outages.

An additional cause of outages in many systems using geostationary satellites is the sun. The sun is a wideband noise source at microwave frequencies, with a radiation temperature around 6000 K. The actual temperature varies considerably with solar activity, a term that encompasses solar flares and sunspots [18]. The sun appears to rise in the east and move in a parabolic arc across the sky, to set in the west. At two periods in the year, usually around the Spring and Fall equinox, the sun's track will intersect the position of a geostationary satellite, as seen from an earth station. When the sun appears to be behind the satellite, the antenna temperature rises so far that communication becomes impossible. The time of day and months of the year when this occurs depend on the azimuth and elevation look angles of the satellite.

The antennas used at earth stations typically have 3-dB beamwidths of 0.5° or less. The sun subtends an angle of 0.5°, as seen from earth, and will completely fill the antenna beam if it passes directly behind the satellite. The intense noise power radiated by the sun raises the antenna noise temperature to several thousand Kelvins, and lowers the receiver (C/N) below the operating threshold. The number of days in the year on which complete outage occurs depends mainly on the earth station beamwidth; large earth stations with narrow beams will suffer an outage whenever part of the sun's disk fills the beam. In smaller earth stations with broader beams, only partial filling of antenna beam occurs, so the (C/N) is reduced to a lesser degree, but on a greater number of days in the year.

A quick estimate of the duration of a sun outage can be obtained for a large earth station antenna as follows. The sun moves in azimuth at approximately 15°/h, and subtends an angle of 0.5°. An antenna with a 3-dB beamwidth of 0.1° will suffer a noise temperature increase of at least 1000 K whenever the 3-dB beamwidth intersects the sun. (This will produce an outage in a system designed with a 100 K system noise temperature and 10-dB margin.) The 3-dB beamwidth of 0.1° fully intersects the sun for about 1.6 min, giving a complete outage of this duration twice per year. For a satellite located due south of the earth station, the sun increases its elevation angle at midday by 0.4°/day at the equinoxes, so the outage would occur on only one or two days in Spring and Fall.

The total time lost to sun outages in a year is not great, but sun outages always occur during the working day when traffic is heaviest. Because the time of the outage is predictable, traffic can be moved to other circuits in advance of the outage; this is a costly and inconvenient procedure for the operator of a large earth station carrying hundreds of telephone or data circuits, but cannot be avoided if continuous service is to be offered to users of those circuits.

Hypothetical Reference Circuit

Following the practice established in the design of long distance terrestrial microwave links, where many repeater stations each add a small amount of noise to the signal as it progresses along the chain of repeaters, CCIR has defined a "hypothetical reference circuit" from the input of the modulator in the transmitter to the output of the demodulator at the receiver [19]. The reference circuit uses a standardized test signal level of 0 dBm at its input and provides a 0 dBm signal at the output, accompanied by noise added by the many stages of amplification and interference picked up in the radio paths. Since noise powers from different sources are assumed to be completely uncorrelated, we must add the individual contributions, in watts. (If we added in decibels, we would be multiplying the powers together.)

The required (S/N) in a telephony channel is typically 50 dB, so the noise power at the output of the receiver demodulator will be 50 dB below 0 dBm or 10^{-8} W. Equation (4.32) can be rewritten using power ratios rather than dB.

$$\frac{S}{N_{out}} = \frac{C \times \text{FM improvement}}{N_{in}} \qquad (4.33)$$

Equating S to 0 dBm and requiring (S/N) = 50 dB, makes $N_{in} = N_{out} = 10^{-8}$ W = 10,000 pW.

Thus we can design the system for a total noise power of 10,000 pW referenced to a test signal output of 0 dBm. The actual noise power at the input of the receiving system will not be 10,000 pW, but this represents a normalized value that it is convenient to use. The normalized carrier level will be (C/N) decibels greater than 10,000 pW, since the receiving system has the same gain for the carrier as it does for the noise.

In practice, we can use any reference level for the calculation of the total noise power in a receiver and do not require an analog modulation system for the approach to be valid. An example for INTELSAT IV-A series will be used to illustrate the procedure. In Section 4.3 we found that the received carrier power of the 36-MHz bandwidth signal had to be -116.6 dBW in clear air to provide a (C/N) of 18 dB. The corresponding noise power, referenced to the LNA input, was -134.6 dBW. Since we only computed thermal noise for the earth station, the full 10,000 pW cannot be allocated to this source; let us allow 5000 pW for earth station thermal noise. Equating 5000 pW reference noise power to -134.6 dBW actual noise power requires a gain factor of 51.6 dB. We must therefore decrease all reference noise powers by 51.6 dB to obtain their true value, in this example.

Calculation of Noise Power Budget

Factors that contribute to noise in an earth station receiving channel include the receiver thermal noise, losses in waveguides and waveguide components, sky noise, interference entering the earth station receiving antenna, interference entering the satellite receiving antenna, thermal noise generated in the satellite, and inter-modulation noise throughout the system. This list is not exhaustive; many other sources of noise may exist in a particular system.

In frequency reuse systems using separate transmission on orthogonal polarizations to double the capacity of a system, crosstalk between the two channels causes a reduction in effective (C/N). The process is complicated, because often the nature of the signals in the two channels makes them partially correlated and the interference in one channel due to *crosstalk* from the other cannot be considered noiselike. The calculation of the effective (C/N) in such cases requires a knowledge of the depolarization statistics of the path. This topic is discussed further in Chapter 8.

For INTELSAT IV-A, the major contributions are listed in Table 4.7 for a typical global beam application [9]. The actual value is the noise power which would be measured at the LNA input due to that source of noise. In the case of

Table 4.7
Noise Power Budget for FDM/FM/FDMA INTELSAT IV-A Link.

Symbol	Noise Source	Normalized Contribution (pW)	System Noise Power (dBW)[a]
N_{ts}	Satellite transponder thermal noise	1,130	-140.5
N_{ims}	Satellite transponder intermod noise	2,160	-137.7
N_{te}	Earth station system thermal noise	4,210	-134.9
N_{eo}	Equipment noise	1,500	-139.3
N_i	Interference noise	1,000	-141.0
		10,000 pW	-131.1 dBW

[a] System noise power values refer to the example discussed in Section 4.6.
Source: Reprinted from K. Miya, *Satellite Communications Engineering*, Lattice Publishing Co., Tokyo, Japan, 1975, with permission.

equipment noise, the actual value represents noise referred to the LNA input, since some noise will be generated after the LNA. Sources of equipment noise are circuit elements such as mixers, which inject local oscillator noise into the signal channel and demodulators, which are inherently noisy devices.

Note that the addition of extra sources of noise to the earth station thermal noise has increased the noise power to -131.1 dBW and would reduce our (C/N) to 14.2 dB in clear air in the example in Section 4.3. The value of 10,000 pW is an extreme when all sources of noise in Table 4.7 are present simultaneously.

The calculation of uplink thermal noise and satellite intermodulation contributions can readily be carried out by reference to the (C/N) in the transponder. Let the carrier-to-noise ratio at the earth station be $(C/N)_e$. Then

$$C_e = N_e + \left(\frac{C}{N}\right)_e \text{ dBW} \tag{4.34}$$

where C_e and N_e are the carrier and noise power levels at the input to the LNA at the earth station.

The various sources of noises that make up N_e, as listed in Table 4.7, are all referenced to C_e, for example

$$N_{ts} = C_e - \left(\frac{C_e}{N_{ts}}\right) \text{dBW} \tag{4.35}$$

Since the path between the satellite transponder output port and the earth station LNA input port is assumed to be linear in its behavior the $(C/N)_{ts}$ ratio in Eq. (4.35) must also be the same at the transponder output. If the transponder is a linear amplifier (with frequency conversion), often called a *transparent transponder*, we can refer the output (C/N) to the input of the transponder LNA stage. Thus Eq. (4.35) can be generalized to give the noise power contribution at the earth station reference point due to the (C/N) in the satellite

$$N_{ts} = C_e - \left(\frac{C}{N}\right)_{ts} \text{ dBW} \tag{4.36}$$

where N_{ts} is the noise power contribution at the earth station due to noise sources in the transponder, which results in a (C/N) in the transponder of $(C/N)_{ts}$.

When only transponder and earth station carrier-to-noise ratios need to be added, the formula

$$\frac{C}{N} = \frac{1}{\dfrac{1}{(C/N_{ts})} + \dfrac{1}{(C/N_{es})}} \tag{4.37}$$

can be used. (C/N) values must be ratios, not in decibels.

Example 4.6.1

The G/T ratio for the INTELSAT IV-A global beam is given as -17.6 dBK^{-1} in Table 4.4. An earth station transmits a single carrier with an RF bandwidth of 7.5 MHz and a EIRP of 80.6 dBW at 6 GHz. The receiving earth station has a system noise temperature of 70 K and is to contribute 4210 pW to the system noise budget. Assuming that the transponder is linear, calculate the thermal noise contribution from the transponder if the earth station receiver (C/N) is 18 dB due to thermal noise alone.

The receiver at the earth station will use a bandwidth of 7.5 MHz for this carrier, so all noise calculations can be carried out using this bandwidth. The thermal noise power at the input to the earth station LNA, in a bandwidth of 7.5 MHz, is given by

$$N_{te} = kT_sB = -228.6 + 18.4 + 68.8 \text{ dBW}$$
$$= -141.4 \text{ dBW}$$

This represents 4210 pW in the noise power budget.

Hence the carrier level at the input to the earth station LNA is C_e; from Eq. (4.34)

$$C_e = N_{te} + \left(\frac{C}{N}\right)_e = -123.4 \text{ dBW}$$

At the input to the satellite transponder, the received power is P_r. We can calculate P_r/T_{sat} from the information given

$$\frac{P_r}{T_{\text{sat}}} = \text{EIRP} - \text{path loss} + \left(\frac{G}{T}\right)_{\text{sat}} \text{ dBK}^{-1}$$

We need the $(C/N)_{ts}$ ratio for the transponder, and since $N_{ts} = kT_{\text{sat}}B$

$$\left(\frac{C}{N}\right)_{ts} = \frac{P_r}{kBT_{\text{sat}}} = \text{EIRP} - \text{path loss} + \left(\frac{C}{T}\right)_{\text{sat}} - k - B \text{ dB}$$
$$= 80.6 - 200 - 17.6 + 228.6 - 68.8 \text{ dB}$$
$$= 22.8 \text{ dB}$$

Using Eq. (4.36) we can calculate the noise power contribution of the transponder thermal noise to the total noise power at the earth station LNA input

$$N_{ts} = C_e - \left(\frac{C}{N}\right)_{ts} = -120.4 - 26.0 \text{ dBW}$$
$$= -146.2 \text{ dBW}$$

Since the thermal noise N_{te} from the earth station of -141.4 dBW is represented by 4210 pW in the noise power budget, the contribution N_{ts} will be 4.8 dB lower, or 1394 pW. Table 4.7 gives a figure of 1130 pW for this contribution: the actual values for EIRP and transponder backoff vary considerably with carrier bandwidth, and the satellite thermal noise contribution is not, in practice, the same for all carriers. Using Eq. (4.37) we have

$$\left(\frac{C}{N}\right) = \frac{1}{\dfrac{1}{63.1} + \dfrac{1}{190.5}} = 47.4 \text{ or } 16.8 \text{ dB}$$

Referenced to a carrier level of -123.4 dBW, the noise power is -140.2 dBW, and this represents 5550 pW in the noise power budget. Thus the transponder has added 1340 pW, using Eq. (4.37). The difference between this figure and the 1394 pW calculated earlier is due to round-off error in dB values.

In the example above, the transponder was assumed to be linear. If the transponder is operated in a nonlinear mode, as might be the case in a TDMA system where there is no intermodulation problem and the transponder output HPA can be driven into saturation, the noise contribution from the transponder may be reduced. In this case, the procedure given above holds, provided the carrier-to-noise ratio $(C/N)_{ts}$ is calculated at the output of the transponder.

Design of Satellite Links to Achieve a Specified Performance

In the previous section, the calculation of the noise power budget for a satellite link was discussed. Because propagation factors influence both carrier level, by way of attenuation, and noise power, by way of increase in sky noise temperature and interference, (C/N) for any given system is not constant. The specification of (C/N) must be made dependent on the statistics of the propagation path, and the link must be designed with sufficient margin in the (C/N) in clear air that it can meet the specified (C/N) during propagation disturbances, for the specified fraction of time.

In both analog and digital radio systems using FM and PSK modulation techniques, the (S/N) or BER in the baseband channel is dependent on the (C/N) at the demodulator input, not on the absolute carrier level. This is because both frequency and phase modulation vary the phase angle of the carrier wave while its amplitude remains constant. In the receiver, a limiter is invariably used before the demodulator to prevent amplitude variations, especially transient noise peaks, from reaching the demodulator. Most demodulators are sensitive to amplitude variation to some degree and will transfer AM noise to the baseband channel unless a limiter is employed.

When the (C/N) is above the threshold of the demodulator, the use of a limiter before the demodulator does not affect the (C/N) of the signal; with both FM and PSK it is the phase deviations in the RF signal caused by added noise that contribute to the noise in the baseband channel after demodulation. The limiter has no effect on the phase of the signal, as measured at the zero crossings, so the (C/N) of the signal remains essentially unchanged by the limiting operation. When a constant amplitude FM signal is demodulated, a constant amplitude baseband signal is obtained. When noise is added to the FM signal, noise appears in the baseband output of the demodulator. With wideband FM, the demodulator suppresses the noise, provided the input (C/N) is above threshold, giving a (S/N) improvement over the (C/N), as discussed in Chapter 5. A similar effect is observed with PSK modulation when a limiter is used in the receiver. The BER rises as the (C/N) of the PSK signal falls, but the signal output from the demodulator maintains a constant amplitude.

Provided the (C/N) of the signal input to the demodulator is above threshold, the output (S/N) or BER is directly proportional to the (C/N). In analyzing the effect of (C/N) changes on the performance of a baseband channel, the practice is to normalize the carrier to its clear air value and increase the noise power as the (C/N) falls. This allows addition of noise power from various sources at any convenient point in the receiver, by reference to the constant carrier level. In this way it is possible to take account of those noise sources that give a constant power level independent of the carrier, such as receiver front end noise, and those whose power level varies as the carrier level changes.

In analog telephony systems, the performance of the link is specified in terms of maximum allowable noise power in a particular telephone channel carrying a 0 dBm test signal. For international circuits using Intelsat satellites, the CCIR specifies the following minimum performance [16]. The noise power in the noisiest channel of an FDM baseband shall conform to the following limits:

1. Noise power not to exceed 10,000 pW for more than 20 percent of any month, in any 1-min period, averaged over 1 min. Noise power not to exceed 50,000 pW for more than 1 min for more than 0.3 percent of any month, in any 1 min, averaged over 1 min.

2. The carrier must remain above threshold for 99.99 percent of the time.

Figure 4.12 shows diagrammatically how the specification fits the performance of a typical 6/4 GHz link. The curve for total noise power in the test channel must

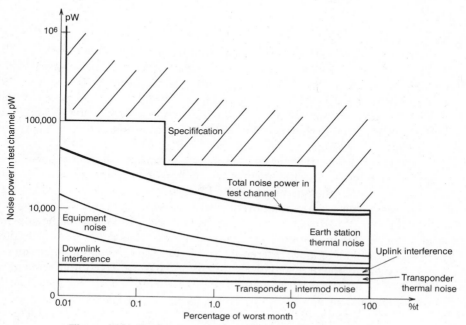

Figure 4.12   Performance and specification of a 6/4 GHz link.

lie below the shaded area of the specification. For attenuation on the downlink, the noise received from the satellite falls as the carrier is attenuated by rain, and therefore does not contribute to any change in the earth station (C/N). Noise at the earth station caused by thermal, equipment, and interference sources increases relative to a constant carrier level, giving the total noise power shown in the figure. To determine whether a particular system meets the specification requires repeated calculation of the noise power in the receiver so that curves similar to those shown in Figure 4.12 can be plotted. The following example illustrates the procedure.

Example 4.6.2
A 6/4 GHz communication system uses an INTELSAT IV-A satellite and 30-m antennas at the earth stations to provide 252 FDM telephone channels between a European capital and the United States. At the earth station in the United States the 4-GHz downlink propagation statistics give the following carrier attenuation and antenna noise temperature values as a function of percent of time of the worst month in an average year.

Condition	Percent Time (%)	Carrier Attenuation (dB)	Antenna Temperature (K)
Clear sky	80	0	50
Moderate rain	0.3	2	95
Heavy rain	0.01	4	135

The clear air (C/N) is 18 dB, the threshold for the FM detector is 11 dB, and the noise budget at the receiving station allocates noise power as in Table 4.7 in clear air. In clear air the system noise temperature is 70 K, due to the addition of the LNA contribution of 20 K. Calculate the noise power in the test channel and compare it to the specification in Figure 4.12. Calculate the (C/N) at the input to the earth station LNA and determine whether it is above threshold for 99.99 percent of the time.

We will assume that the system noise temperature rises in proportion to the antenna temperature. (If there is attenuation due to lossy waveguide components between the antenna and the LNA, this will not be true. Chapter 9 discusses the detailed calculation of system noise temperature.)

Reference to Table 4.7 gives a received noise contribution from the satellite of 3290 pW due to thermal noise and intermodulation effects in the satellite. If we assume half of the interference noise power is due to uplink interference, the fixed part of the receiver noise power is 3790 pW. In clear air the remaining 6210 pW is due to receiver thermal noise, downlink interference, and earth station equipment noise. This part rises as the carrier level falls.

The receiving system noise temperature increases because of the increase in sky noise temperature. This increase must be computed first, before the effect of attenuation of the carrier is applied. The table below gives values at the 80 percent, 0.3 percent, and 0.01 percent time points.

Percent Time (%)	Noise Temp. (K)	Earth Station Thermal Noise Power (pW)	Receiver Contribution (pW)	Carrier Attenuation (dB)	Constant Noise Power (pW)	Total Noise Power (pW)
80	70	4,210	6,210	0	3,790	10,000
0.3	115	6,916	8,916	2	3,790	17,920
0.01	155	9,322	11,322	4	3,790	32,230

Relative to a constant carrier level, the noise in our channel has increased from 10,000 pW in clear air to 32,230 pW for 0.01 percent of the worst month. This represents a 5.1 dB reduction in (C/N), from 18 dB in clear air to 12.9 dB for 0.01 percent of the worst month. The (C/N) is still 1.9 dB above threshold, indicating that the system will function satisfactorily for more than 99.99 percent of the year. The 1.9 dB margin is useful for protection against equipment degradation and other system problems.

4.7 SUMMARY

This chapter has set out the procedures for calculation of received power from a satellite and noise power in a receiver. Together, these figures give the (C/N) for the receiving system. The specification of a system will always require a minimum (C/N) in the receiver, below which the link is considered inoperable. The design of a link to achieve that minimum (C/N) requires repeated application of the link and noise power equations to give (C/N) for clear air conditions with acceptable bandwidth and antenna dimensions.

When the clear air value of (C/N) has been calculated, the propagation path statistics need to be studied to determine how much margin is required to meet worst-case conditions. Although the calculation presented in Example 4.6.2 was for a downlink, which is usually the most critical, the same procedure can be applied to the uplink to determine the effect on (C/N) in the transponder, and thus at the receiving earth station. Fading of both uplink and downlink simultaneously is unlikely for 6/4 and 14/11 GHz systems and can safely be ignored when computing link statistics. At 30/20 GHz the possibility cannot be ignored and the joint effect has to be calculated.

No attempt has been made to derive an optimization procedure for the design of the "best" system within a frequency band and (C/N) specification. There are too many variables in the system, including the cost of antennas, receivers, and other components to produce a single optimization procedure. The designer must go through several trial design procedures and compare the resulting systems to determine which one best suits the particular application.

REFERENCES

1. J. Dicks and M. Brown, Jr., "INTELSAT IV-A Transmission System Design," *Comsat Technical Review.* **5**, 73–103 (1975).

2. *Proceedings of the World Administrative Radio Conference,* 1979. I.T.U., Geneva, Switzerland.

3. *Microwave Systems News,* **7**, Special Issue on Satellite Communication Systems (March 1977).

4. S. Silver, Ed., *Microwave Antenna Theory and Design,* Vol. 12, MIT Radiation Lab Series, 1947. (Republished by Peter Perigrinus, Stevenage, Herts, U.K., 1984.)

5. W. L. Stutzman and G. A. Thiele, *Antenna Theory and Design,* John Wiley & Sons, New York, 1981.

6. H. L. Krauss, C. W. Bostian, and F. H. Raab, *Solid State Radio Engineering,* John Wiley & Sons, New York, 1980.

7. *Standard A Performance Characteristics of Earth Stations in the INTELSAT IV, IVA, and V Systems Having G/T of 40.7 dB/K,* (BG-28-72E Rev. 1), Intelsat, Washington, DC, December 15, 1982.

8. International Radio Consultative Committee (CCIR), *Recommendations and Reports of the CCIR, 1978,* Vol. IV, International Telecommunications Union, Geneva, Switzerland, 1978.

9. K. Miya, *Satellite Communications Engineering,* Lattice Publishing Co., Tokyo, Japan, 1975.

10. K. Jonnalagadda and L. Schiff, "Improvements in Capacity of Analog Voice Multiplex Systems Carried by Satellite," *Proceedings of the IEEE,* **72**, 1537–1547 (November 1984).

11. K. S. Shanmugam, *Digital and Analog Communication Systems,* John Wiley & Sons, New York, 1979, pp. 356–360.

12. H. Taub and D. L. Schilling, *Principles of Communication Systems,* McGraw-Hill, New York, 1971.

13. M. L. Gustafson, "The KLM Sky Eye 10 Receiver," *Satellite Television,* **2**, 66–70 (November 1984).

14. P. Branch and A. Da Silva Curiel, "Inmarsat and Mobile Satellite Communications," Part I, *Telecommunications,* (March 1984).

15. *Technical Requirements for Inmarsat Standard-A Ship Earth Stations* (Issue 2), Inmarsat, 40 Melton Street, London, England (February 1983).

16. International Radio Consultative Committee (CCIR), *Recommendations and Reports of the CCIR, 1970,* Vol. IV, Recommendation No. 465–1, International Telecommunications Union, Geneva, Switzerland, 1970.

17. *FCC Regulations, No. 25.209,* FCC, Washington, DC, August 1983.

18. J. D. Kraus, *Radio Astronomy,* Cygnus-Quasar Books, Powell, OH, 1982. (Originally published by McGraw-Hill, New York, 1966.)

19. *Reference Data for Radio Engineers* (6th ed.,), Howard W. Sams and Co., Indianapolis, 1975.

PROBLEMS

1. A satellite carrying an 11.7-GHz continuous wave (CW) beacon transmitter is located in geosynchronous orbit 38,000 km from an earth station. The beacon's output power is 200 mW, and it feeds an antenna with an 18.9-dB gain toward the earth station. The earth station receiving antenna is 12 ft in diameter and has an aperture efficiency of 50 percent.
  a. Calculate the satellite EIRP in W, dBW, and dBm.
  b. Calculate the receiving antenna gain in dB.
  c. Calculate the path loss in dB.
  d. Calculate the received signal power in W, mW, and dBm.

2. If the overall system noise temperature of the earth station in Problem 1 is 1250 K, determine
  a. The earth station G/T in dBK^{-1}.
  b. The received noise power in a 100-Hz noise bandwidth in W and in dBm.
  c. The received carrier-to-noise ratio in dB in a 100-Hz noise bandwidth.

3. An earth station receiving antenna for an INTELSAT V tracking beacon delivers -119 dBm carrier power at the antenna's output flange. The antenna noise temperature is 68 K. Following the antenna is a waveguide with 1 dB of loss and a physical temperature of 295 K. The output of the waveguide is connected to a GaAsFET amplifier with a U.S. standard noise figure of 4 dB and a gain of 25 dB. Following the amplifier is a mixer with a U.S. standard noise figure of 12 dB. Calculate the following:
  a. The GaAsFET amplifier noise temperature.
  b. The mixer noise temperature.
  c. The waveguide effective input noise temperature.
  d. The overall noise temperature of the receiver, referred to the antenna output port.
  e. The overall noise temperature of the receiver, referred to the GaAsFET input port.
  f. The receiver carrier-to-noise ratio for a 200-Hz bandwidth.

4. A hypothetical satellite network requires participating stations to have a minimum G/T given by

$$G/T = 40.7 + 20 \log_{10}(f/4) \ \mathrm{dBK}^{-1}$$

where the value used for frequency f is in GHz. Assume operation at 4 GHz with a terminal consisting of an antenna followed by a waveguide (physical temperature 300 K) with 0.6-dB loss and a parametric amplifier with a 1.0 dB U.S. standard noise figure and 15-dB gain. The paramp drives a mixer-preamp with a 7-dB noise figure and 30-dB gain. After the mixer-preamp is an IF receiver with a 12-dB noise figure. Calculate the antenna diameter in meters necessary to meet the G/T specification. Assume that the antenna noise temperature is 15 K independent of diameter and that the aperture efficiency is 65 percent.

5. A satellite transmits frequency-modulated TV in a 36-MHz bandwidth with a 34-dBW carrier EIRP. Assuming a 40,000-km path length and a center frequency (carrier frequency) of 3700 MHz, determine the (C/N) in decibels for all combinations of these parameters:

   a. Antenna diameters—3, 4.5, and 6 m (all have 55 percent aperture efficiency)
   b. Antenna noise temperature—90 K
   c. Receiver noise temperatures—120, 150, and 170 K. Neglect waveguide losses and any waveguide noise effects.

6. A receiver operating at 2800 MHz is shown in block diagram form in Figure P.6. Calculate its G/T in dBK^{-1} referred to the output port of the antenna.

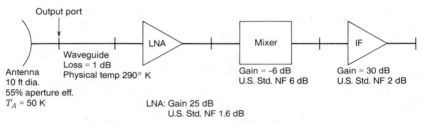

Figure P.6. The receiving system for Problem 6.

7. A valuable tool for law-enforcement agencies would be a "Dick Tracy" two-way wrist radio that would allow a police officer to communicate with his or her base from anywhere in the United States. As an approximation to such a radio, consider a 900-MHz system involving a synchronous satellite at 100° W longitude. Assume that the user is a police officer in Blacksburg, Virginia (37° N latitude and 80° W longitude), using a helmet-mounted quarter-wave whip antenna with a gain (ratio, not decibels) of 1.585. The officer's transmitter can deliver 1 W of RF to the antenna. The satellite's receiver has a U.S. standard noise figure of 2 dB, and its antenna has a noise temperature of 300 K. The uplink must be designed to provide a 20-dB (C/N) over a 3-kHz bandwidth. For this system, determine the following.

   a. The signal power in dBW required at the input to the satellite receiver.
   b. The diameter D of the satellite receiving antenna in meters assuming that the antenna is a parabolic reflector with a 55 percent aperture efficiency.
   c. The approximate area on the ground that the system will cover without repointing the antenna. You may assume that the satellite antenna gain is constant over a cone that is $70\lambda/D$ degrees wide.

8. The INTELSAT V 14 GHz West Spot transponder receiver has a G/T of 3.3 dBK^{-1} and a saturation flux density of -80.3 dBW/m^2. The 11-GHz transmitter part of this transponder has a saturated EIRP of 44.4 dBW. The transponder bandwidth is 72 MHz, centered at 14,205 MHz up and 11,155 MHz down. The satellite is used to set up a 72-MHz bandwidth link between two earth stations. Assume that both uplink and downlink path lengths are 40,000 km. Both transmitting and receiving antennas use 4-m diameter dishes, and the receiving earth station has an overall noise temperature (includes the antenna noise temperature) of 120 K. Both antennas have aperture efficiencies of 60 percent. The link operates with a 4-dB backoff on satellite input and on satellite output. Determine the following:

a. The uplink EIRP and uplink transmitter output power in W and dBW.
b. The uplink carrier-to-noise ratio $(C/N)_{ts}$ in dB.
c. The downlink carrier-to-noise ratio $(C/N)_{te}$ in dB.
d. The overall carrier-to-noise ratio $(C/N)_e$ in dB.

9. A Ku-band satellite carries a number of narrow bandwidth transponders to permit communications between small earth stations. The major parameters of the satellite are given in Table P.9.

a. Assume that the receive and transmit antennas on the spacecraft have rectangular shapes. Determine the dimensions of each antenna, in meters, and the gain of each antenna.
b. Calculate the flux density at the center of the coverage area, in dBW/m^2, when the satellite transponder is fully saturated, that is, output power is 20 W. What is the flux density at the edge of the coverage zone?
c. Find the G/T for an earth station at the edge of the coverage zone to achieve a (C/N) of 22 dB in a bandwidth of 5 MHz when the transponder is fully saturated by a single carrier.

Table P.9
Major Parameters of the Satellite System

Parameter	Uplink		Downlink
Frequency	14.90 GHz		11.30 GHz
3-dB contour of zone beam antenna	$3.6° \times 2.4°$		$3.6° \times 2.4°$
Antenna overall efficiency	65%		65%
Transponder O/P power (fully saturated)			20 W
Transponder input noise temp.	520 K		
Transponder gain; bandwidth		125 dB; 5MHz	
Pointing accuracy		$\pm0.1°$	
Propagation Path Data			
Range to satellite		39,000 km	
Path loss, including clear air atmospheric loss	208 dB		206 dB

d. Find the G/T for an earth station at the edge of the coverage zone to achieve a (C/N) of 10 dB in a bandwidth of 50 kHz when 20 carriers access the transponder (with equal power) simultaneously, and the transponder output power is backed off by 5 dB.

e. Find the minimum earth station EIRP to fully saturate the transponder output with a single carrier. Assume linear operation up to saturation.

f. Find the earth station EIRP to fully load the transponder output for 5 dB of output backoff when 20 identical stations transmit in an FDMA mode, each earth station producing equal power at the transponder input.

10. The characteristics of a digital service transponder to be flown on a satellite are as follows:

Frequencies: 14.9 GHz up
                11.3 GHz down

Antennas: 1.5° beamwidth (3 dB) up and down

Transponder saturated output power: 20 W

Transponder input noise temperature: 500 K

Transponder gain, up to saturation: 120 dB

Transponder bandwidth: 50 MHz

Path loss at 14.9 GHz in clear air: 208 dB

Path loss at 11.3 GHz in clear air: 206 dB

The parameters of the earth station are as follows:

Antenna diameter 3 m

Antenna aperture efficiency: 65% at 11.3 GHz
                              60% at 14.9 GHz

Ohmic loss between antenna and LNA: 1 dB

Receiver IF bandwidth: 50 MHz

a. For an earth station at the edge of the coverage zone of the satellite antenna, calculate the transmitter power required to just saturate the transponder output amplifier.

b. Find the earth station G/T required to give a (C/N) of 18 dB in the earth station IF amplifier. Ignore any noise contributed by the satellite transponder. Allow a pointing and miscellaneous loss figure of 0.5 dB at the earth station, and calculate the values for a station at the edge of the coverage zone. Hence determine the earth station noise temperature.

c. Calculate the (C/N) in the transponder when its output is just saturated. Hence find the (C/N) in the earth station receiver IF amplifier when transponder noise is included.

11. A satellite called "Cardinal I" provides a direct broadcast television service to the state of Virginia, with a beam 1° wide at the 3 dB points. The satellite is positioned at 60° W in the geostationary orbit. Cardinal I has an up-link at a frequency of 30 GHz from a central transmitting station in Richmond, Va. The downlink frequency is 42 GHz, and Cardinal I carries two transponders, each capable of relaying one television channel. The domestic receiving antenna is a parabolic dish of 0.8-m diameter, with an overall efficiency of 60 percent. However, a radome is used to prevent serious buildup of snow during winter, and the radome causes a 1.0-dB loss when its surface is wet. The receiver front end is mounted directly behind the antenna feed and has a U.S. standard noise figure of 7 dB at an ambient temperature of 17°C.

The television video signal at the transmitter is frequency modulated onto a 30-GHz carrier for transmission to the satellite. The video bandwidth is 4.2 MHz and the RF bandwidth of each channel is 30 MHz.

   a.   Calculate the G/T ratio for the receiving terminal.
   b.   Calculate the (C/N) for the 30 MHz IF bandwidth of the receiver when the satellite transmitter has an output power of 1 W. Assume no losses in the atmosphere, accurate pointing of the satellite, 50 percent efficiency for the transmitting antenna of Cardinal I, and a receiver located in the center of the 1° coverage zone.
   c.   Calculate the (C/N) for a receiver located at the edge of the coverage zone, with 0.33° pointing error in the receiving antenna, 1-dB clear-air atmospheric loss, and attenuation due to heavy rain of 10 dB in the atmosphere.
   d.   The FM threshold of the demodulator in the domestic receiver is 13 dB. Calculate the transponder output power required to achieve this (C/N) in one TV channel under the conditions in (c) above.

   12.   A regional satellite communication system using the 6/4 GHz band has the following parameters:

Satellite: Transponder gain—variable in range 85 to 100 dB
            Transponder bandwidth—36 MHz
            Transponder peak output power—6.3 W
            Antenna gain: Receive (6 GHz) 22 dB
                          Transmit (4 GHz) 20 dB
Earth Stations: Antenna gain: Receive (4 GHz) 60.0 dB
                              Transmit (6 GHz) 61.3 dB
                System noise temperature (4 GHz receive)—100 K
                Uplink path loss 200 dB
                Downlink path loss 196 dB

   a.   Four identical ground stations share one transponder in an FDMA mode. The allocated channel capacities are:

   stations 1 and 2: 132 channels/10 MHz RF bandwidth
   stations 3 and 4: 24 channels/5 MHz bandwidth

   In the FDMA mode the transponder is operated with 5-dB output backoff to minimize intermodulation distortion.

Find the transmitter power required at each earth station, if the transponder gain is 90 dB. Ignore any reduction in signal level due to satellite antenna pattern. (Power in the transponder must be shared in proportion to the bandwidth of each carrier.)

b. Calculate the (C/N) at a receiving earth station located on the 3-dB contour of the satellite's transmitting antenna beam, for the 10-MHz and the 5-MHz bandwidth signals.

5

MODULATION AND MULTIPLEXING TECHNIQUES FOR SATELLITE LINKS

Communications satellites carry telephone, television (TV), and data signals. Obviously data are always transmitted digitially, but telephone signals may be analog or digital. Digital television is used for teleconferencing, but entertainment TV is still analog. A satellite link will normally relay many signals from a single earth station; these must be separated to avoid interfering with each other. This separation is called *multiplexing,* and its most common forms are *frequency division multiplexing* (FDM) and *time division multiplexing* (TDM). In the first case the signals pass through the transponder on different frequencies; in the second they enter it at different times. Theoretically either multiplexing technique could be used with analog or digital modulation, but TDM is easier to implement with digital modulation and FDM is more convenient with analog modulation. Since choices of multiplexing and modulation techniques cannot be made separately, this chapter treats the two topics together. The next chapter discusses the related problem of multiple access, which is how to allow signals from different earth stations to use the same satellite without interference.

Many books have been written on modulation, demodulation, and multiplexing, and we lack the space to treat these topics in complete detail here. This chapter will review the characteristics of the signals commonly carried by satellites and stress those aspects of modulation and demodulation that are important to satellite link engineering.

5.1 ANALOG TELEPHONE TRANSMISSION

While digital modulation has some inherent advantages over analog frequency modulation (FM) for telephone signals, much of the early investment in the Intelsat

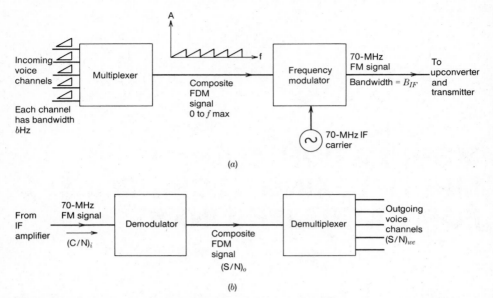

(a)

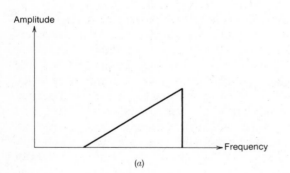

(b)

Figure 5.1 Transmitting and receiving ends of a typical FDM system. (a) At uplink earth station. (b) At downlink earth station.

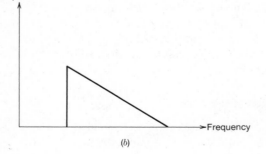

Figure 5.2 Representations of the spectrum of a telephone system baseband voice signal. (a) Normal spectrum. (b) Inverted spectrum.

network was for FDM/FM telephone systems, and these are still widely used. In this section we will discuss analog multiplex FDM/FM system design and characteristics in detail. We will also present two other analog schemes, *single-channel-per-carrier* (*SCPC*) and *companded single sideband* (*CSSB*), both of which offer advantages over conventional FDM/FM multiplex transmission and over digital modulation under some circumstances.

Satellite FDM/FM analog telephone links resemble the terrestrial microwave point-to-point links that have carried most long distance telephone traffic since the early 1950s. Figure 5.1 sketches a typical system. In it a *multiplexer* takes the *baseband* signals from many individual telephone conversations, translates them to adjacent channels in the RF spectrum, and combines them. Essentially, the multiplexer "stacks" the individual channels in nonoverlapping spectral bands. The resulting composite FDM signal frequency-modulates an IF carrier (usually at 70 MHz) to create an FM (frequency-modulation) multiplex signal. The IF carrier is converted to the appropriate uplink frequency, amplified, and transmitted to the satellite. At the satellite the signal is amplified, downconverted to the downlink band, and retransmitted. At the receiving earth station the downlink signal is amplified and downconverted to IF. The frequency-modulated IF signal drives an FM *demodulator*, which recovers multiplex signals with the voice channels stacked in frequency. Then a *demultiplexer* uses product detectors and filters to translate each channel back to baseband.

Baseband Voice Signals

A baseband voice signal is the voltage generated by an individual telephone set. While its detailed characteristics depend upon the speaker, the Bell System treats it as having a flat spectrum extending from 300 to 3100 Hz. The CCITT recommends 300 to 3400 Hz, but some designs assume a 0 to 3000 Hz spectrum. Here we will follow Intelsat practice and use the CCITT 300 to 3400 Hz spectrum.

Schematically the spectrum of a baseband voice signal is often represented by the triangle shown in Figure 5.2; in a normal spectrum the peak of the triangle is to the reader's right and in an inverted spectrum (one in which the order of frequencies has been reversed) the peak is to the reader's left. The spectrum is not really triangular; this is just a convenient symbol.

The amplitude of a voice signal in a communications link depends on where and how it is measured. In telephone engineering practice (see reference 1, p. 23ff), signal powers are expressed in terms of *transmission level*—that is, their decibel levels with respect to a reference point. At the reference point, the signal power in *dBm* is indicated by the unit *dBm0*; the 0 stands for the zero transmission level point or the test point. Thus, a −2-dBm0 signal is one that produces an average power of −2 dBm at the reference point. A suitable power meter placed at the −5-dB transmission level point would measure the absolute power in the −2-dBm0 signal as −7 dBm.

When telephone engineering began, the test point was accessible and meters could be connected to it. The Bell System standardized the transmission level at the outgoing side of the toll transmission switch as 2 dB below the test level point or -2 dB0 (-2 dB with respect to the zero test level point). With later switch-boards, the test point lost its accessibility and disappeared. But the -2 dB standard transmission level at the toll transmission switch remained, and transmission levels are defined from this reference exactly as if the zero test level point still existed.

Under Bell System standards that prevailed for many years, the long-term average power carried by a single voice channel in a telephone system was taken to be -18 dBm0 (CCIR assumes -15 dBm0). The peak instantaneous power in the channel is about 18 dB higher or 0 dBm0. Thus, telephone equipment is often adjusted by applying a 1-kHz tone at 0 dBm0 to the system to simulate peak power on one channel. This is called the *test tone*. We will return to it later in our discussion of multiplexing.

The reader should be aware that the original Bell and CCIR values of -18 dBm0 and -15 dBm0 for the average power level in a single telephone channel are very conservative and that many carriers (including Bell) use other values in some applications. RCA bases its designs on -22 dBm0 while the Bell System uses -19.8 dBm0 for terrestrial FDM/FM links and -22 dBm0 for some satellite links [2]. In general, the number of voice channels that a transponder can carry varies inversely with the average power level per channel [3].

Voice Signal Multiplexing

The process of shifting analog voice channels in frequency and combining them for transmission is called frequency division multiplexing (FDM). The procedure is hierarchial; individual channels are combined into groups; the groups are combined into larger groups; the larger groups are combined into still larger groups, and so on. The names of the groups and their internal channel arrangements vary between administrations and countries. In this section we will use terminology largely drawn from reference 1 for the Bell System and from reference 4 (p.43) for Intelsat.

The first step in voice channel multiplexing is to combine 12 baseband signals into a *basic group* (often called simply a *group*) extending from 60 kHz to 108 kHz. The channels are stacked one above the other at 4-kHz intervals. The stacking is done by double-sideband suppressed-carrier (DSBSC) amplitude modulating each voice channel onto an appropriate carrier, filtering out the upper sideband, and saving and summing the lower sidebands. The result is a single-sideband suppressed-carrier (SSBSC) signal. See reference 5 for a detailed explanation of SSBSC and DSBSC techniques. Figure 5.3 illustrates the process. The carrier frequency in kHz of the nth channel is given by $112 - 4n$; thus channel 1 is at the top of the spectrum and channel 12 is at the bottom. Since each channel occupies only 3.1 kHz, there are 0.9 kHz *guardbands* between channels. These prevent interference and simplify the filtering process when the baseband signals are recovered at the receiver. Selecting the lower sideband in the modulation process inverts the spectra of the channels, but they will be inverted again and put back in the right order at the receiver.

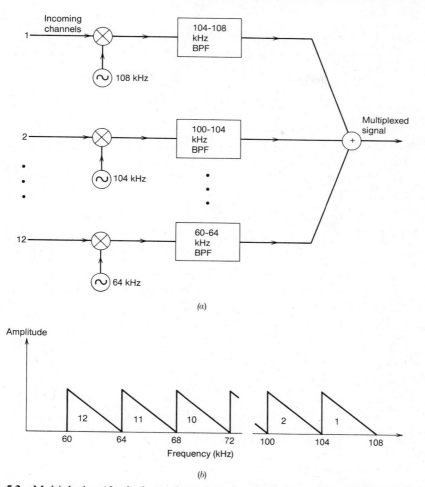

(a)

(b)

Figure 5.3 Multiplexing 12 telephone channels to form a basic group. (a) The basic hardware. (b) The spectrum of the multiplex signal showing individual channel spectra.

While a single basic group could be transmitted by itself, most satellite and terrestrial microwave links carry significantly more channels. Bell terrestrial systems use a rigid hierarchy of channel combinations extending from the 12-channel basic group through the 60-channel *basic supergroup* to 600-channel *basic master-groups* and beyond. The largest named combination in the Intelsat hierarchy is the basic supergroup, five groups stacked in a 240-kHz band. The stacking is done by SSBSC modulating the individual groups onto appropriate carriers and summing the resulting lower sidebands. Figure 5.4 illustrates generation of a 312 to 552 kHz supergroup. The carriers are space 48 kHz apart; that for group 5 is highest at 612 kHz. The spectra of the individual groups are inverted when the basic super-group is formed. But the individual channel spectra in the groups are themselves

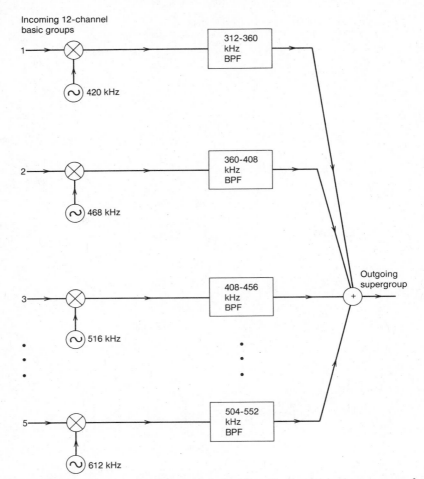

Figure 5.4   Schematic hardware for multiplexing five 12-channel basic groups to form a Bell System basic supergroup.

inverted; the second inversion puts them back in the original order and the individual channel signals in the supergroup are frequency-shifted versions of the original baseband signals. Supergroups are normally separated by 12-kHz guard bands.

Since transponder bandwidth is usually limited, the arrangement of channels in the Intelsat system follows a more flexible format than that used by Bell. In INTELSAT V, for example, earth stations are allowed 12, 24, 36, 48, 60, 72, 96, 132, 192, 252, 312, 432, 492, 552, 612, 792, 972, 1092, 1332, or 1872 voice channels. All of these numbers are divisible by 12 (the number of channels in a basic group), but the channel arrangement varies with the number of channels in order to make efficient use of the transponder. For example, 132 channels are multiplexed by combining a basic group in the 12 to 60 kHz band with one basic supergroup from 60 to 300 kHz and a second basic supergroup from 312 to 552 kHz. Table 5.1 [4:pp. 59ff] summarizes the channel combinations available on the INTELSAT IV through VI spacecraft and lists some of the associated system specifications. The baseband spectrum between 0 and 12 kHz is reserved for an intra-system channel called the *order wire*, which carries housekeeping information, and for the energy-dispersal signal discussed later.

The amplitude of a multiplexed telephone signal is a random function of time whose characteristics depend upon N, the number of channels. For N greater than or equal to 24, the signal amplitude is usually represented by a Gaussian probability distribution with zero mean and rms value σ. The probability $p(v)$ that the instantaneous voltage has the value v is given by

$$p(v) = \frac{1}{\sigma\sqrt{2\pi}} e^{-(v^2/2\sigma^2)} \tag{5.1}$$

For design purposes, the rms value σ is usually taken as that which will result in 1 mW (0 dBm) total power at the impedance level of the system.

Usually the amplitude of a multiplexed voice signal with $N > 24$ is hard limited to lie between ± 3.16 times the rms value. This causes clipping for something less than 0.2 percent of the time and introduces negligible distortion. Thus it is common practice to equate the peak value of the signal to 3.16 times the rms values. For $N < 24$ the multiplexed spectrum is more "peaky" and an 8.5 peak-to-rms ratio is often used.

Frequency Modulation (FM) Theory

To date frequency modulation is the only form of analog modulation widely used on satellite links. In exchange for a wide bandwidth and poor spectral efficiency, it offers considerable signal-to-noise ratio (S/N) improvement. This means that the signal-to-noise ratio at the output of an FM detector is much larger than the carrier-to-noise ratio (C/N) at the detector input, provided that the input (C/N) is above a threshold value that is characteristic of that detector. Since satellite links have been power limited rather than bandwidth limited and have had to operate with low (C/N) levels, FM has been the analog modulation of choice.

For a detailed discussion of the theory and practice of frequency modulation the reader should consult reference 5. Here we will summarize by quoting from reference 5 that "Frequency modulation results when the deviation Δf of the instantaneous frequency f from the carrier frequency f_c is directly proportional to the instantaneous amplitude of the modulating voltage." An FM modulator is characterized by a maximum frequency deviation $\Delta\omega$, which occurs when the modulating voltage reaches its maximum value. For modulation by a single-frequency sinusoid at radian frequency ω_{mod} whose peak amplitude produces maximum deviation, the expression for an FM waveform $v(t)$ with carrier frequency ω_c is given by

$$v(t) = A \cos\left[\omega_c t + \left(\frac{\Delta\omega}{\omega_{\text{mod}}}\right) \sin(\omega_{\text{mod}} t)\right] \tag{5.2}$$

The ratio $\Delta\omega/\omega_{\text{mod}}$ is called the *FM modulation index m*. This FM waveform can be conveniently represented in terms of m by

$$v(t) = A \cos(\omega_c t + m \sin \omega_{\text{mod}} t) \tag{5.3}$$

Because the FM waveform of Eq. (5.3) takes the form of a cosine of a sine, its spectrum is not obvious. But it can be expanded in an infinite series of discrete components as

$$v(t) = A\{J_0(m) \cos \omega_c t + \sum_{n=1}^{\infty} J_n(m)[\cos(\omega_c + n\omega_{\text{mod}})t + (-1)^n \cos(\omega_c - n\omega_{\text{nod}})t]\} \tag{5.4}$$

where the $J_0, J_1, \ldots J_n$ are Bessel functions of the first kind and order 0, 1, $\ldots n$. In theory the spectrum of even a single-frequency sinusoidally modulated FM signal has an infinite number of side frequencies and requires infinite bandwidth; in practice the signal must be filtered to reduce its bandwidth for transmission. An approximate value for the required bandwidth B is given by an equation known as Carson's rule:

$$B = 2f_{\text{mod}}(m + 1) = 2(\Delta f + f_{\text{mod}}) \tag{5.5}$$

where B, Δf (the maximum frequency deviation of the modulator), and f_{mod} (the modulating frequency) are all in Hz. Satellite link designers normally use Carson's rule to calculate bandwidth. The energy associated with sidebands outside the bandwidth B is small, and very little distortion of the modulating waveform occurs when the FM signal is passed through a filter with bandwidth B. An FM signal that is intentionally transmitted through a transponder or to a receiver whose bandwidth is significantly less than the Carson's rule bandwidth is said to be *over-deviated* [2]. We will discuss overdeviation in Chapter 6.

The spectrum of an FM waveform modulated by a real signal is much more complicated than that for a single-frequency sinusoid. But in this case the required bandwidth may still be estimated by Carson's rule if f_{mod} is replaced by f_{max}, the maximum modulating frequency.

$$B = 2(\Delta f + f_{\text{max}}) \tag{5.6}$$

FM Detection Theory: (S/N) Improvement

An FM detector produces an output voltage whose value is proportional to the difference between the instantaneous frequency of the incoming signal and a reference frequency sometimes called the rest frequency. The reference frequency corresponds to the carrier frequency of the previous section. Under normal conditions the detector output is a replica of the modulating waveform that was applied to the carrier before transmission.

As indicated previously, the bandwidth of a frequency-modulated waveform with wideband FM is much greater than the bandwidth of the modulating waveform. Hence the bandwidth of the input signal to an FM detector is much greater than the bandwidth of the output signal. The bandwidth compression provided by the detector is accompanied by an improvement in signal-to-noise ratio (S/N), provided that the input carrier-to-noise ratio is sufficiently large. In other words, the postdetection signal-to-noise ratio $(S/N)_o$ can be considerably larger (by perhaps 20 dB) than the input carrier-to-noise ratio $(C/N)_i$. Since this process is very important to the design and operation of analog FM satellite links and since it is often poorly or incorrectly described in the literature, we will describe it in some detail. For a full development, see pp. 298ff of reference 6.[1]

Assume that an incoming FM signal has an rms amplitude A, occupies an IF bandwidth B_{IF}, and is sinusoidally modulated to have an rms frequency deviation Δf_{rms}. Let the single-sided rms noise power spectral density in the IF bandwidth be η W/Hz so that the noise power at the detector input is ηB_{IF}. For a carrier voltage amplitude A, the average carrier power at the detector input is $A^2/2$ and the incoming carrier-to-noise ratio $(C/N)_i$ is then

$$(C/N)_i = \frac{A^2}{2\eta B_{IF}} \tag{5.7}$$

Note that $(C/N)_i$ is the same as the overall carrier-to-noise ratio $(C/N)_e$ of the previous chapter.

Let the transfer characteristic of the demodulator be K. This means that a frequency deviation Δf on the incoming carrier produces $K \Delta f$ volts at the demodulator output. The rms signal power available at the demodulator output is then proportional to $(K \Delta f_{rms})^2$. If the *output* frequency response of the demodulator extends from f_1 to f_2 Hz, then the noise power output N is given by

$$N = 2\eta(K/A)^2 \int_{f_1}^{f_2} f^2 df = 2\eta \left(\frac{K}{A}\right)^2 \frac{(f_2^3 - f_1^3)}{3} \tag{5.8}$$

Combining Eqs. (5.7) and (5.8) we find that the output signal-to-noise ratio $(S/N)_o$ is

$$(S/N)_o = \frac{K^2(\Delta f_{rms})^2}{2\eta(K/A)^2 \dfrac{(f_2^3 - f_1^3)}{3}}$$

$$= (C/N)_i \left[\frac{3B_{IF}(\Delta f_{rms})^2}{(f_2^3 - f_1^3)} \right] \tag{5.9}$$

---

[1] The reader who wishes to compare our results with those of reference 6 should note that, in accordance with common satellite communications practice, we compute the noise input noise power in the occupied IF bandwidth while reference 6 computes the noise power in the bandwidth that the modulating waveform would occupy at baseband. Our K is related to the α of reference 6 by $K = 2\pi\alpha$.

Recognizing that for sinusoidal modulation the rms deviation Δf_{rms} is related to the peak deviation Δf_{peak} by a factor of $2^{\frac{1}{2}}$, we may rewrite Eq. (5.9) as

$$(S/N)_o = (C/N)_i \left(\frac{3}{2}\right) \frac{B_{IF}(\Delta f_{peak})^2}{(f_2^3 - f_1^3)} \tag{5.10}$$

Remember that Eq. (5.10) is true only for a single-frequency sinusoidally modulated FM signal on a link whose overall $(C/N)_i$ value exceeds a threshold that is typically 10 dB.

For a single (i.e., nonmultiplexed) but nonsinusoidal modulating signal like that used in TV or SCPC telephony, we may use Eq. (5.10) to estimate the FM improvement. To do this, we assume that the spectrum of the modulating waveform extends from 0 to f_{max} Hz. These limits define the output frequency range of the detector; hence, $f_2 = f_{max}$ and $f_1 = 0$. Thus

$$(S/N)_o = (C/N)_i \frac{3}{2} \frac{B_{IF}}{f_{max}} \left(\frac{\Delta f_{peak}}{f_{max}}\right)^2 \tag{5.11}$$

Writing the modulation index m as

$$m = \frac{\Delta f_{peak}}{f_{max}} \tag{5.12}$$

and using Carson's rule

$$B_{IF} = 2f_{max}(1 + m) \tag{5.13}$$

we can express Eq. (5.11) as

$$(S/N)_o = (C/N)_i \times 3(1 + m)m^2 \tag{5.14}$$

For large m, $3(1 + m)m^2 \simeq 3\,m^3$, while for $m \ll 1$, $3(1 + m)m^2 \simeq 3\,m^2$ and Eq. (5.14) matches the equation given on page 23-11 of reference 7. Both results are correct but they refer to different situations. Such simplified versions of Eq. (5.14) appear frequently in the literature, and the reader should be sure they apply to the problem at hand before using them.

Frequency Modulation with Multiplexed Telephone Signals

The signal-to-noise ratio described by Eq. (5.14) exists at the output of the FM demodulator and describes the ratio of the total power in the multiplexed telephone channels to the total thermal noise power. Let us now consider the FM detector output signal-to-noise ratio for a single telephone channel located at the high-frequency end of a multiplex signal. This is the (S/N) at the output of the demultiplexer. If the channel bandwidth is b Hz, then the noise power output of interest is that between $f_{max} - b$ and f_{max}. Hence the term $f_2^3 - f_1^3$ in Eq. (5.9) becomes $f_{max}^3 - (f_{max} - b)^3$. Since $f_{max} \gg b$, $(f_{max} - b)^3 \simeq f_{max}^3(1 - 3b/f_{max}) =$

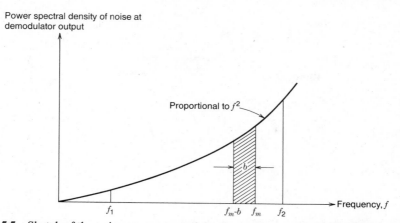

Figure 5.5 Sketch of the noise power spectral density at the output of an FM demodulator. For the narrow band voice channel indicated by the shaded region, the noise power N may be calculated by multiplying the noise spectral density at f_m by b. This leads to Eq. (5.15). For wideband channels the noise power must be calculated by integrating from f_1 to f_m.

$f_{max}^3 - 3bf_{max}^2$ and Eq. (5.9) becomes

$$(S/N)_{wc} = (C/N)_i \left[\frac{3B_{IF}(\Delta f_{rms})^2}{(3bf_{max}^2)} \right] = (C/N)_i \left(\frac{B_{IF}}{b} \right) \left(\frac{\Delta f_{rms}}{f_{max}} \right)^2 \qquad (5.15)$$

where the subscript wc means worst channel. See Figure 5.5. Note that b is the bandwidth of one baseband voice channel (nominally 3100 Hz), and $(C/N)_i$ is the overall carrier-to-noise ratio of the link.

Equation (5.15) describes the ratio of signal power to thermal noise power in a telephone channel at the upper end of a multiplexed baseband signal. But the frequency response of neither the human ear nor a telephone receiver is flat, and a telephone listener will respond differently to noise in different parts of the audio spectrum. Some of the noise that is present in bandwidth b will be unnoticed, and the effective signal-to-noise ratio will be higher than that given by Eq. (5.15) by a *weighting factor*. Its value depends on the frequency response of the telephone receiver and of the user's ear. The Bell System uses what is called *C-message weighting*, while CCITT and common satellite practice use *psophometric weighting*. We will adopt the latter and use the symbol p for the psophometric weighting factor. The numerical value of p is 1.78; this corresponds to 2.5 dB [8].

Noise at the high-frequency end of the input spectrum to an FM detector is demodulated with greater output than noise at the low end. (See reference 5, pp. 298 and 322–325.) Figure 5.6a sketches this effect. The rising noise at the detector output above some arbitrary frequency f_d can be suppressed as shown in Figure 5.6a if a filter with the characteristic of Figure 5.6b is installed as indicated in Figure 5.6c. This *deemphasis* filter will also reduce the high-frequency content of the modulating signal, but this problem can be eliminated if a *preemphasis* filter is

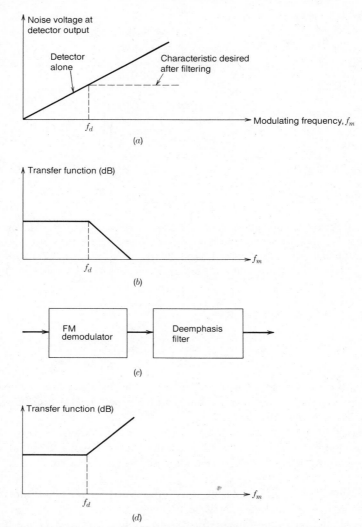

Figure 5.6 Preemphasis. (*a*) Noise voltage at FM detector output. (*b*) Deemphasis filter characteristic. (*c*) Location of deemphasis filter. (*d*) Preemphasis filter characteristic.

inserted at the transmitter ahead of the modulator. The preemphasis and deemphasis filters have inverse characteristics for frequencies above f_d. In practice the characteristics shown need to be maintained only up to the highest baseband frequency present.

This filtering process is called preemphasis although it includes both a preemphasis and a deemphasis filter. It improves the overall (S/N) of the demodulated signal. The degree of improvement depends on the filters and the modulating waveforms used. In CCIR standard satellite and microwave links, preemphasis improves the output signal-to-noise ratio of a telephone system by a factor of 2.5 (4 dB) over that given by Eq. (5.15) [9]. Other values apply to SCPC and TV transmission. Generally the more nonuniform the modulation spectral density (i.e., the more energy below f_d in Figure 5.6), the larger the preemphasis improvement.

CCIR Recommendation 464 [8, pp. 77ff] describes in detail the preemphasis standards for FDM/FM telephone systems. If the highest baseband frequency is f_{max}, the preemphasis filter should provide minimum attenuation at $f_r = 1.25 f_{max}$. Taking attenuation at baseband frequency f_r as the 0 dB reference, the filter attenuation A at any other baseband frequency f should be

$$A = 10 \log_{10} \left\{ 1 + \frac{6.90}{1 + 5.25/(f_r/f - f/f_r)^2} \right\} \text{dB} \tag{5.16}$$

Reference 8 provides several circuits for preemphasis and deemphasis filters.

Preemphasis improvement and the improvement due to psophometric weighting are independent of each other; hence the right side of Eq. (5.15) may be multiplied by p and by w to yield the psophometrically weighted signal-to-noise ratio on the worst multiplexed telephone channel at the output of an FM link using preemphasis and having an overall carrier-to-noise ratio of $(C/N)_i$.

$$(S/N)_{wc} = (C/N)_i \left(\frac{B_{IF}}{b} \right) \left(\frac{\Delta f_{rms}}{f_{max}} \right)^2 pw \tag{5.17}$$

In decibel form

$$(S/N)_{wc} = (C/N)_i + 10 \log_{10} \left(\frac{B_{IF}}{b} \right) + 20 \log_{10} \left(\frac{\Delta f_{rms}}{f_{max}} \right) + P + W \text{ dB} \tag{5.18}$$

where P is 2.5 dB and W is 4 dB.

Bandwidth Calculation for FDM/FM Telephone Signals

We derived Eq. (5.18) and its predecessors assuming a sinusoidally modulated waveform with an rms frequency deviation Δf_{rms} and requiring a transmission or IF bandwidth B_{IF}. In this section we will relate these quantities to the number of channels N carried by a multiplexed telephone signal and to the available transponder bandwidth.

For link performance calculations, the rms frequency deviation Δf_{rms} that should be used is the *rms test-tone deviation*. This is the rms carrier deviation that a single 1-kHz 0-dBm sinewave called the test tone would produce when supplied to the modulator input, and it represents a standardized test signal in one telephone channel. Putting this another way, the transmitter is designed and adjusted to produce this rms carrier frequency deviation when the modulator input signal is a standard 1-kHz 0-dBm test tone. The rms test-tone deviation is related to the rms deviation that a multiplexed telephone signal will cause by the *loading factor*,

l. For N voice channels, l is given by reference 9 as

$$20 \log_{10}(l) = L = \begin{cases} -15 + 10 \log_{10}(N), & N > 240 \\ -1 + 4 \log_{10}(N), & 12 \le N \le 240 \end{cases} \qquad (5.19)$$

The product $l \Delta f_{rms}$ is called the *rms multicarrier deviation*.

The ratio of the peak frequency deviation Δf_p to the rms multicarrier deviation $l \Delta f_{rms}$ is given by the *peak factor*, g. For a large number of channels (typically $N > 24$), g is taken as 3.16 (corresponding to 10 dB) and for small numbers of channels (typically $N < 24$), a value of 6.5 (18.8 dB) may be used [9]. If necessary, the incoming voice signals may be amplitude limited to force a true peak-to-rms ratio of 3.16 on the multiplex signal. Thus for an analog FDM/FM telephone link

$$\Delta f_p = lg \, \Delta f_{rms} \qquad (5.20)$$

The use of a high peak-to-mean ratio results in a low rms frequency deviation and consequently a low (S/N) improvement factor for an average baseband level. Amplitude limiting will cause distortion on large signal peaks, but it allows a higher average frequency deviation and a greater (S/N) improvement. In SCPC systems where the peak-to-mean ratio of the single telephone signal is large, companding is often used to reduce the peak-to-mean ratio of the signal before it is applied to an FM modulator. We will discuss companding later.

The 3.16 factor used in calculating the peak deviation of the carrier frequency represents the 0.1 percent extreme of a Gaussian distribution of signal voltages. For 0.2 percent of the time, the signal voltage will exceed 3.16 times the rms value, assuming a Gaussian probability distribution. Because we have to restrict the bandwidth of our RF signal to avoid interference with adjacent channels, the FM modulator at the transmitter will have to be preceded by a limiter that prevents large peaks of signal from overdeviating the carrier frequency. This distorts the multiplexed signal, but for $N > 24$ the effects of limiting are small.

The maximum modulating frequency, f_{max}, depends upon the multiplexing scheme used—that is, on the number of channels multiplexed and how they are organized into basic groups, basic supergroups, and so on. When the standards of a satellite system are established, f_{max} is tabulated for the allowed values of N. Table 5.1 contains the f_{max} values for INTELSAT IV through VI. If f_{max} is not known or if a new satellite link is being designed, a good estimate to use for f_{max} in kHz is 4.2 N [9].

For a minimum required worst channel $(S/N)_{wc}$ (typically about 50 dB) and an overall $(C/N)_i$ fixed by the link power budget, a satellite systems engineer may trade off values of N, B_{IF}, and Δf_{rms}. The number of channels, N, determines B_{IF} through Carson's rule by

$$B_{IF} = 2(lg \, \Delta f_{rms} + f_{max}) \qquad (5.21)$$

where l and f_{max} depend on N. The minimum $(S/N)_{wc}$ and B_{IF} in turn determine the required value of $(\Delta f_{rms}/f_{max})$ in Eq. (5.17). A satisfactory solution requires that B_{IF} not exceed the allocated transponder bandwidth and that the rms test tone deviation Δf_{rms} be achievable by the modulator. After discussing minimum $(S/N)_o$ requirements, we will present an example B_{IF} calculation that illustrates the interdependence of all the variables involved.

Telephone Performance Specifications

While U.S. engineers tend to think of system performance requirements in terms of decibel signal-to-noise ratios, international practice often expresses system specifications in terms of absolute channel noise levels measured in picowatts (psophometrically weighted), abbreviated *pWp*, or in dB above a 1 pWp reference level, abbreviated dBp. (Unfortunately the dBp abbreviation is used both for weighted and for unweighted picowatts.) Picowatts are particularly useful when noise power contributions from several sources must be combined. Decibel powers cannot be added directly.

To convert between picowatts and dBp and milliwatts and dBm, it is necessary first to remember that a psophometric weighting filter reduces the power level of white noise by 2.5 dB and second that 0 dBp (unweighted) corresponds to -90 dBm (unweighted). If P is an absolute power level to be expressed in different units, then

$$P \text{ in dBp (unweighted)} = 10 \log_{10}(P \text{ in pWp}) + 2.5 \qquad (5.22)$$

and

$$P \text{ in dBm (unweighted)} = P \text{ in dBp} - 90 \qquad (5.23)$$

Assuming a standard 0 dBm signal level, then

$$(S/N) \text{ unweighted} = -P(\text{dBm}) = 87.5 - 10 \log_{10}(P \text{ in pWp}) \qquad (5.24)$$

$$(S/N) \text{ weighted} = 90 - 10 \log_{10}(P \text{ in pWp}) \qquad (5.25)$$

Thus a 7500-pWp channel noise level corresponds to a weighted (S/N) of 51.25 dB and an unweighted (S/N) of 48.75 dB. Typical satellite link designs allow 7500 to 10,000 pWp total thermal noise for the space segment (up and down links including the intermodulation noise added at the spacecraft). The Intelsat specification for the INTELSAT IV, IV-A, and V space segments is 8000 pWp [4].

Practical Examples

In this section we will apply the equations that describe FDM/FM analog telephone transmission to several examples involving the INTELSAT V spacecraft. The numbers describing the satellite are taken from Table 5.1.

Example 5.1.1

An INTELSAT V transponder using a global beam achieves a 17.8-dB $(C/N)_i$ at an earth station. The transponder carries 972 channels on a single carrier; the FDM/FM signal fully occupies a 36-MHz bandwidth in the transponder.

Table 5.1a
INTELSAT IV-A, V, V-A, and VI Transmission Parameters (Regular FDM/FM Carriers)

Carrier Capacity (Number of Channels)	Top Baseband Frequency (kHz)	Allocated Satellite BW Unit (MHz)	Occupied Bandwidth (MHz)	Deviation (rms) for 0-dBm0 Test Tone (kHz)	Multichannel rms Deviation (kHz)	Carrier-to-Total Noise Temperature Ratio at Operating Point $(8000 + 200\ pW0p)$ (dBW/K)	Carrier-to-Noise Ratio in Occupied BW (dB)
n	f_m	b_s	b_0	f_r	f_{mc}	(C/T)	(C/N)
12	60	1.25	1.125	109	159	−154.7	13.4
24	108	2.5	2.00	164	275	−153.0	12.7
36	156	2.5	2.25	168	307	−150.0	15.1
48	204	2.5	2.25	151	292	−146.7	18.4
60	252	2.5	2.25	136	276	−144.0	21.1
60	252	5.0	4.0	270	546	−149.9	12.7
72	300	5.0	4.5	294	616	−149.1	13.0
96	408	5.0	4.5	263	584	−145.5	16.6
132	552	5.0	4.4	223	529	−141.4	20.7
96	408	7.5	5.9	360	799	−148.2	12.7
132	552	7.5	6.75	376	891	−145.9	14.4
192	804	7.5	6.4	297	758	−140.6	19.9
132	552	10.0	7.5	430	1020	−147.1	12.7
192	804	10.0	9.0	457	1167	−144.4	14.7
252	1052	10.0	8.5	358	1009	−139.9	19.4

252	1052	15.0	12.4	577	1627	−144.1	13.6
312	1300	15.0	13.5	546	1716	−141.7	15.6
372	1548	15.0	13.5	480	1645	−138.9	18.4
432	1796	15.0	13.0	401	1479	−136.2	21.2
432	1796	17.5	15.75	517	1919	−138.5	18.2
432	1796	20.0	18.0	616	2279	−139.9	16.1
492	2044	20.0	18.0	558	2200	−137.8	18.2
552	2292	20.0	18.0	508	2121	−136.0	20.0
432	1796	25.0	20.7	729	2688	−141.4	14.1
492	2044	25.0	22.5	738	2911	−140.3	14.8
552	2292	25.0	22.5	678	2833	−138.5	16.6
612	2540	25.0	22.5	626	2755	−136.9	18.1
792	3284	36.0	32.4	816	4085	−137.0	16.5
972	4028	36.0	32.4	694	3849	−133.8	19.7
972	4028	36.0	36.0	802	4417	−135.2	17.8
1092	4892	36.0	36.0	701	4118	−132.4	20.7

Source: (Reprinted with permission of the International Telecommunications Satellite Organization from *Standard A Performance Characteristics of Earth Stations in the INTELSAT IV, IV-A, and V Systems Having a G/T of 40.7 dB/K* (BG-28-72E Rev. 1), Intelsat, Washington, DC, December 15, 1982.)

Table 5.1b
INTELSAT IV-A, V, V-A, and VI Transmission Parameters (High-density FDM/FM Carriers)

Carrier Capacity (Number of Channels)	Top Baseband Frequency (kHz)	Allocated Satellite BW Unit (MHz)	Occupied Bandwidth (MHz)	Deviation (rms) for 0-dBm0 Test Tone (kHz)	Multichannel rms Deviation (kHz)	Carrier-to-Total Noise Temperature Ratio at Operating Point $(8000 + 200\ pW0p)$ (dBW/K)	Carrier-to-Noise Ratio in Occupied BW (dB)
n	f_m	b_s	b_0	f_r	f_{mc}	(C/T)	(C/N)
72	300	2.5	2.25	125	261	−141.7	23.4
192	804	5.0	4.5	180	459	−136.3	25.8
252	1052	7.5	6.75	260	733	−137.1	23.2
312	1300	10.0	9.0	320	1005	−137.1	22.0
492	2044	15.0	13.5	377	1488	−134.4	22.9
612	2540	20.0	17.8	454	1996	−134.2	21.9
792	3284	20.0	18.0	356	1784	−129.9	26.2
792	3284	25.0	22.4	499	2494	−132.8	22.3
972	4028	25.0	22.5	410	2274	−129.4	25.7
1332	5884	36.0	36.0	591	3834	−129.3	23.8

Source: (Reprinted with permission of the International Telecommunications Satellite Organization from *Standard A Performance Characteristics of Earth Stations in the INTELSAT IV, IV-A, and V Systems Having a G/T of 40.7 dB/K* (BG-28-72E Rev. 1), Intelsat, Washington, DC, December 15, 1982.)

If the weighted (S/N) on the top baseband channel is 51.5 dB, find the rms test-tone deviation and the rms multicarrier deviation that must be used. Compare these with the tabulated values.

First we will illustrate the procedure to follow if the multiplexing scheme is not known. Estimating f_{max} as $4200N = 4.082$ MHz and substituting into Eq. (5.18) we have

$$51.0 = 17.8 + 10 \log_{10}\left(\frac{36 \times 10^6}{3.1 \times 10^3}\right) + 20 \log_{10}\left(\frac{\Delta f_{rms}}{4.082 \times 10^6}\right) + 6.5$$

Solving,

$$51.0 - 17.8 - 40.6 - 6.5 = -13.9 = 20 \log_{10}\left(\frac{\Delta f_{rms}}{4.082 \times 10^6}\right)$$

Hence $\Delta f_{rms} = 778$ kHz is the rms test-tone deviation.

Under the loading rule of Eq. (5.19), $L = -15 + 10 \log_{10}(972) = 14.88$ and $l = 10^{(14.88/20)} = 5.55$. Thus the rms multicarrier deviation is $l \Delta f_{rms} = 5.55 \times 778$ kHz $= 4.32$ MHz.

We may check this answer by computing the *Carson's rule bandwidth* $B = 2(3.16 \times 4.32$ MHz $+ 4.082$ MHz$) = 35.5$ MHz, which is close to the 36 MHz allowed.

These are slightly different from the published values because the true value of f_{max} (determined by the multiplexing hierarchy) is 4.028 MHz. Using this value of f_{max} we find Δf_{rms} to be 813 kHz and the occupied bandwidth is 36.6 MHz. The published Δf_{rms} is 802 kHz; this leads to an occupied bandwidth of 36.2 MHz and a weighted (S/N) of 50.9 dB.

Example 5.1.2

A single carrier that will occupy (when modulated) 9 MHz of an INTELSAT V transponder can produce a $(C/N)_i$ of 14.7 dB at a standard earth station using the satellite's global beam. Assuming an 8000-pWp space segment noise allocation, how many telephone channels can the transponder carry?

This example illustrates the kind of analysis a systems engineer would perform to determine telephone channel allocations for a proposed spacecraft. It requires an iterative solution.

First, by Eq. (5.24), 8000 pWp corresponds to a weighted (S/N) of 51.0 dB. Substituting this value, the 14.7-dB (C/N), 3.1-kHz channel bandwidth, and IF bandwidth equal to the 9-MHz occupied bandwidth into Eq. (5.18), we obtain

$$51.0 = 14.7 + 10 \log_{10}\left(\frac{9 \times 10^6}{3.1 \times 10^3}\right) + 20 \log_{10}\left(\frac{\Delta f_{rms}}{f_{max}}\right) + 4 + 2.5$$

and

$$20 \log_{10}\left(\frac{\Delta f_{rms}}{f_{max}}\right) = -4.83$$

or

$$\left(\frac{\Delta f_{rms}}{f_{max}}\right) = 0.57$$

For N channels f_{max} in Hz is approximately $4200N$. Putting all these numbers, a g value of 3.16, and l for $N < 240$ from Eq. (5.19) into Eq. (5.21), we obtain

$$9 \times 10^6 = 2\{10^{[(-1 + 4 \log_{10} N)/20]} \times 3.16 \times 0.57 \times 4200N + 4200N\}$$

This reduces to

$$1071.43 = N[1.61 \times 10^{(0.2 \log_{10} N)} + 1]$$

Substituting a few values we find that $N = 191.2$ solves the equation. The tabulated value for INTELSAT V is 192. Solving this problem required a preliminary assumption that $N < 240$. Suppose instead we had assumed $N > 240$ when getting l from Eq. (5.18). The equation to be solved for N then would have become

$$9 \times 10^6 = 2\{10^{[(-15 + 10 \log_{10} N)/20]} \times 3.16 \times 0.57 \times 4200N + 4200N\}$$

or

$$1071.43 = N[0.320 \times 10^{0.5 \log_{10} N} + 1]$$

The solution to this equation is approximately $N = 195.6$ and it violates the hypothesis that $N > 240$. At this point in the process the incorrect initial assumption becomes apparent.

Analog FM SCPC Systems

Single-channel-per-carrier (SCPC) systems avoid the voice signal FDM/FM multiplexing process and instead transmit each telephone channel on its own carrier. Details on the characteristics of FM SCPC systems are given in reference 10, and this section is based on that reference. SCPC reduces the cost of earth terminals on so-called light routes that handle only a few channels because it eliminates expensive multiplexing and demultiplexing equipment. In addition, an SCPC system is easy to reconfigure to meet changing traffic conditions, and thus it is compatible with the demand assignment (DA) schemes discussed in Chapter 6. Further, the carrier for an SCPC channel must be transmitted only when the link is active. In an FDM/FM system the carrier is always present and consuming transponder power. Energy dispersal (Section 5.3) makes this power consumption essentially independent of the channel loading. Since each link in an SCPC system will be active for less than half the time under fully loaded conditions, SCPC offers a saving in transponder power over FDM/FM. But SCPC requires more bandwidth than FM/FDM for the same number of channels, and it is not an economical way to move large amounts of traffic over a fixed route between two earth terminals. Single sideband (SSB) transmission, on the other hand, may be viewed as an SCPC system that uses bandwidth more efficiently than FDM/FM. We will discuss SSB in a later section.

The (S/N) behavior of an SCPC link is described by Eq. (5.14) modified for appropriate preemphasis and weighting. Using 6.3 dB for preemphasis improvement and 2.5 dB for psophometric noise weighting, Ferguson [10] has derived Eq. (5.26) for an analog FM SCPC system.

$$(S/N)_o = (C/N_0)_i - 95.4 + 20 \log_{10}(\Delta f_p) \text{ dB} \tag{5.26}$$

Here N_0 is the noise power density in watts per hertz and Δf_p is the *peak* test-tone deviation in hertz. Note that Eq. (5.26) uses the overall *carrier-to-noise power density ratio* (C/N_0) instead of the overall (C/N). N_0 is the noise power in watts divided by the IF bandwidth in Hz. In decibels

$$(C/N) = (C/N_0) - 10 \log_{10}(B_{IF}) \text{ dB} \tag{5.27}$$

Further improvement is possible through companding, a process in which the dynamic range of speech is compressed before transmission and expanded after detection. Companding distorts a voice waveform in a way that increases the average power by bringing the level of the soft portions close to that of the loud; we will have more to say about it in the following section. Ferguson [10] quotes a companding improvement of 17 dB for SCPC systems, changing Eq. (5.26) to

$$(S/N)_o = (C/N_0)_i - 78.4 + 20 \log_{10}(\Delta f_p) \tag{5.28}$$

RCA achieves a 16-dB companding improvement in its SATCOM system [3].

Companded Single Sideband (CSSB) [2]

Earlier in this chapter we described the process by which individual voice channels are "stacked" in frequency by single sideband suppressed carrier (SSB) modulation and then added to form a multiplexed telephone signal. The multiplexed telephone signal contains a frequency-shifted replica of each outgoing voice channel. With guardbands, each channel occupies 4 kHz, and the individual channels can be recovered independently of each other from the multiplexed signal by an appropriate combination of multipliers and filters.

Now suppose that instead of the 70-MHz frequency modulator used in FDM/FM, another SSB modulator shifted the multiplexed signal up to IF. The resulting uplink signal would be a replica of the incoming voice channels, appropriately stacked in frequency. At the downlink station, each channel would be immediately accessible at RF and at IF. More important, the occupied bandwidth of the uplink and downlink signals would be exactly 4 kHz times the number of channels. Thus this SSB modulation scheme uses bandwidth much more efficiently than FDM/FM (and more efficiently than digital modulation), and it offers the theoretical possibility of sending 9000 channels through a single 36-MHz transponder. Practical considerations, however, reduce this number to about 6000 channels.

Besides bandwidth efficiency, the other advantage that CSSB offers over FM is graceful failure. If the (C/N) at the input to an FM demodulator falls below threshold, the link will immediately stop working. A CSSB receiver has no threshold, and its output (S/N) degrades in proportion to the input (S/N) without any sudden failure.

Since an SSB signal of the type considered here lacks a carrier, it cannot be described in terms of a carrier-to-noise ratio. Instead a signal-to-noise ratio (S/N) is used that is simply the ratio of the received signal power to the received noise power. If all the voice channels feeding a SSB modulator are at their peak levels, then the transmitter output will be at its rated power level, and this power will be divided evenly between the voice channels. Since SSB demodulation (i.e., returning the original voice channels to baseband) provides no improvement in (S/N), the output (S/N) for each channel will be essentially the same as the overall (S/N) at RF, and this will be essentially the same as the overall (C/N) that the same earth stations and transponder could deliver using FDM/FM. This falls far short of the 48 to 50 dB required for telephone channels, and it rules out ordinary SSB for multiplexed voice channel transmission.

If the dynamic range of each voice channel is compressed before multiplexing and expanding after demultiplexing, the channels are said to have been companded. Companding reduces the required (S/N) at RF to a level significantly below that required at baseband and permits satisfactory operation with approximately the same power levels and antenna sizes as for FDM/FM. We will present the equations that govern the capacity and performance of CSSB links in the next chapter.

5.2 ANALOG TELEVISION TRANSMISSION

While telephone signals represent the bulk of communications satellite traffic, satellite technology has had a more dramatic effect on the television industry than on the telephone system. The first commercial satellites made live coverage of international news and sporting events possible, and in the United States these were soon followed by domestic satellites for network program distribution. Since anyone with an earth station and access to a transponder could originate programs, the cost of entering the U.S. television market fell drastically and a number of commercial, educational, and religious organizations began to offer programs via satellite for distribution over cable TV systems in competition with the established networks. Radio amateurs and other electronic experimenters were able to construct equipment capable of receiving these transmissions. Manufacturers began to offer home satellite TV reception equipment for sale to the general public and pressure grew on the U.S. government to allow unrestricted TV broadcasting by satellite. At the time this text was written neither the technology nor the regulatory environment of satellite TV transmission had reached steady state, but clearly TV was one of the most active parts of the satellite communications industry. In this section we will outline television modulation and demodulation; a later chapter will discuss network TV distribution and home satellite reception.

Television Signals

While a number of television transmission standards exist worldwide, the two in most common use are the North American and Japanese 525 line/60 Hz NTSC system and the European 625 line/50 Hz PAL system. These are also called CCIR systems M and B, respectively. In this text we will emphasize the NTSC system.

The video signal of a monochrome (black and white) TV transmission carries an analog representation of the brightness (i.e., the amount of white light) in the picture along a series of horizontal scanning lines. This is called the *luminance* signal. Along with the luminance signal, synchronization pulses are transmitted so that the TV receiver can recreate the scanning process of the camera.

Historically, monochrome TV developed before color, and in the United States color TV was designed so that the color information could be added to monochrome transmissions without degrading the performance of existing black-and-white receivers. Any color may be created by an appropriate combination of red, green, and blue light. Color TV could be transmitted by transmitting the color components of each picture separately, but this scheme would require excessive bandwidth. Instead, three linear combinations of the three components are transmitted and the component values themselves are recovered at the receiver.

The TV camera generates voltage levels corresponding to the red, green, and blue light at each point in the picture. We will identify these voltage levels by the letters R, G, and B. A monochrome receiver would respond to the amount of white light at a point in the picture; this is the *luminance*, Y, and is related to the color voltage levels by

$$Y = 0.30R + 0.59G + 0.11B \qquad (5.29)$$

The luminance signal is transmitted so that monochrome receivers can receive a color image in black and white.

For color reconstruction, two other independent linear combinations of R, G, and B must be transmitted along with Y so that all of the color components can be recovered. These are called the I and Q signals, given by

$$I = 0.60R - 0.29G - 0.32B \qquad (5.30)$$

$$Q = 0.21R - 0.52G + 0.31B \qquad (5.31)$$

The letters I and Q stand for in-phase and quadrature, and together the I and Q signals (when decoded with the luminance signal) carry the *chrominance* information about the color at each point in the picture.

The I and Q signals modulate a color (or chrominance) subcarrier in such a way that the amplitude of the resulting chrominance signal determines the *saturation* (degree of purity) of the color at a point and the phase of the chrominance signal determines the *hue* (perceived shade) of the color. From the amplitude and phase of the chrominance signal, a TV receiver determines the shade of the color and the amount of white light to add. From the luminance signal it determines how bright the color should be.

In terrestrial broadcasting the luminance (Y) signal, filtered to occupy the band from 0 to 4.2 MHz, modulates a "picture" carrier with a vestigial sideband (VSB) modulator. The upper sideband is transmitted in full; the lower sideband is partially removed. The resulting VSB signal is all that needs to be transmitted for the video portion of monochrome television.

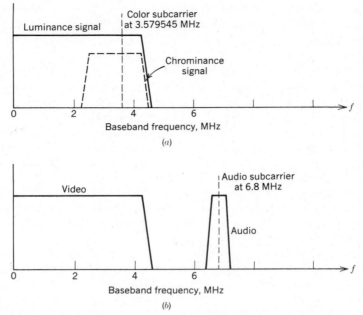

Figure 5.7 Spectra of baseband TV signals. (*a*) Baseband video signal. (*b*) The composite (video plus audio) TV signal as transmitted by US domestic satellites.

The chrominance information is transmitted by a *color subcarrier* at 3.579545 MHz (hereafter abbreviated as 3.58). This value was chosen because it places the chrominance signal at a relatively empty part of the luminance spectrum and minimizes color interference with black-and-white reception. Both the *I* and *Q* signals modulate the color subcarrier through double-balanced mixers to generate double sideband suppressed carrier (DSBSC) signals. The subcarrier is phase shifted by 90° before it enters the *Q* modulator. Thus, both *I* and *Q* components may be recovered at the receiver. Figure 5.7*a* presents a sketch of the spectrum of the baseband video signal.

The baseband audio signal extends from 50 Hz to 15 kHz. It frequency modulates an audio subcarrier and the resulting FM waveform is added to the video baseband signal. This leads to the composite TV signal of Figure 5.7*b*; it consists of the baseband video signal below an FM modulated audio subcarrier. In U.S. domestic systems an audio subcarrier frequency of 6.8 MHz is standard; 6.2 MHz is also used.

In terrestrial broadcasting, the audio and video signals are combined and shifted in frequency to an appropriate part of the VHF or UHF band for transmission. The radiated signal is a complex combination of FM (the sound), VSB (the luminance), and quadrature DSBSC (the chrominance). It occupies a 6-MHz bandwidth. For satellite transmission the baseband video signal (luminance and chrominance), frequency modulates a video carrier and the two audio signals frequency modulate two audio carriers. The details of the video modulation depend on the transponder bandwidth available. Typical values for network TV are a peak deviation Δf_p of 10.75 MHz and a maximum video modulating frequency f_V of 4.2 MHz. By Eq. (5.5) this requires a 29.9 MHz transponder bandwidth. Television signals are often overdeviated, trading the larger improvement in video (S/N) that results for the smaller degradation in picture quality associated with truncating some of the sidebands [3].

A TV signal from a satellite is quite different from a broadcast TV signal. Converters that allow reception of satellite television transmission on conventional home receivers must demodulate the incoming FM signals, recover the baseband video and audio channels, and remodulate the audio and video onto a locally generated carrier using the same modulation scheme as a broadcast TV transmitter.

Signal-to-Noise Ratio Calculation for Satellite TV Links

Equations (5.11) and (5.14) are two equivalent formulas that relate the signal-to-noise ratio at the output of an FM demodulator to the overall carrier-to-noise ratio at the input. As derived, they compare total signal power to total noise power and the (S/N) values they predict can be improved by preemphasis and should be weighted to account for the non-uniform response of the eye to white noise in the video bandwidth. The preemphasis factor is called p and the weighting factor is q. Their decibel values, P and Q, add in the decibel versions of Eqs. (5.11) and (5.14):

$$(S/N)_V = (C/N)_i + 1.76 + 10\log_{10}\left(\frac{B_{IF}}{f_V}\right) + 20\log_{10}\left(\frac{\Delta f_{peak}}{f_V}\right) + P + Q \text{ dB} \quad (5.32)$$

$$(S/N)_V = (C/N)_i + 10\log_{10}\left[3m^2(1+m)\right] + P + Q \text{ dB} \quad (5.33)$$

Here f_V is the maximum video modulating frequency (4.2 MHz by U.S. standards), and the modulation index m is $\Delta f_{peak}/f_V$. Substituting a typical value of 10.75 MHz for Δf_{peak} into these equations we obtain

$$(S/N)_V = (C/N)_i + 18.5 + P + Q \text{ dB} \quad (5.34)$$

The values used for P and Q depend on the noise characteristics of particular TV systems and on the subjective response of individual viewers to the noise included in a television picture. Numbers for $P + Q$ ranging from roughly 18 to 26 dB are quoted in the literature. These lead to overall improvements in signal-to-noise ratio ranging from 36.5 to 44.5 dB.

5.3 ENERGY DISPERSAL

The satellite telephone and television transmission systems described in the previous two sections both employ frequency modulation. When an input modulating signal is absent, an FM transmitter radiates all of its power at the carrier frequency. With modulation the average carrier power is spread over a large bandwidth. The larger the amplitude of the modulating signal, the smaller is the average transmitted spectral power density in watts per hertz of bandwidth.

To minimize interference with terrestrial microwave systems sharing the same frequencies, most administrations restrict the maximum spectral power density that a satellite may radiate toward the earth. Minimum spectral power density occurs with maximum modulation amplitude; this condition is called *full loading* in telephone practice. Typically for an Intelsat spacecraft the radiated power per 4 kHz of bandwidth must not rise more than 2 dB above its value for full loading.

The process of controlling the radiated spectral density is called *energy dispersal*. It is accomplished at the uplink earth station by adding a symmetric triangular voltage waveform called the *dispersal signal* to the modulating waveform before modulation. The dispersal waveform is removed at the downlink earth station. In television transmission the dispersal signal has a constant amplitude and currently a frequency of 30 Hz for the U.S. NTSC system. The frequency is scheduled to be changed to 60 Hz. The amplitude depends on the spacecraft and modulation used; for example with INTELSAT V the dispersal signal must provide between 1 and 2 MHz peak-to-peak carrier frequency deviation [4].

For telephone transmission the amplitude of the dispersal signal is dynamically adjusted to keep the radiated spectral density within the required bounds. The frequency of the dispersal signal depends on the system; Intelsat requires values between 20 and 150 Hz. In general the amplitude of the dispersal signal for multichannel telephone transmission is determined by finding the frequency shift (peak deviation) ΔF that the dispersal waveform must cause. If ΔF and the FM modulator characteristics are known, then the dispersal signal amplitude may be computed. Reference 11 provides a detailed derivation of the equations for calculating ΔF; we will summarize those results and give a numerical example.

Consider an FDM/FM analog telephone signal modulated onto a carrier whose power is C W and whose loading produces rms multichannel deviation d Hz. Assume that the resulting spectrum is Gaussian and that its power density $W(f)$ may be expressed in terms of frequency f by Eq. (5.35) in which ΔF is the difference between the (unmodulated) carrier frequency f_c and f.

$$W(f) = \left(\frac{C}{d\sqrt{2\pi}}\right) \exp\left[-(\Delta F)^2/(2d^2)\right] \qquad (5.35)$$

At full loading the density is $W_{\min}(f)$, given by

$$W_{\min}(f) = \left[\frac{C}{d\sqrt{2\pi}}\right] \exp\left[-(\Delta F)^2/(2d_m^2)\right] \qquad (5.36)$$

where d_m is the fully loaded rms multichannel deviation. At no loading the density $W_{\max}$ is determined by the deviation $\Delta F_{\max}$, which the maximum dispersal waveform causes. Assuming that the triangular dispersal signal simply spreads the carrier power C uniformly over a band that extends ΔF Hz on either side of the carrier frequency, then

$$W_{\max} = \frac{C}{2\,\Delta F_{\max}} \qquad (5.37)$$

The usual practice is to choose $W_{\max} = W_{\min}(0)$, which leads to the result

$$\Delta F_{\max} = \frac{C}{2W_{\min}(0)} = d_m \left(\frac{\pi}{2}\right)^{\frac{1}{2}} \tag{5.38}$$

The rms multicarrier deviation d_m equals the loading factor of Eq. (5.18) multiplied by the rms test-tone deviation Δf_{rms} of the link. Thus from d_m we may calculate the maximum deviation $F_{\max}$ that the dispersal waveform must be capable of producing.

For less than full loading $(d < d_m)$ the dispersal waveform added to the incoming multiplexed telephone signal acting by itself would produce a deviation $\Delta F < F_{\max}$. To calculate ΔF we must solve the integral equation derived in reference 11:

$$\frac{1}{\sqrt{2\pi}} \frac{\Delta F}{d_m} = \frac{1}{\sqrt{2\pi}} \int_0^{\Delta F/d_m \, d_m/d} e^{-x^2/2} \, dx \tag{5.39}$$

Equation (5.39) is written in this form so that the integral may be evaluated from readily available tables. Solving Eq. (5.39) requires a trial-and-error process. The equation actually specifies the maximum ΔF required for a particular loading; Intelsat will permit the spectral density to go 2 dB above the value corresponding to ΔF calculated from Eq. (5.39).

Example 5.3.1

According to Miya's tabulated data [11], a 60-channel telephone link carried by INTELSAT III had an rms test-tone deviation Δf_{rms} of 410 kHz and a fully loaded rms multicarrier deviation d_m of 830 kHz. Find $F_{\max}$ and the value of ΔF that the dispersal signal must produce (i.e., the ΔF the signal would produce if it were acting by itself) at 75 percent loading.

By Eq. (5.39), $\Delta F_{\max} = d_m(\pi/2)^{\frac{1}{2}} = 830(\pi/2)^{\frac{1}{2}} = 1040.25$ kHz. Assume that the deviation d at 75 percent loading is $0.75 d_m$ or 622.5 kHz. Writing $\Delta F/d_m$ as u in Eq. (5.39), we have

$$\frac{u}{\sqrt{2\pi}} = \frac{1}{\sqrt{2\pi}} \int_0^{1.33u} e^{-x^2/2} dx$$

The right-hand side of this equation is the area under the Gaussian probability density function between 0 and $1.33u$. We will represent it by Area $(1.33u)$. It is conveniently tabulated in *C.R.C. Standard Mathematical Tables* and similar publications. The equation to be solved then becomes

$$\frac{u}{\sqrt{2\pi}} = \text{Area } (1.33u)$$

A trial-and-error solution yields $u = 1.05$. Thus, $\Delta F/d_m = 1.05$ and $\Delta F = 871.5$ kHz.

5.4 DIGITAL TRANSMISSION

Digital modulation is the obvious choice for satellite transmission of signals that originate in digital form and that are used by digital devices. Familiar examples are data transmissions between computers, printed text, communications between remote terminals and computers, and the like. Such analog signals as telephone channels and television may be put into digital form for transmission and then converted back to analog form for routing to the end user. While this process may be costly in terms of bandwidth, it usually offers improved noise performance and increased immunity to interference. Digital transmission lends itself naturally to time division multiplexing (TDM) and time division multiple access (TDMA); both of these techniques allow one signal to use a transponder at a time and thus avoid intermodulation problems. Finally, analog signals that are transmitted digitally can share channels with digital data; all digital signals are to be handled in the same way, and their content is immaterial. Thus a digital satellite link can carry a mix of telephone and data signals that varies with traffic demand.

There are basically two problems in satellite digital transmission: (1) how to get incoming analog signals into digital form and then back again, and (2) how to transmit and receive digital signals efficiently—whatever their origin and destination. Sections 5.5 and 5.6 will discuss these topics in more detail.

Baseband Digital Signals

We will represent baseband digital signals as serially transmitted logical ones and zeroes. While in computer circuitry a logical zero may be represented by a low voltage (nominally zero) and a logical one may be represented by a high voltage (say 5 V)—or vice versa—this arrangement is inconvenient for transmission over any significant distance and is not used. To understand why, imagine a transmission line carrying a bit stream encoded this way and containing approximately equal numbers of ones and zeroes. About half the time the line voltage will be 5 V and about half the time it will be 0 V; hence the line signal will have a 2.5-V DC component. All circuits that carry this signal must have a frequency response that extends to DC, and this is difficult to achieve since many communication circuits contain transformers. To avoid this problem, digital modulators usually accept their input in a *polar non-return-to-zero (NRZ)* format: logical ones and zeroes are transmitted as plus or minus a stated value. Thus a one might be transmitted as $+1$ V and zero might be transmitted as -1 V. Zero volts is not transmitted except as a transient value. Throughout this text we will assume a polar NRZ format for data signals unless we explicitly state otherwise.

Baseband Transmission of Digital Data

A random sequence of rectangular binary pulses has a power spectral density

$$G(f) = T_b \left[\frac{\sin(\pi f T_b)}{\pi f T_b} \right]^2 \tag{5.40}$$

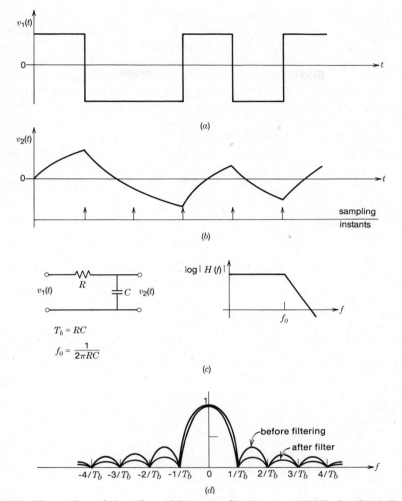

Figure 5.8 Illustration of the effect of low pass filtering on a NRZ signal. (*a*) Random NRZ polar pulse train. (*b*) Waveform output from an RC filter with $T_b = RC$. (*c*) RC filter and its transfer function $|H(f)|$. (*d*) Spectrum of bandlimited NRZ pulse train.

where T_b is the duration of the pulse [12]. This spectrum is illustrated in Figure 5.8*d*. The familiar $\sin x/x$ shape shows that energy exists at all frequencies; to retain the rectangular pulse shape would require an infinite transmission bandwidth. Practical communication systems are always bandwidth limited; not only is infinite bandwidth not available, interference considerations in radio links dictate that a communication system should use the smallest possible bandwidth, and this is usually one of the design criteria of a communication system.

If we take the random pulse train shown in Figure 5.8*a* and bandlimit it by passing it through a low-pass filter, the pulse shape will be altered. As an example, consider the effect of passing the rectangular pulse train through a single RC section, representing a very simple low pass filter. The resulting waveform, shown in Figure 5.8*b*, has been delayed and pulses are "smeared" in time—the decaying pulse from one transition extends into the next pulse interval. Where the pulse pattern is 10 or 01, the amplitude of the second pulse at the sampling instant shown in Figure 5.8 has been reduced by the presence of a delayed portion of the preceding pulse. This is called *intersymbol interference* (*ISI*) and is likely to occur whenever a digital signal is passed through a bandlimiting filter. When noise is added to the waveform, ISI increases the likelihood that the receiver will detect a bit incorrectly, causing a bit error. In a baseband system, ISI can be avoided by an appropriate

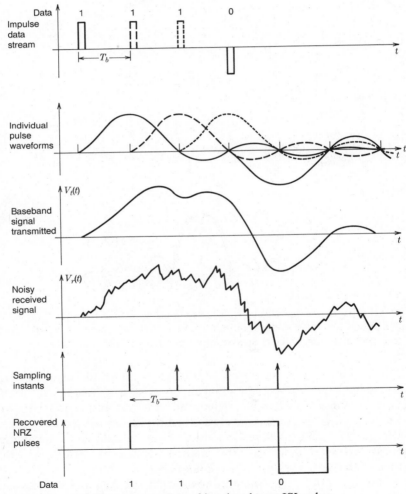

Figure 5.9   Transmission and reception of baseband zero-ISI pulses.

choice of low-pass filter. Nyquist [13] proposed a technique that can theoretically produce zero ISI, now known as the Nyquist criterion. The objective is to create in the receiver a pulse that resembles the sin x/x shape, crossing the axis at intervals of T_b, where T_b is the bit period. The receiver samples the incoming wave at intervals of T_b, as shown in Figure 5.9, so that at the instant one pulse is sampled, the "tails" from all preceding pulses have zero value. Thus previous pulses cause zero intersymbol interference (zero ISI) at each sampling instant.

Filters that produce the required zero ISI waveform in the receiver can be synthesized in several ways. The filter proposed by Nyquist was the "raised cosine" filter, which has a frequency characteristic given by

$$V_r(f) = \begin{cases} T_b, & |f| \le \dfrac{R_b}{2}(1-\alpha) \\[2em] T_b \cos^2 \left\{ \dfrac{\pi}{2\alpha R_b} \left[\dfrac{|f| - R_b(1-\alpha)}{2} \right] \right\}, & \\[1em] & \dfrac{R_b}{2}(1-\alpha) < |f| < \dfrac{R_b}{2}(1+\alpha) \\[2em] 0, & |f| \ge \dfrac{R_b}{2}(1+\alpha) \end{cases} \tag{5.41}$$

where $0 < \alpha < 1$ and $R_b = 1/T_b$ is the bit rate in bits/second. The pulse shape generated when the filter is driven by an impulse, $\delta(t)$, is $v_r(t)$, the required zero ISI waveform. The waveform $v_r(t)$ is obtained as the inverse Fourier transform of the output from the Nyquist filter, which is simply the spectrum of the input pulse multiplied by the frequency response of the filter.

$$v_r(t) = F^{-1}[V_r(f) \times S(f)] \tag{5.42}$$

where $F^{-1}[\ \ ]$ indicates the inverse Fourier transform and $S(f)$ is the spectrum of the input pulse. If we use an impulse $s(t)$ as the input signal, $S(f) = 1$ and then

$$v_r(t) = F^{-1}[V_r(f)] \tag{5.43}$$

For the raised cosine filter with $V_r(f)$ given by Eq. (5.43)

$$v_r(t) = \left[\frac{\cos \pi \alpha R_b t}{1 - (2\alpha R_b t)^2} \right] \left[\frac{\sin \pi R_b t}{\pi R_b t} \right] \tag{5.44}$$

Figure 5.10 shows the shape of several raised cosine filter characteristics and the corresponding waveforms generated by the impulse response of these filters. The case of $\alpha = 0$ in Eq. (5.41) yields a filter with a bandwidth of $R_b/2$, the minimum bandwidth through which a bit rate R_b can be transmitted while still satisfying the zero ISI condition. Such a filter is not realizable in practice, since we cannot have an infinitely rapid attenuation slope at one frequency. Practical filters use values of α between 0.2 and 1.0. Figure 5.11 shows a baseband digital link with typical waveforms, and the corresponding spectra.

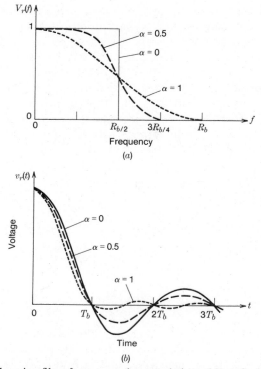

Figure 5.10 Raised-cosine filter frequency characteristic and impulse response. (*a*) Raised-cosine filter transfer characteristics. (*b*) Corresponding impulse responses.

Bandpass Transmission of Digital Data

In a radio frequency communication system that transmits digital data, a parameter of the RF wave must be varied, or modulated, to carry the baseband information. The most popular choice of modulation for a digital satellite communication system is phase shift keying (PSK), as described in the following section. Bandpass (or radio frequency) transmission of digital data differs from baseband transmission only because modulation of an RF wave is required: the receiver demodulates the modulated RF wave to recover the baseband data stream. Thus intersymbol interference will occur at the receiver due to bandlimiting of the modulated waveform unless filters that satisfy the Nyquist criterion are used.

An additional constraint usually exists with radio communication systems. The bandwidth occupied by a transmission is specified to avoid interference with other transmissions at adjacent frequencies: the output of a transmitter must have a carefully controlled spectrum that reduces out-of-band signals to a low level. Figure 5.12 shows the spectrum of a binary PSK ($BPSK$) signal generated from a random train of binary digits. The slow decay of the spectrum beyond $f_c \pm R_b$ results from the sudden phase reversals of the PSK waveform.

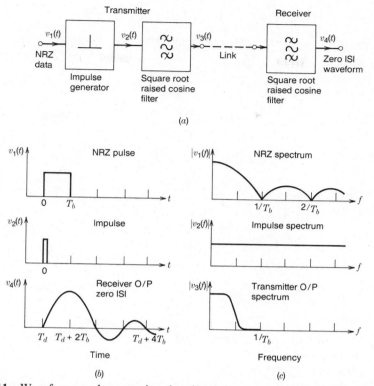

Figure 5.11 Waveforms and spectra in a baseband data system with raised-cosine filters. (*a*) System block diagram. (*b*) Waveforms. (*c*) Spectra.

In many data transmission systems the baseband waveform used has the NRZ format. Nyquist filters produce zero ISI waveforms only when driven by an impulse, as shown by Eqs. (5.42) and (5.43). If the filter is driven by a NRZ waveform, the spectrum of the driving pulse has a sin x/x shape, and the spectrum of the filter output will be $V_r(f)$ sin x/x [14]. To obtain zero ISI at the receiver, we must supply a signal with a spectrum $V_r(f)$, which can be achieved by using an equalizer with a frequency characteristic given by $x/\sin x$. The arrangement is illustrated in Figure 5.12*a*. The raised cosine filter cuts off at $f_c \pm f_0$ where $f_0 \leq 1/T_b$, so the $x/\sin x$ equalizer operates only within the central lobe of the sin x/x function. At $f = 1/T_b$, $x/\sin x$ goes to infinity, so α must be less than 1 for this system to work. In practice, RF filters with raised cosine shaping use α around 0.4, so the maximum gain at the edge of the equalizer band is 8.7 dB in this case.

The discussion of filter characteristics and signal spectra thus far has ignored the phase response of the filters and the resulting phase spectra of the waveforms. It can readily be shown [12] that the phase response of all filters and equalizers must be linear with frequency for the zero ISI condition to be met. Achieving a

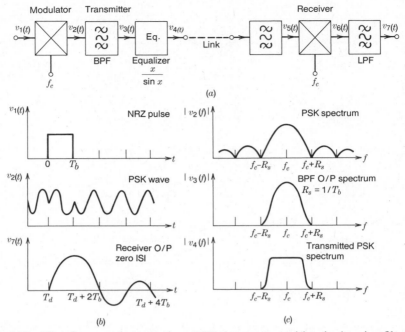

Figure 5.12 Waveforms and spectra in a PSK data system with raised-cosine filters and $x/\sin x$ equalization. (*a*) Block diagram of one channel of a QPSK system. (Equivalent to a BPSK system.) (*b*) Waveforms. (*c*) Spectra.

linear phase response throughout a communication system can be difficult in practice.

Transmission of QPSK Signals Through a Bandlimited Channel

QPSK (quadrature phase shift keying) is the most popular choice of modulation technique for use in satellite communication links carrying digital data. It will be described in more detail in Section 5.5, but basically a digital data stream is taken two bits at a time and used to generate one of four possible phase states of the transmitted carrier. If the data rate is R_b bits/s, the symbol rate for the QPSK carrier is $R_b/2$ bits/s $= R_s$ symbols/s.

In order to recover the symbol stream with zero ISI, we must shape the transmitted spectrum such that after demodulation a single symbol creates a zero ISI waveform at the output of the demodulator. Then sampling of the symbol stream can be achieved with zero intersymbol interference. In practice, a QPSK system has two demodulators, one for each pair of symbols (phase states) in the QPSK carrier. We shall consider only one channel in looking at ISI.

Figure 5.12*a* shows a typical arrangement for one half of a QPSK transmit–receive link. The other half is identical except that the carrier used for modulation and demodulation is shifted in phase by 90°. Since the carriers in the two channels have a 90° phase difference, the channels are identified as *in-phase* (I) and *quadrature* (Q).

The data presented to the QPSK modulators is in NRZ format and causes a jump in carrier phase at each symbol transition. The input data rate to the demodulator is $R_s = R_b/2$, giving the QPSK spectrum shown in Figure 5.13a. The

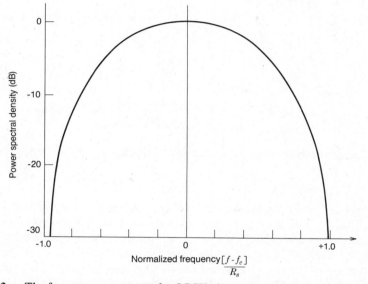

Figure 5.13a The frequency spectrum of a QPSK signal. Frequency is relative to the carrier and normalized to symbol rate R_s. Only the central lobe is shown.

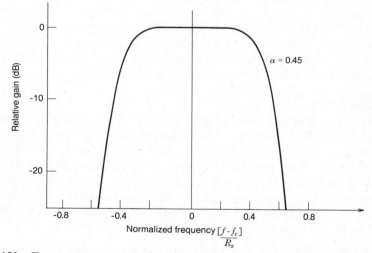

Figure 5.13b Frequency response of a bandpass square root raised-cosine filter with $\alpha = 0.45$.

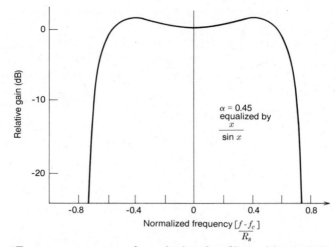

Figure 5.13c Frequency response of a raised-cosine filter with $\alpha = 0.45$, equalized by $x/\sin x$.

central lobe of the spectrum extends from $(f_c - R_s)$ to $(f_c + R_s)$, giving a band occupancy of $2R_s$. The spectrum must be narrowed for transmission via a radio channel, and this is achieved by use of a bandpass filter meeting the zero ISI criterion, for example, a raised cosine filter. The bandpass raised cosine filter is a transformation of the low-pass raised cosine filter and has a response $|H(f)| = 1/2$ at frequencies $(f_c - R_s/2)$ and $(f_c + R_s/2)$. The frequencies at which $H(f)$ falls to zero are determined by the roll-off factor α in Eq. (5.41). Matched filter operation of the link requires that the raised cosine filter response be split between the transmit end and the receive end of the link. Thus a square-root raised cosine response filter is required after the modulator and before the demodulator. Finally, because we are using NRZ pulses rather than impulses, we need an $x/\sin x$ equalizer with $x = [\pi(f_c - f)]/R_s$ to equalize the spectrum from the modulator. The frequency responses of some typical filters are shown in Figures 5.13b, c.

Example 5.4.1
A data stream at 240 Mbps is to be sent by QPSK on a microwave carrier. The receiver IF frequency is 240 MHz. Find the RF bandwidth needed to transmit the QPSK signal when raised cosine filters with $\alpha = 0.45$ are used.

The 240 Mbps signal is divided into two 120 Msps symbol streams and applied to I and Q channel modulators fed by an IF carrier, with a 90° phase difference. The resulting spectrum from each modulator has a width of 240 MHz between zeros of the central lobe of the PSK spectrum. The I and Q signals are added and applied to an $x/\sin x$ equalizer with

$$x = \frac{\pi(f_c - f)}{R_s}$$

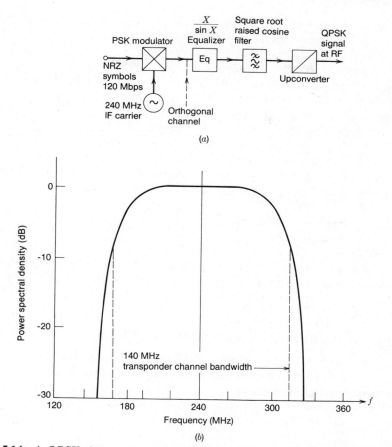

Figure 5.14 A QPSK data system with a NRZ format and $R_s = 120$ Msps symbol rate. IF frequency is 240 MHz, (a) Transmit end block diagram for model of 120 Msps symbol rate QPSK link. Only one channel is shown. (b) Transmitted QPSK spectrum after processing by square-root raised cosine filter and $x/\sin x$ equalization. Filter roll-off factor α is 0.45.

extending to ± 87 MHz from the carrier. The maximum gain of $x/\sin x$ at ± 87 MHz from the carrier is 9.53 dB. Figure 5.14a shows a block diagram of one half of the transmit end of the QPSK link.

The equalized spectrum is applied to the square-root raised cosine filter. The response of this filter is 3 dB down at $f_s \pm R_s/2$, that is, at $f_s \pm 60$ MHz. In practice, one filter combining the square-root raised cosine and $x/\sin x$ responses is used. The combined response of this single filter is shown in Figure 5.13c. Thus in the IF amplifier of the receiver, the signal spectrum is 6 dB down at 180 MHz and 300 MHz. The low-pass raised cosine filter with $\alpha = 0.45$ and $R_s = 120$ Msps has $|H(f)| = 0$ for $|f| > R_s/2 + \alpha R_s/2$, so the bandpass filter will have zero response for $f < f_c - (R_s/2)(1 + \alpha)$ and for $f > f_c + (R_s/2)(1 + \alpha)$, that is, below 153 MHz and above 327 MHz. Figure 5.14b shows the transmitted QPSK spectrum centered on the IF carrier.

If we examine the spectrum of the QPSK signal at the receiver, we find that the 3-dB bandwidth is 120 MHz and the total frequency band containing all of the signal energy is 174 MHz. A typical satellite transponder for such a signal would have a 3-dB bandwidth of 140 MHz. Beyond 140 MHz the spectrum of the QPSK signal would be attenuated by the transponder filter, leading to some spectral distortion of the receiver signal and consequent ISI in the demodulated waveform. However, the energy contained in the QPSK spectrum beyond ± 70 MHz from the carrier is small, and the ISI caused by the transponder filter is minimal.

Practical filters invariably cause some ISI because it is impossible to realize the raised cosine characteristic exactly. Appendix A.2 presents some spectra for PSK signals using Chebyshev and Butterworth filters. Typically, an extra 2 dB of carrier power must be provided to achieve a 10^{-6} BER, compared to the theoretical power level needed for this error rate, in a carefully filtered QPSK link. The extra power is sometimes called *implementation margin*.

5.5 DIGITAL MODULATION AND DEMODULATION

In this section we will review methods for digital transmission used on or proposed for current satellite links. We will not attempt to summarize the extensive literature of digital communications in general.

Terminology

While any feature of a signal—amplitude, frequency, or phase—may be digitally modulated, phase modulation is almost universally used for satellites. For historical reasons, digital phase modulation is frequently called *phase shift keying*, abbreviated PSK. An M-phase PSK modulator puts the phase of a carrier into one of M states according to the value of a modulating voltage. Two-state or biphase PSK is usually called BPSK, and four-state or quadriphase PSK is termed QPSK. Other numbers of states and some combinations of amplitude and phase modulation are possible and are employed in terrestrial links, but satellite users have been reluctant to adopt anything besides BPSK or QPSK. An important reason for this is the high values of (C/N) required for acceptable bit error rates—typically > 26 dB. Any type of PSK can be *direct* or *differential*, depending on whether it is the state of the modulating voltage or the *change* in state of the modulating voltage that determines the transmitted phase.

Whether direct or differential, a PSK modulator causes the phase of a carrier waveform to go to one of a finite set of values. The transition time plus the time spent at the desired phase constitute a fixed time interval called the *symbol period*; the transmitted waveform during the interval is called a *symbol*. The set of all symbols for a particular modulation type is called its *alphabet*. Thus BPSK has a two-symbol alphabet and QPSK has a four-symbol alphabet.

In the digital modulation process, a stream of incoming binary digits (bits) determines which symbol of the M available in the alphabet will be transmitted. Mathematically, N_b bits are required to specify which of M possible symbols is being transmitted where N_b and M are related by

$$N_b = \log_2(M) \tag{5.45}$$

As defined by this equation, N_b is the number of *bits per symbol* for the M-PSK modulation scheme. Standard practice is to make M a power of 2 so that N_b will be an integer.

Modulation and Coding

The boundary between digital modulation and digital encoding is not well defined. In encoding for forward error correction (*FEC*), redundant bits are added to an incoming bit stream so that errors in transmission may be detected and corrected at the other end of the link. When the redundant bits are added at baseband and the composite (information bits plus redundant bits) bit stream is used to phase modulate a carrier and produce the transmitted symbols, then the division between modulation and encoding is obvious. But the modulator itself may be designed to add redundant bits during the modulation process, making encoding and modulation inseparable. In this section we will ignore encoding for FEC and concentrate strictly on the modulation process for turning an incoming bit stream into RF symbols. We will assume that any FEC encoding is done ahead of the modulator by the methods to be presented in a later chapter.

It is unfortunate that differential phase modulation is frequently called differential encoding, since it is a characteristic of the modulation and demodulation equipment and plays no role in coding as it is usually understood. Differential encoding would more properly be called differential modulation, and we will discuss it after we have presented direct modulation.

Bit and Symbol Error Rates

The figure of merit for a digital radio link is its *bit error rate* (*BER*), also called the *bit error probability* (*PB*). Mathematically this is the probability that a bit sent over the link will be received *incorrectly* (i.e., that a 1 will be read as a 0 or vice versa) or, alternatively, the fraction of a large number of transmitted bits that will be received incorrectly. Like a probability, it is usually stated as a single number—for example 1×10^{-4} or .0001. The BER plays the same role as an indicator of quality in a digital communication system that the signal-to-noise ratio plays in an analog link.

Physically a bit error occurs because a *symbol error* has occurred. At some point in the link noise has corrupted the transmitted symbol so that the decision circuitry at the receiver cannot identify it correctly. For example, the carrier phase may have been transmitted as $+90°$ but additive noise may have changed the received carrier phase to $-90°$. If one symbol carries N_b bits and if differential modulation is not used, then a single symbol error may cause $1, 2, \ldots N_b$ bit errors. With differential modulation, an error on one symbol will cause the symbol that follows to be misinterpreted, and the number of bit errors per symbol error may exceed N_b, the number of bits per symbol.

Symbol errors arise from thermal noise, from external interference, and from intersymbol interference. If only thermal noise is considered, then the *symbol error rate (SER)* or *symbol error probability (PE)* may be calculated unambiguously from (E_s/N_0), the energy per symbol in joules divided by the noise density in W/Hz, measured in the IF bandwidth at the demodulator input. The higher the value of (E_s/N_0), the lower will be the SER. (E_s/N_0) may be determined from the input value of (C/N), expressed as a ratio.

Assume that C W of carrier power are transmitted during one symbol interval T_s. The energy received during that symbol period is E_s, where

$$E_s = CT_s = \frac{C}{R_s} \tag{5.46}$$

where R_s is the symbol rate in symbols/second. The noise-density N_0 is the received noise power N divided by the IF bandwidth at the demodulator input

$$N_0 = N/B \tag{5.47}$$

Combining the last two equations we have

$$\frac{E_s}{E_0} = \frac{C}{N} \cdot \frac{B}{R_s} \tag{5.48}$$

The square-root Nyquist cosine filter discussed in the preceding section has a noise bandwidth B equal to the symbol rate R_s. Thus a receiver designed with filters of this type to achieve zero ISI also has $BT_s = 1$ and $E_s/N_0 = C/N$. Practical filters such as Butterworth or Chebychev come close to the shape of the square root raised cosine filter, giving BT products close to unity.

While for BPSK bit and symbol errors are the same thing, for modulation schemes with $M > 2$, the relation between the bit error rate and the symbol error rate is not consistently defined in the literature. Equation (5.49), derived in reference 15, is based on the probability that a particular bit carried by a symbol is in error, given that the symbol itself is in error.

$$\text{PB} = \tfrac{1}{2} \frac{PE}{1 - 2^{-N_b}} \tag{5.49}$$

Here PB and PE are the corresponding values of bit and symbol error probability, and N_b is the number of bits per symbol. For QPSK, $N = 2$ and $PE = 1.5$ PB. Another approach, derived in reference 16 and more frequently quoted in the literature than Eq. (5.49), assumes that a large block of data is to be transmitted either in the form of serial bits or in symbols of M-ary PSK. Each symbol carries

2^M bits. The derivation equates the probability that the word will be received correctly in BPSK with the probability that it will be received correctly in the M-ary system. For PE much less than 1, this equality leads to Eq. (5.50):

$$PB = \frac{\ln 2}{\ln M} PE = \frac{PE}{\log_2 M} \qquad (5.50)$$

For QPSK, $M = 4$ and Eq. (5.50) yields $PE = 2\,PB$.

The reader may find it confusing that, while Eqs. (5.49) and (5.50) yield slightly different results for the relation between PB and PE, both show the bit error rate PB to be less than the symbol error rate PE for QPSK. This happens because each QPSK symbol carries two bits. When a given symbol is sent, three symbol errors are possible (i.e., a 00 may be detected as an 01, 10, or 11), but two of these cause only a single bit error, and one will leave a given bit unchanged. Thus if we look at one particular bit out of the two carried by the symbol, the probability that a symbol error will change that bit is about two-thirds—that is, two of the three possible symbol errors that can be made will change the bit. Hence we would expect the bit error rate probability PB to be two-thirds the symbol error probability PE, and this is what Eq. (5.49) yields.

Binary Phase Shift Keying (BPSK)

In binary phase shift keying, an incoming bipolar bit stream $u(t)$ sets the phase of a carrier to plus or minus 90° ($\pi/2$ rad). Thus, if u_i is the ith bit, then the transmitted carrier v_c is given by

$$v_c = V \cos\left(\omega_c t - u_i \frac{\pi}{2}\right) \qquad (5.51)$$

where V is an arbitrary amplitude frequently set to 1. Since u_i must have a value of ± 1, a logical one is transmitted by setting the phase to $-\pi/2$ rad and a logical zero (baseband -1) is transmitted with a $+\pi/2$ phase. Using trigonometric identities, we may rewrite Eq. (5.51) as

$$v_c = V u_i \sin(\omega_c t) \qquad (5.52)$$

and we see that BPSK resembles amplitude modulation in which the modulating signal has a value $+1$ or -1 only. This causes the BPSK waveform to have a constant amplitude and an envelope AM detector cannot demodulate it.

To recover u_i the receiver must compare the phase of the received signal with that of a reference voltage that has the same phase as the original unmodulated carrier. This may be done with the simple product detector (mixer) of Figure 5.15 where the output voltage v_o is ideally u_i. At the center of each symbol interval, a decision circuit decides whether v_o is positive or negative and thus determines whether u_i was a $+1$ and represented a one or whether it was a -1 and represented a zero. This technique is called coherent detection since it requires a reference voltage that is phase coherent with the transmitter carrier.

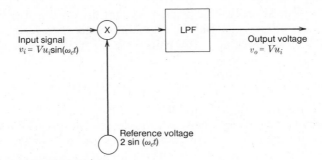

Figure 5.15 A coherent BPSK detector.

The decision circuit will make an error if noise changes the sign of v_o. We may calculate the probability that this will happen and thus the symbol error rate PE by the following argument, based on reference 16. Let the channel noise voltage $n(t)$ be Gaussian distributed with zero mean and rms value σ. Assume that u_i is a -1. At the decision time, v_o is given by

$$v_o = n(t) - V \tag{5.53}$$

If v_o is positive, the decision circuit will interpret u_i as $+1$. That will happen if $n(t) > V$; see Figure 5.16. The probability of $n(t)$ being greater than V is

$$P(N > V) = PE = \tfrac{1}{2}\,\mathrm{erfc}\left(\frac{V}{\sigma\sqrt{2}}\right) \tag{5.54}$$

The complementary error function, abbreviated erfc, is given by

$$\mathrm{erfc}(x) = \frac{2}{\sqrt{\pi}} \int_x^\infty e^{-u^2} \cdot du \tag{5.55}$$

where $u = x/\sigma\sqrt{2}$. See reference 6, pp. 60ff., for a discussion of its properties. Equation (5.45) involves the erfc because the noise voltage has a Gaussian distribution.

One symbol lasts for T_s s. The power in the symbol waveform is $V^2/2R$, where R is the input resistance of the decision circuit. Thus the energy per symbol, E_s, is given by

$$E_s = \frac{V^2}{2R} T_s \tag{5.56}$$

assuming that we have a matched filter in our receiver. Then

$$V = \sqrt{\frac{2RE_s}{T_s}} \tag{5.57}$$

and the noise power is given by σ^2/R. The noise density N_0 is the noise power divided by the bandwidth. If we assume that the bandwidth is $1/T_s$, then N_0 is

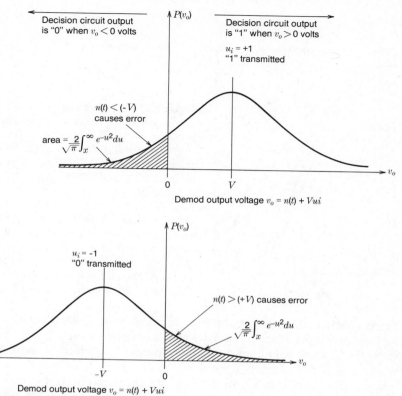

Figure 5.16   Illustration of errors in a binary decision circuit caused by additive Gaussian noise. The threshold is at zero volts.

given by

$$N_0 = \frac{\sigma^2}{R} T_s \tag{5.58}$$

and

$$\sigma = \sqrt{\frac{RN_0}{T_s}} \tag{5.59}$$

Combining Eqs. (5.54), (5.58), and (5.59) yields

$$PE = \tfrac{1}{2}\text{erfc}\left[\frac{\sqrt{2RE_s/T_s}}{\sqrt{RN_0/T_s}} \frac{1}{\sqrt{2}}\right] = \tfrac{1}{2}\text{erfc}\left[\sqrt{\frac{E_s}{N_0}}\right] \tag{5.60}$$

Since for BPSK a bit and a symbol are the same thing, Eq. (5.60) is often written

$$PB = \tfrac{1}{2}\text{erfc}(\sqrt{E_b/N_0}) \tag{5.61}$$

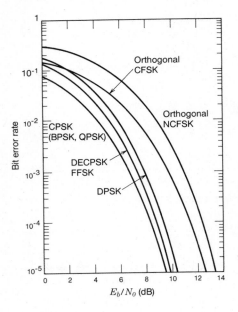

Figure 5.17 Bit error rate versus (E_b/N_0) for a variety of digital modulation and demodulation schemes. CPSK stands for coherent phase shift keying and is the same as the BPSK discussed in the text. DECPSK is differentially encoded PSK, and DPSK is differentially detected PSK. The other curves represent fast frequency shift keying (FFSK), coherently detected frequency shift keying (CFSK), and noncoherently detected frequency shift keying (NCFSK). (Reprinted with permission from V. K. Bhargava, D. Haccoun, R. Matyas, and N. Nuspl, *Digital Communications by Satellite*, John Wiley & Sons, New York, 1981.)

Since coherent detection is the most efficient way of demodulating direct BPSK, Eq. (5.61) is the relation normally used to determine the (E_b/N_0) and hence the (C/N) that a satellite link must maintain to meet a specified bit error rate requirement. Figure 5.17 displays a curve of PB versus (E_b/N_0).

The reference carrier in Figure 5.15 may be generated from the received signal using a carrier recovery circuit, a form of phase locked loop. See Chapter 5 of reference 17 for a detailed discussion of carrier recovery techniques.

Most carrier recovery loops have a 180° phase ambiguity, that is, when the loop is locked the phase of the recovered carrier may differ by 180° from the correct value. This has the effect of interchanging logical ones and zeroes and causes the demodulated bit stream to be the complement of what was transmitted. There are several ways to eliminate the ambiguity; one is to use differential encoding in which adjacent symbols have the same phase if the modulating voltage is a 1 and are 180° out of phase if it is a 0. This may be realized by a binary phase shifter that toggles between 0° phase shift and 180° phase shift each time the modulating bit is a 0. Incoming 1 values have no effect.

Differential modulation is more error prone than direct modulation, since an error on a single bit in a differential system will cause one or more subsequent bits to be interpreted incorrectly. See Section 5.6 of reference 15 for a detailed analysis of errors in differential PSK. Most practical satellite systems avoid differential encoding and check the status of the recovered carrier phase periodically by transmitting a known word. Logic at the receiver looks for this word. If it receives it correctly, then the recovered carrier phase is correct. If it receives the complement of the known word, than the recovered carrier phase is off by 180° and the demodulated data stream should be complemented before it is sent to the end user.

Quadrature Phase Shift Keying (QPSK)

In QPSK the phase ϕ of the carrier is set by the modulator to one of four possible values. We may write the result as

$$v = V\sqrt{2}\cos(\omega_c t - \phi) \tag{5.62}$$

where ϕ takes on the values $\pi/4$, $3\pi/4$, $5\pi/4$, and $7\pi/4$ rad. The factor 2 is for our later convenience. Using trigonometric identities to expand Eq. (5.62) we obtain

$$v = V\sqrt{2}\cos\omega_c t \cos\phi + V\sqrt{2}\sin\omega_c t \sin\phi \tag{5.63}$$

The first term is a BPSK signal in phase with the carrier; it is called the I channel. The second term is a BPSK signal in quadrature with the carrier and is called the Q channel. Thus a QPSK waveform may be generated by combining two BPSK waveforms in quadrature. We may write the result as

$$v = u_I V \cos\omega_c t + u_Q V \sin\omega_c t \tag{5.64}$$

where u_I represents a binary data stream modulating the I channel and u_Q represents a binary data stream modulating the Q channel. On both of these a logical 1 corresponds to u_I or $u_Q = +1$ and a logical 0 corresponds to u_I or $u_Q = -1$. The relationship between u_I, u_Q, and ϕ is given by

$$u_I = \sqrt{2}\cos\phi \tag{5.65}$$

$$u_Q = \sqrt{2}\sin\phi \tag{5.66}$$

and is summarized in Table 5.2. Note that ϕ is conveniently visualized as the phase angle of a phasor whose real component is u_I and whose imaginary component is u_Q. See Figure 5.18.

The bits u_I and u_Q may be selected alternately from one incoming bit stream. For example, u_I may represent the odd-number bits and u_Q the even. In this case one binary data channel enters the QPSK modulator and the outgoing symbol rate is equal to half of the incoming bit rate. Alternatively u_I and u_Q represent binary data channels coming from independent sources, and QPSK may be viewed

Table 5.2
The Relationship Between the Modulating Bit Streams u_I, u_Q and the Phase Angle ϕ of the Modulated QPSK Waveform

u_I	u_Q	ϕ
1	1	$\pi/4$
-1	1	$3\pi/4$
-1	-1	$5\pi/4$
1	-1	$7\pi/4$

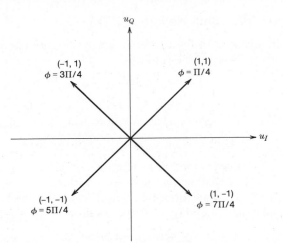

Figure 5.18 Phasor diagram showing the phase of a QPSK waveform modulated with all possible pair combinations of the bits (u_I, u_Q).

as a form of digital multiplexing that combines two BPSK signals with orthogonal carriers. This is the interpretation we first put on Eq. (5.64). When u_I and u_Q come from independent channels, then the incoming bit rate on each of two modulator inputs is equal to the outgoing symbol rate.

QPSK modulators and demodulators are basically dual-channel BPSK modulators and demodulators. One channel processes the u_I bits and uses the reference carrier; the other processes the u_Q bits and uses a 90° phase shifted version of the reference. Figures 5.19 and 5.20 [17] show generalized block diagrams of a QPSK modulator and demodulator. More detailed information is available in reference 18.

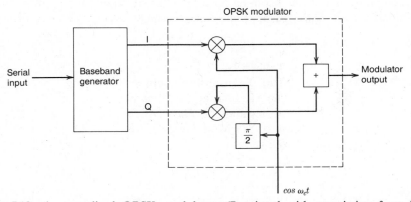

Figure 5.19 A generalized QPSK modulator. (Reprinted with permission from V. K. Bhargava, D. Haccoun, R. Matyas, and N. Nuspl, *Digital Communications by Satellite*, John Wiley & Sons, New York, copyright © 1981.)

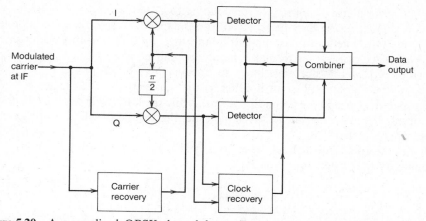

Figure 5.20 A generalized QPSK demodulator. (Reprinted with permission from V. K. Bhargava, D. Haccoun, R. Matyas, and N. Nuspl, *Digital Communications by Satellite*, John Wiley & Sons, New York, copyright © 1981.)

If the transmitted phase angle ϕ takes on the values $\pi/4$, $3\pi/4$, $5\pi/4$ or $7\pi/4$ rad, then a QPSK receiver must simply decide in which quadrant the received phasor signal lies. A decision circuit interprets the phase of all signals that lie in the first quadrant as $\pi/4$; all those in the second quadrant are assumed to have been transmitted with a phase of $3\pi/4$, and so on.

Symbol errors occur when noise pushes the received phasor into the wrong quadrant. Figure 5.21 illustrates the process. In it we assume that the transmitted

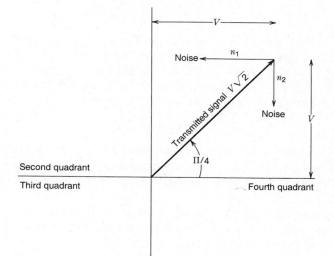

Figure 5.21 Phasor diagram of QPSK signal with phase of $\pi/4$ rad and narrow band noise components n_1 and n_2.

symbol had a phase of $\pi/4$ rad, corresponding to $u_I = 1$ and $u_Q = 1$. Narrow-band white Gaussian noise may be resolved into independent orthogonal components that are respectively in-phase and in quadrature with any arbitrary phasor. We will orient the noise so as to make the component noise phasors n_1 and n_2 point in the directions that are most likely to cause symbol errors.

A symbol error will occur if either n_1 or n_2 exceeds V. The former will place the apparent phase in the second quadrant, while the latter will put it in the fourth quadrant. If both exceed V at the same time, the resultant will go into the third quadrant.

Let the rms noise voltage be σ. This is also the rms value of n_1 and n_2. The probability that n_1 exceeds V is P_A.

$$P_A = \tfrac{1}{2}\operatorname{erfc}\left(\frac{V}{\sigma\sqrt{2}}\right) \tag{5.67}$$

The probability that n_2 exceeds V is P_B. Since n_1 and n_2 have the same rms value, $P_A = P_B$. The probability that n_1 and n_2 simultaneously exceed V is P_A^2. We will assume that this is negligible in comparison to P_A.

An error will occur if either $n_1 > V$ or $n_2 > V$. Since the noise components are independent the probability PE of this happening is

$$PE = 2P_A = \operatorname{erfc}\left(\frac{V}{\sigma\sqrt{2}}\right) \tag{5.68}$$

The rms signal power is proportional to V^2. Following steps similar to those that led to Eq. (5.51), we obtain

$$PE = \operatorname{erfc}\left(\frac{1}{\sqrt{2}}\frac{E_s}{N_0}\right) \tag{5.69}$$

At this point we should emphasize that PE in Eq. (5.68) is the *symbol* error rate for QPSK. Some texts present the same result as if it were a bit error rate, PB. We may calculate PB for QPSK by recognizing that there are two bits per symbol, and hence $E_s = 2E_b$. If we use Eq. (5.50) to relate PB to PE, we find that $PB = 0.5\,PE$. Thus the bit error rate for QPSK is given by

$$PB = \tfrac{1}{2}\operatorname{erfc}(\sqrt{E_b/N_0}) \tag{5.70}$$

Thus QPSK and BPSK have the same bit error rate for the same E_b/N_0 and the error performance of the two modulation systems would seem to be identical; a plot of PB versus (E_b/N_0) for QPSK would be the same as that shown in Figure 5.17.

To compare QPSK and BPSK on an equal basis, assume that we must send R_0 bits per second over a satellite link with fixed bandwidth B and a fixed value of (C/N). For QPSK $R_s = R_0/2$. Hence by Eq. (5.48)

$$\frac{E_s}{N_0\,\text{QPSK}} = \frac{C}{N}\frac{R_s}{B_{\text{QPSK}}} = \frac{CR_0}{N2B} \tag{5.71}$$

For BPSK, $R_s = R_0$ and

$$\frac{E_s}{N_{0\text{ BPSK}}} = \frac{C}{N} \cdot \frac{R_s}{B_{\text{BPSK}}} = \frac{CR_0}{NB} \tag{5.72}$$

By Eq. (5.61)

$$PB_{\text{PSK}} = \tfrac{1}{2}\operatorname{erfc}\left(\frac{R_0}{B}\frac{C}{N}\right) \tag{5.73}$$

and by Eqs. (5.69) and (5.50)

$$PB_{\text{QPSK}} = \tfrac{1}{2}PE_{\text{QPSK}} = \tfrac{1}{2}\operatorname{erfc}\left(\tfrac{1}{2}\frac{R_0}{B}\frac{C}{N}\right) \tag{5.74}$$

Thus QPSK will have a higher BER than BPSK when the two modulation schemes are compared for equal bit rates, bandwidths, and (C/N) values. But QPSK carries twice as much data as BPSK for the same RF bandwidth and using QPSK can double the communications capacity (and revenue-earning power) of a transponder. Some TDMA systems (see Chapter 6) use BPSK in their preamble for rapid and accurate establishment of a link that will subsequently carry QPSK.

Thus far we have discussed direct QPSK. In differential QPSK [18] the carrier phase ϕ of Eq. (5.63) *changes* by an integer multiple of $\pi/2$ rad. The value of the integer depends on the incoming bits u_I and u_Q. Because of the higher error rate associated with differential modulation, this technique has not been widely adopted.

QPSK Variants

We noted after Eq. (5.64) that QPSK may be visualized as the sum of two independent BPSK signals whose carriers are in phase quadrature. In conventional QPSK the bits u_I and u_Q that modulate these carriers both make step changes at the same time. If the bit changes are staggered so that u_I makes step changes at the beginning of each symbol period and u_Q makes step changes at the midpoint of each symbol period, the result is called *staggered QPSK (SQPSK)* or *offset QPSK (OQPSK)*. If instead of steps the bits make sinusoidal transitions between their allowed values of 1 and -1, the result is *minimum shift keying* (MSK) or *fast frequency shift keying* (FFSK). These modulation systems produce spectra that are slightly different from conventional QPSK and that would appear to have some advantages over QPSK for satellite transmission. While they have received considerable academic attention, OQPSK, SQPSK, MSK, and FFSK have not yet been adopted for commercial satellite applications. At the time of writing, the prevailing attitude in the industry seems to be that any theoretical advantages that they might have over conventional QPSK either vanish when these techniques are used over a real transponder or else are so slight as not to justify the added expense that their implementation would require. Because of space limitations and the present lack of practical applications of these QPSK variants, we will not discuss them further in this text. For additional information the reader should consult references 17 and 18.

5.6 DIGITAL TRANSMISSION OF VOICE

The previous sections have discussed techniques for transmitting and receiving digital information via satellite. Now we will turn our attention to the problem of putting analog voice signals into digital form for transmission and returning them to analog form after reception. While the material presented is generally applicable to all analog signals, we will emphasize baseband voice channels because these are virtually the only analog signals sent over commercial satellite links in digital form.

Sampling and Quantizing

The basic processes in digital transmission of analog information are sampling, quantizing, and encoding. The principles underlying sampling are routinely presented in beginning courses in communications theory, and we will not reproduce them here. See Chapter 5 of reference 6 or Chapter 2 of reference 18 for details. The sampling theorem states that a signal may be reconstructed without error from regularly spaced samples taken at a rate f_s (samples/second), which is at least twice the maximum frequency f_m present in the signal. Instead of transmitting the continuous analog signal, we may transmit the samples. For example, voice signals on satellite links are normally filtered to limit their spectra to the range 300 to 3400 Hz. Thus, one voice channel could be transmitted with samples taken at least 6800 times per second or, as it is usually expressed, with a minimum sampling frequency of 6800 Hz. Common telephone system practice is to use a sampling frequency of 8000 Hz. While transmitting the samples requires more bandwidth than transmitting the original waveform, the time between samples of one signal may be used to transmit samples of other signals. This is *time division multiplexing* (TDM), and we will discuss it later in this chapter.

The samples to which the sampling theorem refers are analog pulses whose amplitudes are equal to that of the original waveform at the time of sampling. The original waveform may be reconstructed without error by passing the samples through an ideal low-pass filter whose transfer function is appropriate to the sampling pulse shape. A communications system that samples an input waveform and transmits analog pulses is said to use *pulse amplitude modulation*. Figure 5.22 sketches this process.

Analog pulses are subject to amplitude distortion, and they are also incompatible with conventional baseband digital signals in which pulses take on only one of two possible values. Hence pulse amplitude modulation is not used over satellite links. Instead, the analog samples are *quantized*—resolved into one of a finite number of possible values—and the quantized values are binarily encoded and transmitted digitally. Thus each sample is converted into a digital word that represents the quantization value closest to the original analog sample. *Quantization* may be *uniform* or *nonuniform* depending on whether or not the quantized voltage levels are uniformly or nonuniformly spaced. At the receiver a digital-to-analog (D/A) converter converts each incoming digital word back into an analog sample; these analog samples are filtered and the original input waveform is reconstructed.

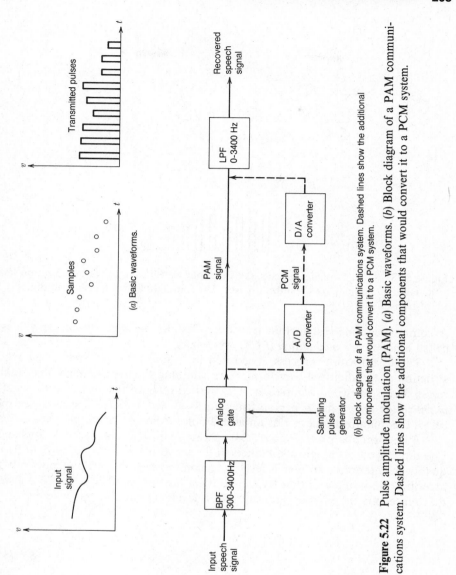

Figure 5.22 Pulse amplitude modulation (PAM). (*a*) Basic waveforms. (*b*) Block diagram of a PAM communications system. Dashed lines show the additional components that would convert it to a PCM system.

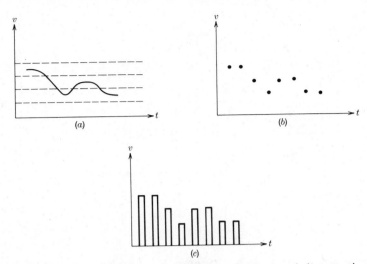

Figure 5.23 The quantizing process. (*a*) The input waveform and the quantization levels. (*b*) Quantized samples. (*c*) Quantized pulses. Their amplitude will be encoded digitally for PCM transmission.

A communications system that transmits digitally encoded quantized values is called a *pulse code modulation (PCM)* system.

The quantization process, illustrated in Figure 5.23, prevents exact reconstruction of the digitized waveform. (The sampling theorem requires that analog rather than quantized samples be transmitted.) The error introduced is called *quantization error*; and a person listening to a reconstructed speech signal perceives the quantization error as an added noise called *quantization noise*.

A uniform quantizer (Figure 5.24) operates with L levels spaced Δ volts apart. The input signal is amplitude limited to lie between $-\Delta(L/2)$ and $+\Delta(L/2)$. The quantizer determines in which level an incoming sample falls and puts out the identification number of that level. This identification number is the digital word that represents the sample. Transmitting L levels requires N_l bits where

$$N_l = \log_2 L \qquad (5.75)$$

or

$$L = 2^{N_l} \qquad (5.76)$$

The levels are normally numbered 0 through $L - 1$. Thus an 8-bit ($N = 8$) PCM system would quantize its incoming samples into one of 256 ($L = 2^8 = 256$) levels numbered 0 through 255. These would be transmitted as binary words ranging from 00000000 (decimal 0) through 11111111 (decimal 255).

If the input signal amplitude is uniformly distributed with an rms value of V_{rms}, the signal-to-noise ratio (S/N) of the reconstructed signal (assuming that *only* quantization noise is present) is given by [19]

$$(\text{S/N}) = 12\left(\frac{V_{rms}}{\Delta}\right)^2 \qquad (5.77)$$

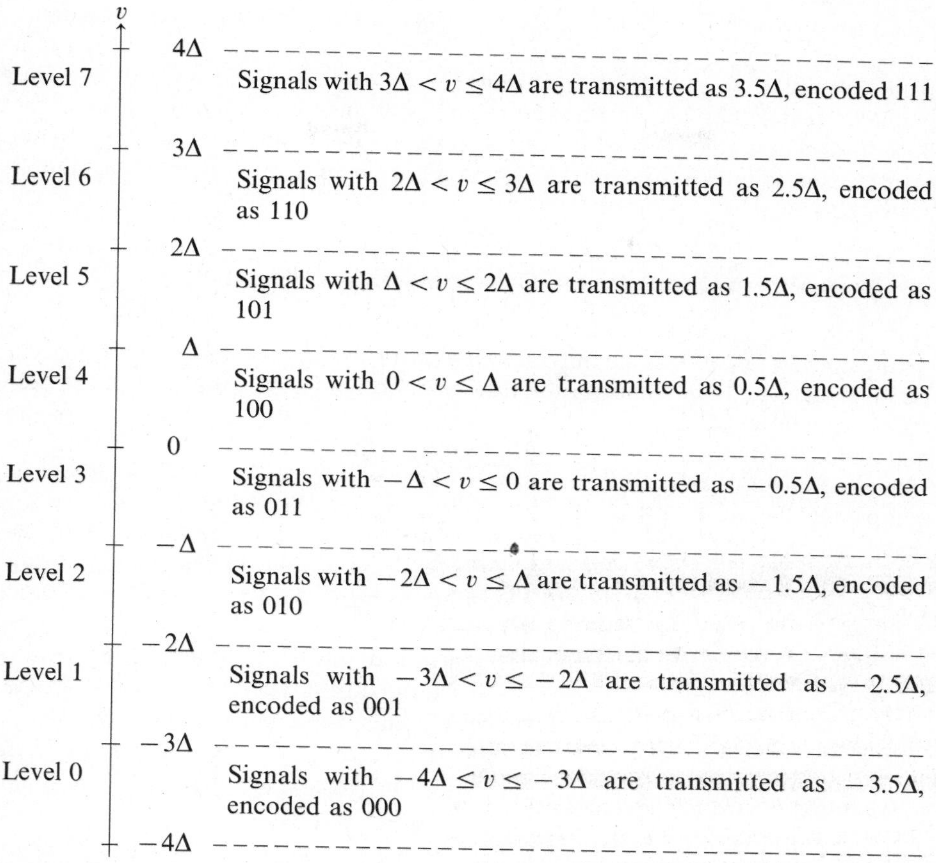

Figure 5.24 Levels and encoding for a uniform 3-bit (8 level) PCM quantizer.

It is common practice (but not a requirement) for uniform quantizers to be designed so that

$$\Delta 2^{N_l} = 8V_{\text{rms}} \tag{5.78}$$

For uniform quantization and a sine wave input

$$(S/N) = \frac{3}{16} L^2 \tag{5.79}$$

or in decibels

$$(S/N) = 6N_l - 7.27 \, \text{dB} \tag{5.80}$$

For a uniformly distributed input

$$(S/N) = L^2 - 1 \simeq L^2 \qquad \text{when} \qquad N \geq 5 \text{ bits} \tag{5.81}$$

In Section 5.1 we noted that a common specification for satellite telephone channels is a weighted (S/N) of 51.25 dB (7500 pWp of noise). Achieving this performance with uniformly quantized PCM would require $N_l = 10$ bits per sample [rounded up from the 9.75 calculated using Eq. (5.80)]. At 8000 samples per second, a single telephone channel would require 80,000 bits per second (80 kbps) and a bandwidth of about 80 kHz for transmission by uniform PCM and BPSK. This is much greater than the 4 kHz required for SSB analog transmission. While the 80 kHz can be reduced to 64 kHz by the techniques of the next section, digital transmission is inherently bandwidth inefficient when a small number of signals are to be sent.

Nonuniform Quantization: Compression and Expansion

Uniform quantization introduces more noise when a signal is small and one quantization interval is large in comparison with the signal than it does when the signal is large and one quantization interval is insignificant.. Improved noise performance can be obtained using *nonuniform quantization* in which the size of the quantization intervals increases in proportion to the signal value being quantized. The same effect can be obtained from a uniform quantizer if the input signal is compressed before quantization. The distortion introduced by the compressor must be removed at the receiver by an expander. The transfer functions of the compressor and expander are complementary, that is, their product is a constant and the amplitude distribution of a signal that has passed through both a compressor and an expander is unchanged.

Companding was first employed on terrestrial telephone systems using analog compressors that had logarithmic transfer functions. These were the so-called mu-law and A-law compressors. Later developments in digital technology allowed

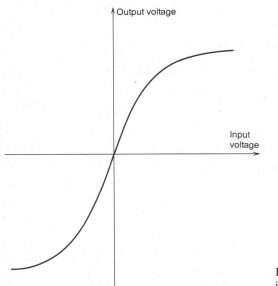

Figure 5.25 The transfer characteristic of a compressor.

digital implementation of the compression and expansion functions and permitted the sampling, compression, quantization, and encoding operations to be combined into one piece of equipment called a *coder*. Common digital compression schemes are the 15-segment coder that provides up to 30 dB improvement over a uniform quantizer with the same number of levels and the 13-segment coder that yields up to 24 dB improvement. (These numbers are taken from reference 1, whose Chapter 28 provides a good discussion of the companding process. See also Chapter 3 of reference 8.) Thus, with a 15-segment coder the 51.25-dB (S/N) requirement of the previous section may be met by the number of bits required by a nonuniform quantizer that delivers a 21.25-dB (S/N) (51.25 minus 30). From Eq. (5.80) this requires $N = 5$ bits. Practical companded systems operate with 7 or 8 bits. Figure 5.25 shows a typical compressor characteristic.

Signal-to-Noise Ratio in PCM Systems

Thermal noise causes bit errors in digital communication links, as discussed in Section 5.5. In a PCM system, the digital data are converted back to a baseband analog signal at the receiver. We need to know the signal-to-noise ratio that corresponds to a given probability of a bit error occurring in the digital data at the receiver. The analysis is straightforward when only one bit error occurs in each PCM word; provided the BER is below 10^{-4} and we have 7 or 8 bits per word, the likelihood of two bit errors occurring in one word is very small. We will assume this to be the case in the analysis that follows.

When a bit is in error in a PCM word, the recovered sample of the baseband analog signal will be at the wrong level. This adds an impulse of amplitude V_n and duration T_s, the period of one sample, to the true analog signal. The bit that is in error may be located in any position in the PCM word. If the least significant bit is in error, V_n is small and equal to Δ, the analog-to-digital converter step size; if it is the most significant bit that is in error, V_n will be large and equal to $2^{N_l - 1}\Delta$. Thus the variance of the error in the analog signal, $(\overline{\Delta m})^2$ is given by [6]

$$(\Delta m)^2 = \frac{1}{N_l} \left[\Delta^2 + (2\Delta)^2 + (4\Delta)^2 + \ldots (2^{N_l - 1}\Delta)^2 \right] \qquad (5.82)$$

The sum of the geometric progression in Eq. (5.82) is

$$(\overline{\Delta m})^2 = \frac{2^{2N_l - 1}\Delta^2}{3N_l} \approx \frac{2^{2N_l}\Delta^2}{3N_l} \qquad (5.83)$$

for $N_l \geq 2$.

Errors occur infrequently for BERs below 10^{-4}, and the mean time between errors is T, where

$$T = \frac{T_s}{N_l PB} \qquad (5.84)$$

and *PB* is the bit error probability.

The two-sided power spectral density of a thermal-noise impulse train is given by [6]

$$G_t(f) = \frac{(\Delta m)^2}{T} = \frac{N_l PB(\Delta m)^2}{T_s} \tag{5.85}$$

To calculate the average noise power in the baseband channel due to noise impulses at the receiver output we must assume a baseband bandwidth. For the square-root raised cosine filter, the equivalent noise bandwidth B is $1/2T_s$, for all values of α. Then the thermal noise after the filter due to random bit errors in the PCM words is

$$N_t = \int_{-B}^{+B} G_t(f)\,df = \int_{-1/2T_s}^{1/2T_s} \left[\frac{2^{2N_l}\Delta^2 N_l PB}{3N_l T_s} \right] df = \frac{2^{2N_l}\Delta^2 PB}{3T_s^2} \tag{5.86}$$

The power in the analog signal, assuming equal probability of any signal voltage, is given by [6]

$$S_0 = \frac{L^2\Delta^2}{T_s^2 12} \tag{5.87}$$

Combining Eqs. (5.86) and (5.87) we find that the signal-to-noise ratio in the baseband channel at the output of the receiver is

$$\frac{S_0}{N_t} = \frac{L^2\Delta^2}{T_s^2 12}\frac{3T_s^2}{2^{2N}\Delta^2 PB} = \frac{1}{4PB} \tag{5.88}$$

since $L^2 = 2^{2N_l}$.

The mean quantization noise power for an equiprobable signal is given by [6]

$$N_q = \frac{\Delta^2}{12T_s^2} \tag{5.89}$$

We can combine thermal noise from Eq. (5.77) with quantization noise from Eq. (5.89) to find the overall PCM output signal-to-noise ratio.

$$\frac{S}{N_{PCM}} = \frac{S_0}{N_t + N_q} = \frac{\dfrac{L^2\Delta^2}{12T_s^2}}{\dfrac{2^{2N_l}\Delta^2 PB}{3T_s^2} + \dfrac{\Delta^2}{12T_s^2}} = \frac{2^{2N_l}}{1 + 4PB2^{2N_l}} \tag{5.90}$$

When PB is small, for example, less than 10^{-8}, the quantization noise will dominate and $(S/N) \simeq 2^{2N_l}$. For $N_l = 8$ bits, this gives S/N = 48 dB. When PB is larger, thermal noise dominates; for example, with $PB = 10^{-4}$ and $N_l = 8$,

$$(S/N) \simeq \frac{1}{4PB} = 34\,\text{dB}$$

Figure 5.26 shows the transition from thermal noise to quantization noise as the predominant noise source as the probability of a bit error decreases, for PCM systems using 7 and 8 bit words, with linear quantization. Clearly, for BERs

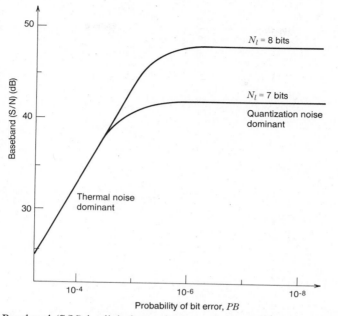

Figure 5.26   Baseband (S/N) in digital speech system using PCM. Signal values between zero and maximum are equally probable.

below 10^{-6}, quantization noise is dominant. Since most PCM links operate with BERs below 10^{-6} most of the time, it is worthwhile using nonlinear encoding (companding) to improve the baseband (S/N) by reducing quantization noise.

Delta Modulation

Delta modulation (*DM*) is a way of digitizing a voice waveform, transmitting the digits, and reconstructing the original analog waveform that avoids the quantizer and the A/D and D/A converters employed in PCM. For details on DM theory and practice the reader should consult reference 20.

In *linear delta modulation (LDM)* a circuit like that of Figure 5.27 determines the difference between an incoming waveform $x(t)$ and an estimate $z(t)$. It calculates an error voltage $e(t)$, where

$$e(t) = x(t) - z(t) \tag{5.91}$$

and a sign quantizer determines the sign of $e(t)$. The quantizer output $Q(t)$ is a positive constant when $e(t)$ is positive (i.e., when the signal is greater than the estimate) and a negative constant when $e(t)$ is negative (i.e., when the signal is smaller than the estimate). A sampling circuit samples $Q(t)$ and generates a positive pulse when $Q(t)$ is positive and a negative pulse when $Q(t)$ is negative. These pulses go to a conventional PSK digital modulator for transmission.

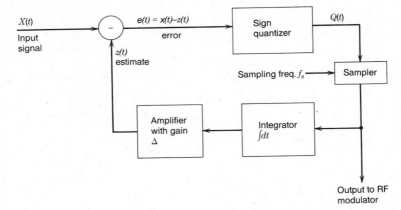

Figure 5.27 A delta modulator.

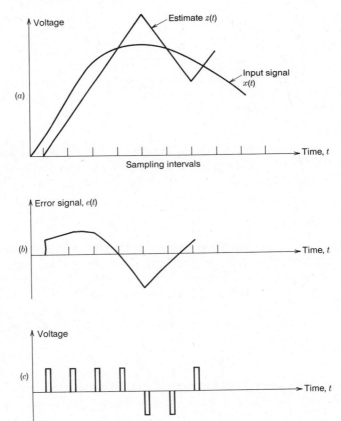

Figure 5.28 Waveforms in the delta modulator of Figure 5.27. (*a*) The input signal $x(t)$ and its estimate $e(t)$. (*b*) The error signal $e(t)$. (*c*) The transmitted pulses. (Based on J. J. Spilker, Jr., *Digital Communications by Satellite*, Prentice-Hall, Inc., Englewood Cliffs, NJ, 1977.)

The waveform reconstruction part of an LDM receiver and the estimator portion of the modulator both use the generated pulses to form the estimate $z(t)$ of $x(t)$. The estimate is made by integrating the pulses (which is equivalent to summing them numerically) and multiplying the result by a step size, Δ. If the pulses are of unit area, then the estimate $z(t)$ increases or decreases in value at a rate equal to the sampling frequency f_s if the estimate was smaller than or larger than the input waveform at the time of the last sample. Note that if input waveform is constant, the estimate will "hunt" around the correct value, alternately overestimating and underestimating it by Δ.

Figure 5.28 illustrates the delta modulation process for a hypothetical input signal. The waveforms in the figure are based on a worst-case assumption that the initial value of the estimate $z(t)$ is zero.

The performance of a DM system depends on the step size in two competing ways. The estimate $z(t)$ can change by only Δ volts at each sampling instant; if the input signal $x(t)$ is changing more rapidly, then the estimate cannot keep up and a condition called *slope overload* occurs. Slope overload can be prevented by making Δ large, but this increases *granularity*, the noise that results when the system hunts around a constant input value. Granularity is minimized by making Δ small. For a particular value of the sampling frequency f_s and a particular input signal, the output (S/N) behavior of a LDM system then depends on Δ as indicated schematically in Figure 5.29 [18]. There is an optimum step size; below it slope overload distortion dominates and above it granular noise dominates.

In linear delta modulation, the step size is fixed at a value that provides performance near the peak of Figure 5.28. But better performance may be achieved through a scheme called *adaptive delta modulation* (*ADM*) in which the value of Δ may be varied during the modulation process. Slope overload is characterized by long strings of pulses with the same algebraic sign. Digital logic in the estimator

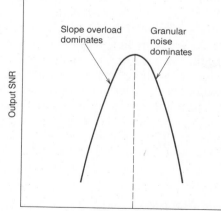

Figure 5.29 Conceptual sketch showing the dependence of (S/N) on step size in a linear delta modulator. (*Source:* J. J. Spilker, Jr., *Digital Communications by Satellite*, copyright © 1977, p. 77. Reprinted by permission of Prentice-Hall, Inc., Englewood Cliffs, NJ.)

circuit watches for this condition and automatically increases the step size after some specified number of same-sign pulses have been sent. Granularity, on the other hand, is associated with long strings of pulses having alternating signs; when this occurs then Δ is made smaller. Thus the step size varies between some specified minimum and maximum; the transmitter and receiver contain identical circuits to determine the step size and they both use the same value for a given sample.

It is difficult to make a definitive comparison between the performance of PCM and DM because DM technology is continually improving. See reference 19 for a detailed discussion of some early comparisons. If the bit rate can be made arbitrarily large, then PCM will outperform DM. But if the systems are compared for successively lower bit rates, then a crossover point will be reached below which DM gives better performance than PCM.

Most commercial satellite links to date use only PCM for digital transmission of speech while DM has been used for military applications or specialized civilian systems like the Space Shuttle in which the simpler equipment of DM and its improved performance advantage at low bit rates (i.e., at narrow bandwidths) over PCM is significant.

5.7 DIGITAL TV AND BANDWIDTH COMPRESSION [21]

While satellite TV for program distribution has retained analog modulation, compressed bandwidth digital television is now offered by at least two carriers (AT&T and SBS) for video teleconferencing. This is a service that offers two-way interactive video and audio for meetings and conferences; see reference 22 for an interesting discussion of the facilities and procedures involved. If the U.S. standard baseband video signal were digitized and transmitted without encoding, it would require a minimum sampling frequency of 8.4 MHz and a bit rate above 50 Mbps. But television pictures are highly redundant, and a much lower bit rate is required to transmit the difference between successive frames than to transmit the frames themselves. Further reduction is possible if the transmitter and receiver use linear predictive encoding [18]. The degree of redundancy in successive TV frames varies inversely with the motion of the main features in the pictures, and a consequence of reducing the bit rate is a deterioration of the system's ability to depict moving objects. This would be objectionable in entertainment TV, but the participants in a teleconference are usually sitting in one place and a technique called motion compensation can be used to improve the presentation of the regular linear motion associated with camera panning. The result is that teleconference TV is now transmitted at bit rates as low as 1.544 Mpbs. (This is the T1 bit rate explained in the next section.)

5.8 TIME DIVISION MULTIPLEXING

In time division multiplexing (TDM) a group of signals take turns using a channel. This contrasts to frequency division multiplexing, presented earlier, where the signals occupy the channel at the same time but on different frequencies. Since digital signals are precisely timed and consist of groups of short pulses with relatively long intervals between them, TDM is the natural way for combining digital signals for transmission.

TDM Terminology: The U.S. T1 24-Channel System

In this section we will describe the U.S. Bell Telephone T1 24-channel TDM system and use it to introduce the terminology of time division multiplexing. While T1 was developed for terrestrial microwave circuits, it appears on FDMA digital satellite links like those operated by RCA Government Communications Services [2]. We present it here as a convenient vehicle for explaining TDM operation. In pure TDMA systems, the multiplexing blends into the multiple access process.

A TDM system transmits a digital word from each channel in turn. Each word is a group of bits that identify the quantization interval of the current sample. The words are organized into *frames*. One frame contains one word from each channel plus some synchronizing information that serves to identify the start of the frame. A frame is then a series of bits numbered sequentially from zero that carry synchronizing information plus the quantized values of one sample from each channel. The bits within a frame are grouped into *slots*. A slot contains all the bits from a common source. The slots within a frame are numbered sequentially from zero. In the Bell T1 system illustrated in Figure 5.30, there are 25 slots, numbered 0 through 24. Slot 0 contains a single bit and carries synchronization information. Slots 1 through 24 each contain 8 bits and carry telephone channels 1 through 24. Thus a T1 frame contains $1 + (8 \times 24)$ or 193 bits.

Standard telephone PCM systems sample at an 8-kHz rate. Said another way, they transmit one sample of each channel every 125 μs. This is the frame interval; the frame rate is the reciprocal of the frame interval and is always equal to the sampling rate. The T1 system must transmit 193 bits in the 125 μs frame interval; hence its bit rate is 193 bits divided by 125 μs or 1.5440 megabits per second (Mbps).

Frame synchronization is established and maintained by transmitting a known bit pattern in slot 0. This bit pattern constitutes what is called the *frame alignment word* (*FAW*). The T1 FAW is 100011011100; it contains 12 bits and requires 12 frames for transmission. The group of frames that transmit the FAW make up a *superframe*. Thus in the first frame of a superframe, slot 0 contains a 1. In the second frame slot 0 contains a 0, and so on.

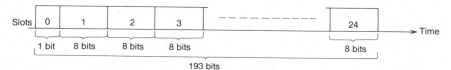

Figure 5.30 Slot organization of one Bell T1 frame.

At this point let us summarize the operation of a Bell T1 digital multiplexer. It receives digitized voice signals from 24 telephone channels, which, for now, we will assume are perfectly synchronized with each other and have exactly the same bit rate. The 8-bit word samples from each channel flow into buffers and wait for the multiplexer to read them out. The multiplexer reads them out and inserts them into outgoing frames as follows. The first frame is transmitted by sending a 1(the first bit of the FAW), then one 8-bit word from channel 1, then one 8 bit word from channel 2, and so forth through one 8-bit word from channel 24. Then a new group of samples flows into the buffers. The multiplexer forms frame 2 by sending a 0 (the second bit of the FAW) followed by the words from each channel. A third group of samples enters and the process continues. When the buffers have been filled and emptied 12 times, one superframe has been sent and the multiplexer begins a new FAW.

At the receiving end of the link a demultiplexer must sort out the bits in each frame and route the appropriate words to each outgoing channel. It must also keep track of the number of the frame (within the superframe) that it is receiving.

We may visualize the demultiplexing process by assuming that the incoming bits are clocked serially into a shift register. At the instant the last bit has entered the register, its contents match the bits in the frame of Figure 5.30. The multiplexer then does a parallel transfer of the bits for each channel into their own individual registers for subsequent serial transmission over their separate paths. At this point the digital channels have been demultiplexed.

The frame alignment bits go into a shift register, which, at the completion of the superframe, should contain the frame alignment word. If it does not, then the multiplexer and demultiplexer are out of sync. When this occurs the demultiplexer seeks to regain alignment; the process is called *reframing*. In it the demultiplexer looks at candidate frame alignment bits until it finds one that is going through the requisite 100011011100 100011011100 100011011100 pattern. Obviously there is a trade-off between the number of frame alignment bits and the time required for reframing. If the entire FAW is transmitted within each frame, then reframing time is much shorter than when the FAW is transmitted with one bit per frame. Typical reframing times are about 50 ms [21], which is sufficiently short not to cause significant degradation of speech in the 24 channels when misalignment occurs.

Along with the information carried in the 24 channels must go the signaling information necessary to route, initiate, and terminate the data channels. In the T1 system this information is transmitted by "robbing" the least significant bit from slots 6 and 12 and using these to form signaling channels A and B, respectively. Thus, channels 6 and 12 are actually carried by a form of 7-bit PCM and channels A and B convey signaling information at an 8 kbps rate.

Other TDM Systems

At the time of writing there seems to be no agreed upon standard international for satellite TDM systems. The CCITT has recommended a standard 1.544 Mbs system (which is slightly different from the T1) and a 2.048 Mbs 30-channel system [23]. For details on their slot and bit organization the reader should consult reference 24.

Channel Synchronization in TDM

Our explanation of the T1 system made the tacit assumption that all 24 incoming PCM channels were synchronized with each other and running at the same bit rate. This condition would hold if the voice channels had reached the originating earth station in analog form and had been digitized by modulators running on a common clock. But if the channels came into the station in digital form, their synchronization would not be guaranteed. They may be resynchronized for TDM transmission by a technique called *pulse stuffing* [1, 18].

In pulse stuffing the incoming words for each channel flow into an elastic buffer. There is one such buffer per channel, and each buffer can hold several words. The multiplexer reads words out of the buffer slightly faster than they come in. Periodically the multiplexer will go to the buffer and find less than a full word remaining. When that happens it inserts a dummy word called a *stuff word* into the frame in place of the word it would have taken from the buffer. At the same time it places a message on the signaling channel that states that a stuff word has been inserted. When the demultiplexer at the other end of the link receives the message it ignores the stuff word. When it is time for the next frame to be sent the buffer will have more than a full word waiting for transmission.

5.9 SUMMARY

Multiplexing is the process of separating the channels transmitted by a single earth station to prevent them from interfering with each other; its most common forms are frequency division multiplexing (FDM) and time division multiplexing (TDM). In the first case the channels are separated in frequency and in the second case they are separated in time.

Most analog telephone channels are transmitted over satellite links using frequency division multiplexing with frequency modulation (FDM/FM). In this method individual voice channels (nominally containing frequencies between 300 and 3400 Hz) are "stacked" in frequency by a multiplexer and the resulting multiplexed telephone signal is used to frequency modulate an uplink carrier. At the downlink earth station, an FM demodulator recovers the multiplexed signal and a demultiplexer recovers the individual channels.

An FM demodulator is characterized by a threshold. Provided that a satellite link's overall carrier-to-noise ratio (C/N) is above this threshold, the signal-to-noise ratio (S/N) of the demultiplexed voice channels will be significantly greater than the incoming (C/N). This effect is called FM improvement. Additional improvement in (S/N) may be obtained through preemphasis and deemphasis. Deemphasis decreases the noise power output of an FM demodulator; preemphasis distorts the multiplexed telephone signal before transmission to compensate for the deemphasis at the downlink earth station.

Analog FDM/FM systems are designed to produce a weighted (S/N) of about 50 dB in the worst telephone channel; this corresponds to about 10,000 picowatts psophometrically weighted (pWp) of noise power. The terms *weighted* and *psophometrically weighted* refer to a procedure for calculating (S/N) that reflects the human ear's nonuniform response to white noise.

Both the bandwidth and the FM improvement of an FDM/FM link depend in a complicated way on the number of voice channels carried, the way in which the voice channels are stacked, and the rms test-tone deviation of the uplink frequency modulator. The last is the rms carrier deviation that a single 1-kHz 0-dBm sinewave called the test tone will produce when supplied to one telephone channel.

Two important alternatives to FDM/FM for analog transmission are FM SCPC (single-channel-per-carrier) and companded single sideband (CSSB) systems. These share a common feature that individual voice channels can be extracted from the downlink radio frequency (RF) signal without demultiplexing all of the unwanted channels. In FM SCPC each voice channel FM modulates its own uplink channel. While requiring more bandwidth per channel than FDM/FM, FM SCPC makes more efficient use of transponder power and is easier to reconfigure to meet changing traffic conditions. It also has economic advantages for low-traffic routes.

In CSSB the individual voice channels are shifted to the uplink frequency by single sideband suppressed carrier modulation. Including guard bands, each signal occupies a 4-kHz bandwidth. Thus CSSB uses bandwidth much more efficiently than competing analog FM and digital modulation schemes, and it provides the same capability as FM SCPC for recovering individual channels at RF. Since SSB demodulation provides no (S/N) improvement, SSB transmission would require prohibitively large uplink and downlink (S/N) values if it were not for companding, a process by which the dynamic range of a voice channel is compressed before transmission and expanded (i.e., restored) after reception. This permits SSB links with (S/N) values on the order of 20 dB at RF to provide subjective (S/N) values of around 50 dB in individual voice channels. Its efficient use of bandwidth makes CSSB more attractive than digital modulation for some applications.

In analog television (TV) transmission by satellite, the baseband video signal and one or two audio subcarriers constitute a composite video signal that frequency modulates an uplink carrier. This system requires a very wide bandwidth for transmission (usually either a full transponder or a half transponder), but it provides the FM improvement necessary to achieve required (S/N) values.

If an uplink FM transmitter is fully modulated (fully "loaded" in the usual terminology), the downlink EIRP is distributed across an entire transponder bandwidth. But if the load decreases and disappears, the downlink EIRP will be concentrated at the center of the transponder bandwidth, and in the absence of uplink modulation it will be radiated at a single frequency. To avoid the interference with terrestrial radio services that this would cause, uplink earth stations are required to add a dispersal waveform to their modulating signal so as to maintain a reasonably constant downlink power spectral density independent of traffic loading. The dispersal waveform is removed by filtering at the downlink earth station.

Digital modulation is obviously the modulation of choice for transmitting digital data. Digitized analog signals may conveniently share a channel with digital data, allowing a link to carry a varying mix of voice and data traffic.

While baseband digital signals are often visualized as rectangular voltage pulses, careful pulse shaping is required to prevent intersymbol interference (ISI) and to permit reasonably distortionless transmission through the limited bandwidth of a transponder. With proper pulse shaping, the symbol rate of a digital link can be made approximately equal to the RF bandwidth.

The common digital modulation schemes used on digital satellite links are binary phase shift keying (BPSK) and quadrature phase shift keying (QPSK). In these an incoming data stream sets the phase of a sinusoidal carrier to one of two (BPSK) or four (QPSK) values. The performance of a BPSK or QPSK link is described by its bit error rate. Digital links are designed to meet bit error rate requirements in the same way as analog links are designed to deliver minimum (S/N) values.

Analog voice signals must be digitized for transmission over a digital link. This involves sampling the signal at a rate that is at least twice the highest frequency present and converting the sample values to digital words. The system that does this is called a quantizer; nonuniform quantizers are analogous to companders in CSSB systems and permit a lower bit rate than uniform quantizers. Standard practice with nonuniform quantization is to sample telephone channels at a 4-kHz rate and transmit each channel at 64 kbps.

Digital signals from different channels are interleaved for transmission through time division multiplexing (TDM). Digitized samples or digitized words from each channel are transmitted in turn; the time interval in which one sample or word from each channel is sent is called a frame. Channels are identified by their position in the frame; individual frames are identified by the presence of synchronization bits that repeat a known pattern.

REFERENCES

1. Technical Staff, Bell Telephone Laboratories, *Transmission Systems for Communications* (5th ed.), Bell Telephone Laboratories, Holmdel, NJ, 1982.
2. K. Jonnalagadda and L. Schiff, "Improvements in Capacity of Analog Voice Mutliplex Systems Carried by Satellite," *Proceedings of the IEEE*, **72**, 1537–1547 (November 1984).
3. W. H. Braun and J. E. Keigler, "RCA Satellite Networks: High Technology and Low User Cost," *Proceedings of the IEEE*, **72**, 1483–1505 (November 1984).
4. *Standard A Performance Characteristics of Earth Stations in the INTELSAT IV, IV-A, and V Systems Having a G/T of 40.7 dB/K* Publication (BG-28-72E Rev. 1), Intelsat, Washington, D.C., December 15, 1982.
5. H. L. Krauss, C. W. Bostian, and F. H. Raab, *Solid State Radio Engineering*, John Wiley & Sons, New York, 1980.
6. H. Taub and D. L. Schilling, *Principles of Communications Systems*, McGraw-Hill, New York, 1971.

7. *Reference Data for Radio Engineers*, (6th ed.), Howard W. Sams and Co., Indianapolis, 1975.
8. *Recommendations and Reports of the CCIR, 1978*, Vol. IV, International Telecommunication Union, Geneva, Switzerland, 1978.
9. H. L. Van Trees, *Satellite Communications*, IEEE Press, New York, 1979.
10. M. E. Ferguson, "Design of FM Single-Channel-per-Carrier Systems," *IEEE Int. Conf. on Commun.*, 1, 12-11–12-16 (June 1975). (Reprinted in [9], pp. 336–341.)
11. K. Miya, Ed., *Satellite Communications Engineering*, Lattice Publishing Co., Tokyo, 1975.
12. K. S. Shamnugam, *Digital and Analog Communication Systems*, John Wiley & Sons, New York, 1979.
13. H. Nyquist, "Certain Topics in Telegraph Transmission Theory," *AIEE Transactions*, 47, 817 (April 1928).
14. K. Feher, *Digital Communications: Satellite/Earth Station Engineering*, Prentice-Hall, Englewood Cliffs, NJ, 1983.
15. W. C. Lindsey and M. K. Simon, *Telecommunication Systems Engineering*, Prentice-Hall, Englewood Cliffs, NJ, 1973.
16. J. R. Pierce and E. C. Posner, *Introduction to Communication Science and Systems*, Plenum Press, New York, 1980.
17. V. K. Bhargava, D. Haccoun, R. Matyas, and N. Nuspl, *Digital Communications by Satellite*, John Wiley & Sons, New York, 1981.
18. J. J. Spilker, Jr., *Digital Communications by Satellite*, Prentice-Hall, Englewood Cliffs, NJ, 1977.
19. N. S. Jayant, "Digital Encoding of Speech Waveforms: PCM, DPCM, and DM Quantizers," *Proceedings of the IEEE*, 62, 611–632 (May 1974). (Reprinted in [9], pp. 86–107.)
20. R. Steele, *Delta Modulation Systems*, John Wiley & Sons, New York, 1975.
21. T. Ishiguro and K. Iinuma, "Television Bandwidth Compression Transmission by Motion-Compensated Interframe Coding," *IEEE Communications Magazine*, 20, 24–30 (November 1982).
22. B. A. Wright, "The Design of Picturephone Meeting Service (PMS) Conference Centers for Video Teleconferencing," *IEEE Communications Magazine*, 21, 30–36 (March 1983).
23. J. Martin, *Communications Satellite Systems*, Prentice-Hall, Englewood Cliffs, NJ, 1978.
24. G. H. Bennett, *Pulse Code Modulation and Digital Transmission*, Marconi Instruments Ltd., St. Albans, Hertfordshire, England, 1978.

PROBLEMS

1. An earth station is to receive 972 multiplexed telephone channels from the INTELSAT V global beam. The rms test-tone deviation specified for the system is 802 kHz and the top baseband frequency is 4028 kHz. Using a peaking factor $g = 3.16$, determine the following.

a. The number of groups and supergroups that are involved.
b. The required receiver bandwidth before FM detection.
c. The overall (C/T) at the receiver input required for operation at a 10-dB (C/N).
d. The overall (C/N) required at the receiver input required for a worst-channel (S/N) of 50 dB.

2. A television signal at baseband extends from 0 to 4.2 MHz. The television signal frequency modulates a 6-GHz transmitter with a peak carrier deviation of 5 MHz. The transmitted signal occupies a bandwidth of 18.40 MHz.

The 6-GHz TV signal will be relayed by a satellite whose antenna noise temperature is 290 K and whose U.S. standard noise figure is 8 dB. Assume that the transponder bandwidth equals the 18.40 MHz signal bandwidth. Successful operation of this system requires a minimum (C/N) of 10 dB at the satellite receiver input.

a. Calculate the required carrier power in dBW at the output of the satellite antenna (i.e., at the input to the transponder).
b. Assuming a satellite antenna gain of 17 dB and a distance from the transmitting station to the satellite of 3.8×10^7 m, calculate the required earth–station EIRP required in dBW.
c. For a 45-dB gain earth station antenna, find the required transmitter power in watts.
d. Assuming that the downlink (C/N) is 11 dB, calculate the overall (C/N) including retransmitted noise.

3. You have rented 36 MHz of bandwidth on an INTELSAT V transponder and your link can deliver an overall carrier-to-noise ratio $(C/N)_o$ to your earth station of 15 dB. You want to run analog telephone FDM/FM/FDMA telephone channels over this downlink. Determine the number of channels that the link can carry if the weighted worst-channel signal-to-noise ratio $(S/N)_{wc}$ is 51.0 dB. In making this calculation use the following equation to relate $(S/N)_{wc}$ to $(C/N)_o$. The equation includes preemphasis and weighting.

$$(S/N)_{wc} = (C/N)_o + 6.5 + 10 \log_{10}(B_{IF}/b) + 20 \log_{10}(\Delta f_{rms}/f_m) \text{ dB}$$

Here b is the bandwidth of a single telephone channel (use 3100 Hz), B_{IF} is the occupied (Carson's rule) bandwidth, $f_m = 4200 \times$ number of channels, and Δf_{rms} is the rms test-tone deviation (you have to find it). If the channels are organized into groups and supergroups, how many can be carried?

4. One of Intelsat's goals is a system of directly relaying messages from one satellite to another without going through an intervening earth station. Assume that they intend to use a frequency in the assigned 55 to 65 GHz band where, at present, 10-W output transmitters and receivers with 750 K noise temperatures are available. In this problem you will do some preliminary analysis for such a system using a *10-W transmitter* and a *750 K* overall noise temperature receiver operating at *60 GHz*. The satellite-to-satellite link will use BPSK (for ease of calculation) with a bit rate of *100 megabits per second* and a bit error rate (*BER*) *of*

10^{-5}. In your link calculations, assume that the two satellites are spaced 120° apart in a 24-h (exactly) synchronous orbit and use parabolic reflector antennas.

   a.   Find the linear distance (in kilometers) between the two satellites.

   b.   Assuming identical transmitting and receiving antennas, find the minimum antenna diameter (in meters) necessary to achieve the 10^{-5} BER without leaving any margin for error.

   c.   Suppose that the two antennas are misaligned so that the *sum* of their gains is 3 dB below the peak value. What will the BER be now?

   d.   What diameter should each antenna have to allow a 3-dB error margin per antenna? (In other words ensure that both antennas can be simultaneously mispointed so that the gain of each is down 3 dB from peak and the system still maintains a 10^{-5} BER.)

5.   The following bit steam is to be transmitted, left-most bit first:

01100011

Explain how it would be transmitted using (a) direct BPSK with phases of $\pm \pi/2$ and (b) direct QPSK using phases of $\pi/4$, $3\pi/4$, $5\pi/4$, and $7\pi/4$. For each case define the transmitted symbol stream and sketch the carrier phase that corresponds to each transmitted symbol.

6.   Transponder No. 5 on SBS-1 has its input (uplink) center frequency at 14.221 GHz and its output (downlink) frequency at 11.921 GHz. The transponder bandwidth is 43 MHz. While SBS-1 is intended for digital applications, it can be used for FDM/FM analog telephone as well. In this problem we will consider both applications.

   a.   In a pure TDM mode using QPSK how many telephone channels could the transponder carry? For purposes of this problem, you may assume that a single earth station has full use of the transponder. Explain your reasoning fully.

   b.   Assume that the transponder must carry exactly 912 FDM/FM voice channels. At baseband, they are organized into one basic group, which extends from 12 to 60 kHz followed by an unspecified number of supergroups. Determine the peak frequency deviation Δf_{peak} and the rms multicarrier deviation $l\Delta f_{rms}$ that should be used. Explain each step in your calculations and any assumptions fully.

   c.   At the receiving earth station the weighted signal-to-noise ratio in the worst demultiplexed telephone channel must be 53 dB. Using a 4-dB preemphasis improvement factor and a 2.5-dB psophometric weighting factor, determine the overall (C/N) required at the input to the FM demodulator.

   d.   Between what baseband frequencies will the worst channel be found?

7.   A transponder with an EIRP of 34 dBW transmits uncoded T1 digital telephone "carrier" at 1.5440 megabits per second using direct BPSK. For a free-space path loss of 200 dB, determine the earth station G/T required to receive this transmission with a bit error rate of 1×10^{-7}.

8. This question compares the performance of a single-carrier-per-channel (SCPC) system when different analog and digital modulation techniques are used. A small earth station receives one speech channel from a 14/11 GHz satellite. The RF bandwidth is 45 kHz, and modulation could be either FM or PSK. You are asked to compare the system performance in each case.

In clear air, the measured (C/N) at the input to the earth station LNA is 20 dB. For a few hours per year, rain causes attenuation of the signal and an increase in noise temperature of the system. The statistics of the link are given in Table P.8. Complete the other columns in the table to assess the link performance under rain fade conditions.

- a. Calculate the (C/N) for each step of attenuation in Table P.8, correct to ± 0.1 dB. Use the values given in the table for 0-dB attenuation as the starting point, by considering the reduction in C and the (decibel) increase in N.
- b. The FM demodulator has a threshold at 6 dB, at which the (S/N) improvement factor is reduced by 1 dB. Calculate the (S/N) in the baseband channel for the FM system. If the (C/N) falls below threshold, insert an asterisk in the (S/N) column.
- c. Coherent four-phase PSK (QPSK) could be used to modulate digitally coded speech onto the RF carrier. Assume that PCM is used to code the speech, with an 8-kHz sampling frequency and 8-bit words using linear quantization. Using Figure 5.17 determine approximately the bit error rate on the link assuming that E_b/N_0 is equal to the (C/N) calculated in Part (b).
- d. For cases where the BER is $< 10^{-3}$, estimate the (S/N) in the speech channel using Figure 5.26. Insert an asterisk in the (S/N) column if the BER is above 10^{-3}.
- e. Compare the performance of the FM and QPSK modulation schemes, and comment on the differences. Which system would you recommend and why?

Table P.8
Propagation Statistics and System Performance Data for Problem 8

Time Parameter Is Exceeded (hours/year)	Path Attenuation (dB)	System Noise Temp. (K)	FM System (C/N) (dB)	FM System (S/N) (dB)	QPSK System BER	QPSK System (S/N) (dB)
—	0	250	20		$< 10^{-7}$	
12.0	2	275				
6.5	4	322				
3.0	6	352				
1.8	8	371				
1.0	10	382				
0.5	12	390				

6

MULTIPLE ACCESS

Multiple access is "the ability of a large number of earth stations to simultaneously interconnect their respective voice, data, teletype, facsimile, and television links through a satellite" [1]. To quote a recent survey article, "The multiple-access problem is fundamental to satellite communications because it is the means by which the wide geographic coverage capability and broadcast nature of the satellite channel are exploited. It affects all elements of the system, determines the system capacity and flexibility, and has a strong influence on costs" [2]. The basic problem involved is how to permit a changing group of earth stations to share a satellite in a way that optimizes (1) satellite capacity, (2) spectrum utilization, (3) satellite power, (4) interconnectivity, (5) flexibility, (6) adaptability to different traffic mixes, (7) cost, and (8) user acceptability [1]. Usually all of the elements in this list cannot be optimized and some may have to be traded off against others.

Classically there are three multiple access techniques, illustrated in Figure 6.1. In *frequency division multiple access* (FDMA), all users share the satellite at the same time, but each transmits in its own unique frequency band. This is most commonly employed with analog modulation, where signals are present all the time. In *time division multiple access* (TDMA), the users transmit in turn in their own unique time slots. While transmitting, each occupant has exclusive use of one or more transponders. The intermittent nature of TDMA transmission makes it particularly attractive for digital modulation. In *code division multiple access* (CMDA), many earth stations simultaneously transmit orthogonally coded spread-spectrum signals that occupy the same frequency band. Decoding ("despreading") systems receive the combined transmissions from many stations and recover one of them.

In all three classical multiple access schemes some resource is shared. If the proportion allocated to each earth station is fixed in advance, the system is called *fixed access* (FA) or *preassigned access* (PA). If the resource is allocated as needed in response to changing traffic conditions, the multiple access arrangement is termed *demand access* (DA). Demand access blurs some of the distinction between

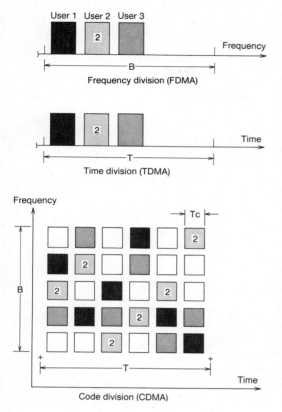

Frequency division (FDMA)

Time division (TDMA)

Code division (CDMA)

Figure 6.1 Multiple-access techniques. Here each "user" corresponds to a one-way communication link between two stations. In frequency division (FDMA), the available bandwidth B is divided among the users and all can transmit simultaneously. In time division (TDMA), only one user transmits at any time and that user can use the entire bandwidth, so the instantaneous data rate is proportional to B. In code division (CDMA) schemes, users can transmit simultaneously and also share the frequency allocation. For the frequency-hopping scheme illustrated, each user transmits with a particular pseudorandom frequency pattern. T_c is normally on the order of a bit time. Such schemes are useful for mobile applications; synchronization among users is not required. (Reprinted with permission from H. L. VanTrees, E. V. Hoverstein, and T. P. McGarty, "Communications Satellites: Looking to the 1980s," *IEEE Spectrum*, **14**, 42–51, (December 1977). Copyright © 1977 IEEE.)

FDMA and TDMA, since stations in a DA FDMA system transmit only when they have traffic. In the sections that follow, our approach will be to present "pure" fixed-access systems first and then discuss demand access.

6.1 FREQUENCY DIVISION MULTIPLE ACCESS (FDMA)

FDM/FM/FDMA

Frequency division multiple access with FM frequency division multiplexing (see Sections 5.1 and 5.2) is abbreviated FDM/FM/FDMA. In it an earth station is premanently assigned a carrier frequency (or several carrier frequencies for a busy station) and a bandwidth around that carrier frequency. The station frequency modulates all of its outgoing traffic—whatever the destination—on that carrier. An originating station's traffic capacity is limited by its allocated bandwidth and the (C/N) that it can achieve on the downlinks. The carrier frequencies and bandwidths assigned to all the earth stations constitute a satellite's frequency plan; Figure 6.2 [3] presents a typical example.

Every station that operates in an FDM/FM/FDMA network must be able to receive at least one carrier from all the stations in the network. Thus most

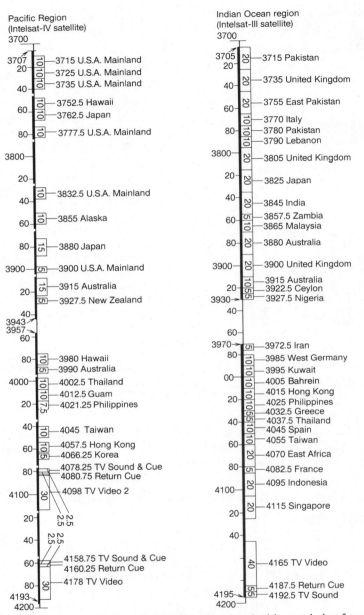

Pacific Region
(Intelsat-IV satellite)

Frequency	Location
3700	
3707	
20	3715 U.S.A. Mainland
	3725 U.S.A. Mainland
40	3735 U.S.A. Mainland
60	3752.5 Hawaii
	3762.5 Japan
80	3777.5 U.S.A. Mainland
3800	
20	
40	3832.5 U.S.A. Mainland
60	3855 Alaska
80	3880 Japan
3900	3900 U.S.A. Mainland
20	3915 Australia
	3927.5 New Zealand
40	
3943	
3957	
60	
80	3980 Hawaii
	3990 Australia
4000	4002.5 Thailand
	4012.5 Guam
20	4021.25 Philippines
40	4045 Taiwan
60	4057.5 Hong Kong
	4066.25 Korea
80	4078.25 TV Sound & Cue
	4080.75 Return Cue
4100	4098 TV Video 2
20	
40	
60	4158.75 TV Sound & Cue
	4160.25 Return Cue
80	4178 TV Video
4193	
4200	

Indian Ocean region
(Intelsat-III satellite)

Frequency	Location
3700	
3705	3715 Pakistan
20	
40	3735 United Kingdom
60	3755 East Pakistan
	3770 Italy
80	3780 Pakistan
	3790 Lebanon
3800	3805 United Kingdom
20	3825 Japan
40	3845 India
60	3857.5 Zambia
	3865 Malaysia
80	3880 Australia
3900	3900 United Kingdom
	3915 Australia
20	3922.5 Ceylon
3930	3927.5 Nigeria
40	
60	
3970	3972.5 Iran
80	3985 West Germany
	3995 Kuwait
00	4005 Bahrein
	4015 Hong Kong
20	4025 Philippines
	4032.5 Greece
40	4037.5 Thailand
	4045 Spain
	4055 Taiwan
60	4070 East Africa
80	4082.5 France
	4095 Indonesia
4100	
20	4115 Singapore
40	
	4165 TV Video
	4187.5 Return Cue
4195	4192.5 TV Sound
4200	

Figure 6.2   Typical frequency plans for FDMA. (Reprinted with permission from K. Miya, *Satellite Communications Engineering*, p. 115. Copyright © Lattice Publishing Co., Tokyo, Japan, 1975.)

FDM/FM/FDMA earth stations have a large number of separate IF receivers and demultiplexers [4].

Satellite FDM/FM/FDMA was patterned after the terrestrial analog telephone microwave and cable transmission systems used in the early days of the Intelsat system. Based on a relatively simple technology of separating analog signals with filters, it lacks the flexibility to take full advantage of the worldwide coverage and potential interconnectivity of a satellite network. In addition, it does not use transponder power efficiently. As a fixed assignment system, FDM/FM/FDMA suffers from a poor "fill factor" (the fraction of the available channels that are actually in use) [1]. Nevertheless many small countries invested heavily in FDM/FM/FDMA; because of this investment and the high quality telephone service that it delivers, FDM/FM/FDMA will remain in use for the foreseeable future [2]. Besides, its proponents argue that FDMA has been making money for years while (they claim) no TDMA system has yet (1984) made a profit.

Calculating the Overall Carrier-to-Noise Ratio on an FDM/FM/FDMA Link

In Chapter 5 we derived the equations for the worst-channel output (S/N) in an FDM/FM multiplexed telephone system and showed how the capacity of a carrier is determined by the available bandwidth and the carrier-to-noise ratio $(C/N)_i$ available at the input to the receiving station's FM demodulator. In this section we will assume that the available bandwidth is fixed and show how $(C/N)_i$ is influenced by the multiple-access scheme. Our presentation follows that of [5].

The overall ratio (*not decibel*) $(C/N)_i$ of a satellite link is given by

$$(C/N)_i = \cfrac{1}{\cfrac{1}{(C/N)_U} + \cfrac{1}{(C/N)_D} + \cfrac{1}{(C/N)_I}} \tag{6.1}$$

Here $(C/N)_U$, $(C/N)_D$, and $(C/N)_I$ are numerical (*not decibel*) carrier-to-noise ratios for the uplink, the downlink, and the intermodulation process.

The decibel value of $(C/N)_U$ is given by

$$(C/N)_U(dB) = F_S + (G/T)_S - 10 \log_{10}\left(\frac{4\pi}{\lambda^2}\right) - 10 \log_{10} k - 10 \log_{10}(B_{IF}) - BO_i \text{ dB}$$

$$\tag{6.2}$$

where F_S is the single-carrier saturation flux density at the beam center that the transmitting earth station can achieve at the satellite in dBW/m². The quantity BO_i is the *input backoff* in dB, that is, the dB difference between the uplink single-carrier saturation flux density that the transmitting earth station could achieve at the satellite and multicarrier flux density that is actually in use. In other words, the flux density at the satellite is given by

$$F(dBW/m^2) = F_S(dBW/m^2) - BO_i(dB) \tag{6.3}$$

Subsequently we will show how the value of BO_i is selected to maximize $(C/N)_i$.

Figure 6.3 [6] shows the input–output characteristic of a traveling wave tube (TWT). The TWT produces maximum output power at what is called the saturation point, but at saturation the TWT is operating in the nonlinear region of its characteristic; the output power is not linearly related to the input power. Thus frequency components appear at the TWT output that were not at the input. Nonlinear operation leads to the creation of intermodulation (IM) products and intermodulation distortion. This is particularly serious when two or more carriers are present; IM products can appear that overlap the spectra of the original incoming modulated carriers. The only way to reduce the IM distortion in a given TWT is to lower the input signal level so that the tube can operate in a more linear region.

The TWT nonlinearity is different for multiple carriers and for single carriers because of the way in which the output power must be divided between all of the output carriers. But the cure for IM is still the same; the input power level must be reduced. For a given carrier, the decibel difference between the single-carrier input power level at saturation and the input power level for that particular carrier in multicarrier FDMA operation is the input backoff BO_i; Figure 6.3 illustrates how it is measured.

Because the TWT is nonlinear, reducing the input power by BO_i dB causes a reduction in the output power that is smaller than BO_i. This is called the output backoff, BO_o. The output backoff influences the overall (C/N_i) in Eq. (6.1) through its influence on $(C/N)_D$.

$(C/N)_D$ is the carrier-to-noise ratio that would be achieved on the downlink if the satellite were radiating a pure signal—that is, if the transponder were not also transmitting incoming and internally generated thermal noise and IM prod-

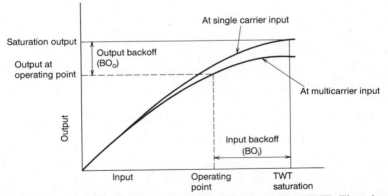

Figure 6.3 Typical transfer characteristic of a traveling wave tube (TWT). The tube delivers maximum output power at the saturation input, but operation at this point on the characteristic causes unacceptable intermodulation distortion. For satisfactory intermodulation performance, the tube is operated with its input BO_i dB below saturation. BO_i is called the input backoff; the decibel reduction in output that it causes is called the output backoff BO_o. (Reprinted with permission from K. Miya, Ed., *Satellite Communications Technology*, p. 224. Copyright © KDD Engineering and Consulting, Inc., (KEC), Tokyo, Japan, 1981.)

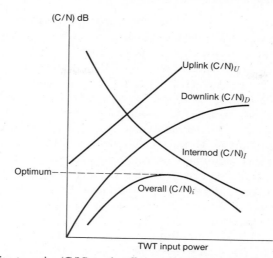

Figure 6.4 Carrier-to-noise (C/N) trade-offs as a function of the input power to the transponder TWT amplifier. Intermodulation distortion increases as the TWT approaches saturation and the downlink (C/N_0) decreases after saturation. These two effects limit the overall (C/N_0) that can be obtained. (Adapted from J. L. Dicks, P. H. Shultze, and C. H. Schmitt, "Systems Planning," *Comsat Technical Review*, **2**, 439–475 (Fall 1972).)

ucts. It is given by

$$(C/N)_D = EIRP_S - L_p + (G/T)_E - 10\log_{10} k - 10\log_{10}(B_{IF}) - BO_o \, dB \quad (6.4)$$

where $EIRP_S$ is the satellite EIRP for saturated single-carrier operation, L_p is the total downlink path loss [i.e., the sum of all of the loss terms in Equation (4.11)], and $(G/T)_E$ is the earth station figure of merit.

The term $(C/N)_I$ is included in Eq. (6.1) to incorporate the effects of IM distortion on the overall carrier-to-noise ratio of the FDM/FM/FMDA link. For a given distribution of carriers in a transponder, $(C/N)_I$ in decibels decreases almost linearly with TWT dB input power. See Figure 6.4 for an illustration. It is often calculated by equations of the form

$$(C/N)_I = (C/N)_{IS} + BO_o \, dB \quad (6.5)$$

where $(C/N)_{IS}$ is the value of $(C/N)_I$ for a given carrier distribution when the TWT is operating at saturation [6]. The values of $(C/N)_{IS}$ used depend on the TWT and the frequency plan involved. The most extensive tabulations (see reference 3) were made for the INTELSAT IV series, designed and tested during a period of high interest in FDM/FM/FDMA technology.

Solid-state power amplifiers (SSPAs) are replacing TWTs in some applications. They provide better linearity and lower intermodulation distortion, but the discussion to follow applies equally well to SSPAs and TWTs.

The trade-offs that determine the optimum value of $(C/N)_i$ are illustrated in Figure 6.4. In interpreting it, imagine that the transmitting earth station's output power is the variable, since the satellite TWT input power is directly proportional to the uplink transmitter power. As the uplink power increases, the $(C/N)_U$ increases linearly without limit. At low power levels, the satellite TWT output increases in proportion to $(C/N)_U$, and therefore so does $(C/N)_D$. But as the TWT approaches saturation, the rate of increase in $(C/N)_D$ drops, and once saturation is passed and overdrive sets in, $(C/N)_D$ decreases with further increases in $(C/N)_U$. The inter-modulation term $(C/N)_I$ decreases monotonically with increasing $(C/N)_U$. At first the decrease is linear, but for large values of $(C/N)_U$, $(C/N)_I$ decreases more rapidly. The result of these trends in $(C/N)_U$, $(C/N)_D$, and $(C/N)_I$, is the curve of $(C/N)_i$, which initially increases with uplink transmitter power, reaches a peak somewhere within the TWTs saturation region, and then decreases. There is a maximum value of $(C/N)_i$ that corresponds to the optimum operating point for the TWT. This point may be as much as 16 dB below saturation [5].

If the available uplink EIRP is large enough, it is possible to use attenuators in the transponder to keep the TWT at the optimum operating point while making the contribution of $(C/N)_U$ to $(C/N)_i$ negligible. This works because the attenuators reduce the incoming carrier and thermal noise by the same amount, maintaining a constant carrier-to-noise density ratio. The INTELSAT IV series provided an attenuation range of 24 dB for this purpose [5].

Measuring and Calculating the Effects of Intermodulation Noise

Intermodulation noise is described by a quantity called the noise power ratio, NPR. This is measured [4] by providing a white noise input whose spectrum approximates a fully loaded multiplexed telephone signal to the FDM/FM/FDMA modulator. A notch filter removes the noise from a single measuring channel before modulation. At the receiver a noise analyzer measures the ratio of the noise level in a nearby loaded channel (called the composite noise power) to the noise level in the empty channel. See Figure 6.5. The NPR in decibels is the difference in these two noise levels, that is

$$\text{NPR} = \text{composite noise power in dBm} - \text{empty channel noise power in dBm} \quad (6.6)$$

The psophometrically weighted signal-to-noise ratio in the worst channel of an N-channel multiplexed telephone system is related to the NPR by [4]

$$(S/N) = \begin{cases} \text{NPR} + 6\log_{10} N + 4.6\,\text{dB} & \text{for} \quad N < 240 \\ \text{NPR} + 18.8\,\text{dB} & \text{for} \quad N > 240 \end{cases} \quad (6.7)$$

The mathematical analysis of intermodulation that leads to expressions for NPR in terms of TWT characteristics and parameters of the modulated carriers entering a transponder is extremely complicated and we will not pursue it in detail here. For descriptions of the philosophies and approaches involved, the reader should consult references 7 and 8. For mathematical details, see references 9 and 10. The practical results are very system-specific; they reduce to tables of $(C/N)_I$ values for use in equations like Eq. (6.5). The basic approach is as follows.

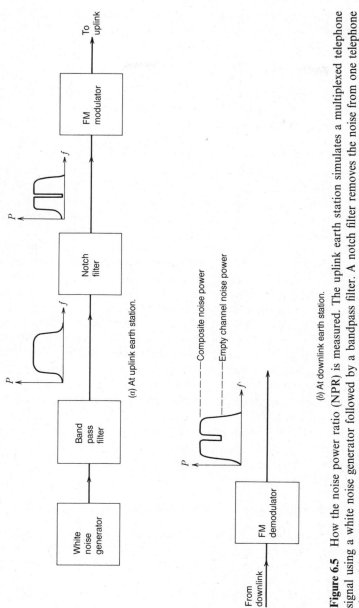

(a) At uplink earth station.

(b) At downlink earth station.

Figure 6.5 How the noise power ratio (NPR) is measured. The uplink earth station simulates a multiplexed telephone signal using a white noise generator followed by a bandpass filter. A notch filter removes the noise from one telephone channel, leaving it empty. Intermodulation noise partially fills in the empty channel. The downlink earth station measures the decibel ratio of the received noise in a filled channel (the composite noise power) to the filled-in noise in the empty channel. (a) At uplink earth station. (b) At downlink earth station.

We may model the transfer function of a general nonlinear device with output voltage e_{out} and input voltage e_{in} by a series expansion [7]

$$e_{out} = \sum_{n=1}^{\infty} a_n e_{in}^n \qquad (6.8)$$

The coefficients a_n are characteristics of the device. The subscript n is called the *order* of the intermodulation products. If the input voltage e_{in} is a group of sinusoids with different frequencies and phases

$$e_{in} = A_1 \cos(\omega_1 t + \phi_1) + A_2 \cos(\omega_2 t + \phi_2) + A_3 \cos(\omega_3 t + \phi_3), \text{ etc.} \quad (6.9)$$

then the first-order terms will be sinusoids at all of the original frequencies $\omega_1, \omega_2, \omega_3 \ldots$, the second-order terms will be sinusoids at all combinations of any two of the original frequencies, $\omega_1 + \omega_2, \omega_2 + \omega_3, \omega_1 + \omega_3 \ldots$, the third-order terms will be sinusoids at all combinations of any three of the original frequencies $\omega_1 + \omega_2 + \omega_3, 2\omega_1 - \omega_2, 2\omega_2 - \omega_1 \ldots$, and so forth.

In a practical system, many of the higher-order terms will fall outside the satellite's bandwidth and they will not affect (C/N). But this is not true for third-order terms. If the frequencies $\omega_1, \omega_2, \omega_3 \ldots$ are all within one transponder bandwidth, then most of the third-order products at frequencies involving the sum of two input frequencies minus a third input frequency (i.e., $\omega_1 + \omega_2 - \omega_3$ or $2\omega_2 - \omega_3$) will also fall within the transponder bandwidth and cause intermodulation distortion and interference. Fifth-order IM products will also have in-band frequencies, but the amplitudes of these and higher-order IM products may be neglected in most practical systems. The third-order products are quite sensitive to the input levels, and a 1-dB change in input amplitude can cause a 3-dB change in third-order intermodulation noise [3]. Hence, IM noise can change sharply with TWT backoff.

Analyzing IM distortion involves calculating the output signal for a realistic combination of input signals and using the results to develop an expression or an algorithm for the NPR. The process is quite difficult and involves considerable computer time.

Overdeviation and Companded FDM/FM/FDMA [11]

Increasing the number of FDM/FM voice channels carried by a transponder increases the maximum modulating frequency, f_{max}. If the rms test-tone deviation Δf_{rms} of Eq. (5.17) is increased in proportion to f_{max}, the FM improvement will remain constant but the Carson's rule bandwidth of Eq. (5.21) will increase. If the FDM/FM/FDMA signal is transmitted through a transponder bandwidth significantly smaller than the Carson's rule bandwidth, the signal is said to be *overdeviated*. Some of the sidebands of an overdeviated signal are truncated by the transponder; the resulting distortion may conveniently be described by an increase in intermodulation distortion and a reduction in $(C/N)_I$. While conventional FDM/FM/FDMA systems require relatively high values of $(C/N)_I$, the companding process described in the sections on CSSB permits FDM/FM/FDMA links to operate with $(C/N)_I$ values that are only 6 or 7 dB below $(C/N)_D$. Thus companding and overdeviation can significantly increase the capacity of an analog FDM/FM/FDMA link; RCA currently uses this technique to carry 2892 channels on each SATCOM transponder.

Practical Limitations of FDM/FM/FDMA

At a fixed carrier-to-noise ratio in the occupied bandwidth, the number of telephone channels that can be accommodated per MHz of transponder bandwidth decreases nonlinearly with the number of carriers sharing the transponder. This has two principal causes: IM noise increases with the number of modulated carriers that are present, and guard bands must be provided to separate the band that each modulated carrier occupies. Even when an FDM/FM/FDMA system operates at its optimum value of $(C/N)_i$, the reduction in performance for multi-carrier operation can be as much as 6 dB over that which would be obtained if the same number of channels were transmitted through the same transponder on a single carrier [12]. Alternatively, the number of channels per MHz that can be carried at a constant $(C/N)_i$ decreases with the occupied bandwidth. For example, at a $(C/N)_i$ of 18.4 dB in the occupied bandwidth, INTELSAT V, V-A, and VI can all accommodate 48 voice channels in an assigned 2.5 MHz of transponder bandwidth. This corresponds to 19.2 channels per MHz. On the other hand, at the same $(C/N)_i$, 372 channels will fit into a 15 MHz transponder allocation [12]. This is 24.8 channels per MHz. These numbers illustrate the main disadvantage of FDM/FM/FDMA besides its inflexibility; particularly when employed for more than two or three accesses per transponder, FDM/FM/FDMA does not use bandwidth efficiently. This situation may be improved by companding; RCA's companded FDM/FM/FDMA system achieves an impressive 80.4 voice channels per MHz with the SATCOM spacecraft [11].

Companded Single Sideband [11]

In companded single sideband transmission, individual voice channels are "stacked" at RF using single sideband suppressed carrier amplitude modulation. Since no carrier is present, CSSB operation is described in terms of (S/N) values instead of (C/N). The uplink and downlink signal-to-noise ratios $(S/N)_U$ and $(S/N)_D$ are calculated for an individual channel by assuming that the transmitter EIRP is divided evenly between the N channels carried and that the noise power is only that present in a single audio channel bandwidth. (Noise in the guardbands is removed by filters as part of the demultiplexing process.) Thus in the notation of Eqs. (6.2) and (6.4)

$$(S/N)_U = F_S + (G/T)_S - 10 \log_{10}\left(\frac{4\pi}{\lambda^2}\right) - 10 \log_{10}(k)$$
$$- \log 10_{10}(N) - 10 \log_{10}(b) - BO_i \text{ dB} \tag{6.10}$$

$$(S/N)_D = EIRP_S - L_p + (G/T)_E - 10 \log_{10}(k)$$
$$- 10 \log_{10}(N) - 10 \log_{10}(b) - BO_o \text{ dB} \tag{6.11}$$

where b is the bandwidth of a single voice channel.[1] Likewise the effects of inter-modulation noise can be incorporated into an $(S/N)_I$ term calculated from an equation like that of Eq. (6.5)

$$(S/N)_I = (S/N)_{IS} + BO_o - 10\log_{10}(N) \text{ dB} \tag{6.12}$$

Reference 13 expresses the intermodulation (S/N) in terms of input backoff and obtains the following equations (in our notation) for the RCA SATCOM family. The first applies to TWTs and the second applies to SSPAs.

$$(S/N)_I = 49.59 - 10\log_{10}(N) + 1.754 \, BO_i \text{ dB} \tag{6.13}$$

$$(S/N)_I = 51.12 - 10\log_{10}(N) + 0.653 \, BO_i \text{ dB} \tag{6.14}$$

The numerical (not decibel) values of $(S/N)_U$, $(S/N)_D$, and $(S/N)_I$ add recipro-cally to give the overall signal-to-noise ratio $(S/N)_i$.

$$(S/N)_i = \cfrac{1}{\cfrac{1}{(S/N)_U} + \cfrac{1}{(S/N)_D} + \cfrac{1}{(S/N)_I}} \tag{6.15}$$

Required (S/N) in a CSSB Voice Channel [14]

The action of the compander makes the subjective (S/N) of a CSSB voice channel significantly higher than the (S/N) calculated from Eqs. (6.10) through (6.15). For this reason, the performance of a CSSB channel is described in terms of a perceived test-tone-to-noise ratio (T_T/N) given by

$$(T_T/N) = (S/N)_i + A + W - (P_S + X) \text{ dB} \tag{6.16}$$

Here A is the improvement in subjective signal-to-noise ratio provided by the compandor, P_S is the average power (dBm0) in the voice channel before com-pression, X is the dB increase in average power provided by the compressor, and W is the 2.5-dB psophometric weighting factor. Design values quoted [14] for these quantities are $A = 16$ dB, $X = 7$ dB, and $P_S = -21$ dBm0.[2]

If the interference environment in which a link must operate is unknown, the value of (T_T/N) must be 51.2 dB. It can be reduced to 50 dB if the interference is included in a $(S/N)_{\text{interference}}$ term. This interference term is included in the reciprocal sum on the right-hand side of Eq. (6.15). Often the interference is separated into a signal-to-cross-polarization-interference ratio (S/X) and a voice channel signal-to-interference ratio (S/I), and Eq. (6.15) is rewritten as

$$(S/N)_i = \cfrac{1}{\cfrac{1}{(S/N)_U} + \cfrac{1}{(S/N)_D} + \cfrac{1}{(S/N)_I} + \cfrac{1}{(S/X)} + \cfrac{1}{(S/I)}} \tag{6.17}$$

[1] The reader may wish to note two minor differences between our presentation and that of Jonnalagadda and Schiff [13]. They take BO_i and BO_o to be negative numbers where we treat them as positive, and they use 3000 Hz for the voice channel bandwidth b where we use the CCITT and Intelsat value of 3100 Hz.

[2] The reader may want to review the discussion of voice signal average power in Chapter 5.

where all quantities are numerical ratios, not decibels [11]. A 50 dB value of (T_T/N) corresponds to an $(S/N)_i$ of 18.7 dB.

Capacity of CSSB Links

The capacity of a CSSB link is determined by the available bandwidth and by the inverse dependence of $(S/N)_U$, $(S/N)_D$, and $(S/N)_I$ on the number of channels, N. With negligible interference, CSSB capacities of 8000 channels per 36 MHz of transponder bandwidth are possible, but under more realistic conditions with (S/I) and (S/X) values around 25 dB, this number drops to 6000 [13]. The last figure corresponds to 166.7 channels per MHz, and thus it still represents significantly more efficient use of bandwidth than is possible with FDM/FM/FDMA. CSSB has the further advantage over FDM/FM/FDMA that its spectral efficiency is not affected by the number of accesses.

Practical Limitations of CSSB

At the present time (late 1984) little can be said definitively about the practical capabilities and limitations of CSSB and the degree to which it will be adopted in the future. While some authors [13] are very optimistic about its performance, others [15] feel that its capacity will be drastically limited by interference and predict that the number of channels that can be carried by a 36-MHz transponder will drop to perhaps 2500 with 2° orbital spacings. This last study also concluded that CSSB might have greater channel capacity than TDMA with QPSK modulation only for earth stations with antennas larger than 18 m. Nevertheless, AT&T is apparently operating 6000 voice channel links over COMSTAR transponders with good results, and a similar system has been proposed for the INTELSAT V family [16].

6.2 TIME DIVISION MULTIPLE ACCESS

In time division multiple access (TDMA) a number of earth stations take turns transmitting bursts through a common transponder. Since all practical TDMA systems are digital, TDMA has all of the advantages over FDM/FM/FDMA and CSSB that digital transmission usually has over analog. TDMA is easy to reconfigure for changing traffic demands, resists noise and interference, mixes voice and data traffic, and so on. But one advantage of TDMA that is essentially unique to satellite systems is that it permits a transponder's TWT to operate at or near saturation and thus it maximizes downlink (C/N). Since only one carrier is in the TWT at a time, there are no intermodulation products to worry about and no backoff is necessary. The principal problem that TWT nonlinearity can cause is increased intersymbol interference, and this can be eliminated by filtering and equalization.

Many of the concepts developed in Chapter 5 for time division multiplexing (TDM) apply without change to TDMA. In TDM digital data streams from many sources are transmitted sequentially in assigned time slots; the slots are organized into frames that also contain synchronization information. A receiving station must first recover the transmitter carrier frequency, then recover the transmitting station clock pulses, and then identify the start of each frame so that it can recover each transmitted channel and route it on to its destination. The principal difference is that in TDM everything comes from the same transmitter, and the clock and carrier frequencies do not change, while in TDMA each frame contains a number of independent transmissions. Each TDMA station has to know when to transmit, and it must be able to recover the carrier and clock for each received burst in time to sort out all wanted baseband channels.

In this section we have tried to present both the theory and the practice of TDMA. The last is difficult to do; while TDMA literature goes back at least 20 years, much of it describes theoretical investigations or one of a small number of experimental systems that were operated to validate the concept. Papers about commercial TDMA systems were often based on FCC filings or proposed network standards that preceded commercial operation by 5 or 10 years. At least one system (SBS) would seem to have evolved somewhat differently than was first envisioned. We have generally based our practical examples on the Intelsat TDMA system that will start operations in 1984 using INTELSAT V and later satellite families; this is well documented in reference 17. Table 6.1, adapted from that

Table 6.1
Primary Modulation Specifications of the Intelsat TDMA System

Modulation Parameters	
Nominal bit rate	120.832 Mbps
Nominal symbol rate	60.416 MBaud
Mode of operation	Burst
Modulation	Four-phase PSK
Demodulation	Coherent
Encoding	Absolute (i.e., no differential encoding)
Carrier and bit timing recovery sequence	48 symbols unmodulated 128 symbols modulated
Unique word length	24 symbols
Phase ambiguity resolution	By unique word detection

Source: Intelsat TDMA/DSI System Specification (TDMA/DSI Traffic Terminals) (BG-42-65E Rev. 2.), Intelsat, Washington, D.C., June 23, 1983. (Reprinted by permission of Intelsat, Washington, D.C. 20008.)

reference, illustrates the primary specifications of that system; we will refer to it as the Intelsat TDMA system.

Bits, Symbols, and Channels

A potential source of confusion in describing TDMA systems that use QPSK modulation is that some specifications are written in terms of bits and others are in terms of symbols. In the material that follows, we will assume that the data to be transmitted start at the uplink earth station and end at the downlink earth station with a single binary data stream. Everything that is transmitted and received can be described by the bits in this stream. If the binary data stream feeds a QPSK modulator, we will follow Intelsat practice and assume that the odd-numbered bits modulate the in-phase carrier (the P channel) and the even-numbered bits modulate the quadrature carrier (the Q channel). Thus each pair of incoming bits (one even bit and one odd bit) defines a transmitted symbol. In some cases we will describe burst and frame structure in terms of the P and Q channels, and in others we will refer to the incoming and outgoing binary streams at the earth stations.

TDMA Frame Structure and Design

Framing and Overhead

Figure 6.6 [18] illustrates the basic problem in TDMA transmissions: a group of earth stations, each a different distance from a satellite, must transmit individual bursts of RF energy in such a way that the bursts arrive at the satellite in a prescribed order. The stations have to adjust their transmissions to compensate

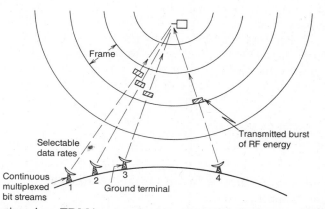

Figure 6.6 Stations in a TDMA network must transmit individual bursts of RF energy in such a way that the bursts arrive at the satellite in a prescribed order. (*Source:* J. J. Spilker, Jr., *Digital Communications by Satellite*, copyright © 1977, p. 266. Reprinted by permission of Prentice-Hall, Inc., Englewood Cliffs, NJ.)

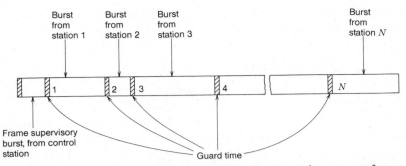

Figure 6.7 The structure of a TDMA frame. Each station transmits once per frame. Each frame contains a supervisory or reference burst. (*Source:* James Martin, *Communications Satellite Systems*, copyright © 1978, p. 248. Reprinted by permission of Prentice-Hall, Inc., Englewood Cliffs, NJ.)

for variations in satellite range, and they must be able to enter or leave the network without disrupting its operations.

These goals are accomplished by organizing TDMA transmissions into frames containing reference bursts that establish absolute time for the network. Each station transmits once per frame so that its burst begins to leave the satellite a specified time interval before or after the start of a reference burst. See Figure 6.7 [19].

Each frame contains one (or two for redundancy) reference burst and a series of traffic bursts. Each traffic burst contains a *preamble*, which provides synchronization (sync) and signaling information and identifies the transmitting station, followed by a group of traffic bits. The traffic bits are the revenue-producing portion of the frame, and the reference bursts and the preamble constitute system overhead. The smaller the overhead, the more efficient a working TDMA system is, but the more difficulty it may have in acquiring and maintaining sync. Some of the trade-offs involved are illustrated by Eq. (6.18) [20], which shows how the number of voice channels n that can be carried by a TDMA transponder is related to the transmission bit rate R, the bit rate V for one voice channel, the number of bursts N in a frame, the number of bits P in each preamble, and the frame period T_F.

$$n = \frac{R - \dfrac{NP}{T_F}}{V} \tag{6.18}$$

For a given number of preamble bits, making the frame time T_F as long as possible maximizes the channel capacity.

The minimum frame time is the 125 μs required by a voice channel sampled at the standard 8-kHz rate. The maximum frame time is arbitrary, provided that each frame contains one sample from each voice channel for each 125 μs of frame duration. Thus the frame time affects the structure of the bursts from the individual

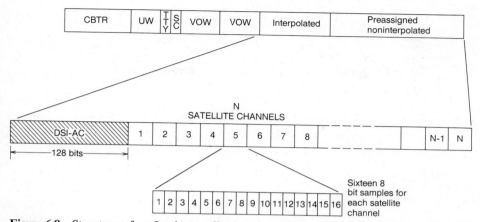

Figure 6.8 Structure of an Intelsat traffic data burst. A satellite channel is a block of 16 consecutive 8-bit samples from one terrestrial channel. The other blocks in the traffic data burst will be explained later in the text. (Reprinted with the permission of the International Telecommunications Satellite Organization from *Intelsat TDMA/DSI System Specification* (*TDMA/DSI Traffic Terminals*) (BG-42-65E Rev. 2), Intelsat, Washington, DC, June 23, 1983.)

stations. In the Intelsat system the frame time is 2 ms, and each traffic burst must contain 16 samples (2 ms divided by 125 μs) from each voice channel being carried by the transmitting station. The bits representing one sample of each such voice channel are called a *terrestrial channel* or *TC*. The 16 TCs that must be sent in each frame for one voice channel are transmitted one after the other in a 128-bit (64-symbol) block called a *satellite channel* or *SC*. Figure 6.8 [17] illustrates this. A TDMA system with a different frame time would have a different SC structure.

For synchronization purposes, frames may be grouped into multiframes or superframes as in TDM. A frame's position in a larger structure may be identified by some feature of the reference burst. We will discuss this point in more detail when we come to TDMA synchronization.

An earth station may transmit into or receive from several transponders; this is called *transponder hopping*. In Intelsat practice all of the transponders through which a given station may transmit or receive must be synchronized. This means that their frames must start and end at the same time. The set of all synchronized transponders is called a *community of transponders*. Figure 6.9 illustrates the frame structure for such a community.

One transponder in a community is designated the *timing reference transponder* (*TRT*) and its first reference burst (RB 1 in Figure 6.9) bridges the *start of frame* (*SOF*) for all members of the community. The reference bursts in all the other transponders start at assigned time offsets, $T1_i$ for the ith transponder, after SOF. Thus by the formal definition of a frame, in all transponders except the TRT some earth stations will transmit before the reference bursts. But note that this is equivalent to transmitting *after* the reference bursts of the previous frame.

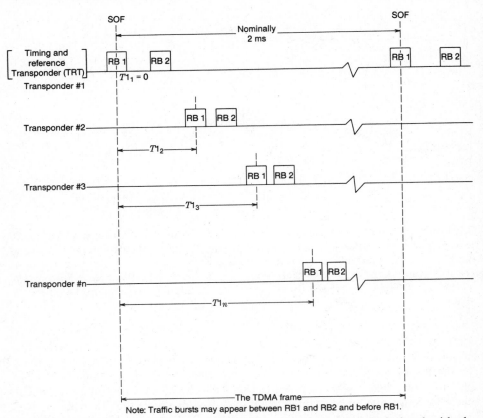

Figure 6.9 The frame structure of a community of transponders. (Reprinted with the permission of the International Telecommunications Satellite Organization from *Intelsat TDMA/DSI System Specification* (*TDMA/DSI Traffic Terminals*) (BG-42-65E Rev. 2), Intelsat, Washington, DC, June 23, 1983.)

Reference Bursts

Each frame must be marked by one or more reference bursts. These are generated from control stations on the ground, but for timing purposes they are treated as if they were originated by the satellite [18]. A reference burst must contain at least three items: carrier and bit timing recovery (*CR/BTR*) information, a unique word (*UW*), and a station identification code (*SIC*) [20]. It will probably contain network housekeeping information as well, and the information in the UW and SIC may be combined. The network housekeeping information may be broken down into a number of separate channels such as teletype (TTY), service channel (SC), voice order wire (VOW), and the control and delay channel (CDC). Figure 6.10 illustrates the structure of a reference burst in the Intelsat system.

The CR/BTR portion of the reference burst serves to synchronize the transmitting and receiving carriers and modems. Typically [21] it consists of a short

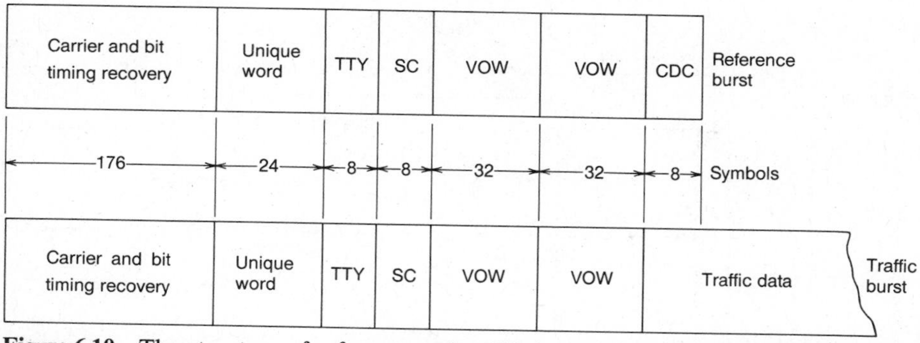

Carrier and bit timing recovery	Unique word	TTY	SC	VOW	VOW	CDC	Reference burst

←——176——→	←—24—→	←8→	←8→	←—32—→	←—32—→	←8→	Symbols

Carrier and bit timing recovery	Unique word	TTY	SC	VOW	VOW	Traffic data	Traffic burst

Figure 6.10 The structure of reference and traffic bursts in the Intelsat TDMA system. TTY, teletype; SC, service channel; VOW, voice order wire; CDC, coordination and delay channel. (Reprinted with the permission of the International Telecommunications Satellite Organization from *Intelsat TDMA/DSI System Specification (TDMA/DSI Traffic Terminals)* (BG-42-65E Rev. 2), Intelsat, Washington, DC, June 23, 1983.)

transmission of unmodulated carrier (equivalent in a direct PSK system to a carrier modulated by a string of logical ones or a string of logical zeros) followed by carrier phase transitions between 0 and π rad at the symbol clock frequency. A CR/BTR sub-burst duration equal to 30 QPSK symbols (60 bits) is widely quoted in the literature, but this does not seem to be followed in practice. Intelsat [17] uses 48 QPSK symbols (96 bits) of all ones on both the P and Q channels followed by 128 QPSK symbols (256 bits) of simultaneous 01010101... on both the P and Q channels. Inmarsat's BPSK system uses a CR/BTR sequence of 50 zeros followed by 29 ones [22].

The Unique Word

The unique word or UW (also called the burst code word or BCW [23]) serves to mark each frame. It plays the same role in TDMA as in TDM and it is chosen by the same criteria. Typical lengths range from 20 [20] to 48 [21] bits. The Inmarsat system, for example uses

$$0111 \quad 1010 \quad 1100 \quad 1101 \quad 0000$$

on its BPSK links [22]. Intelsat uses three 24 QPSK symbol (48 bit) unique words in its reference bursts called UW0, UW1, and UW2, which are given as [17]

UW0: P 0111 1000 1001 0111 1000 1001
$\quad\quad\;\;$ Q 0111 1000 1001 0111 1000 1001

UW1: P 0111 1000 1001 1000 0111 0110
$\quad\quad\;\;$ Q 0111 1000 1001 0111 1000 1001

UW2: P 0111 1000 1001 0111 1000 1001
$\quad\quad\;\;$ Q 0111 1000 1001 1000 0111 0110

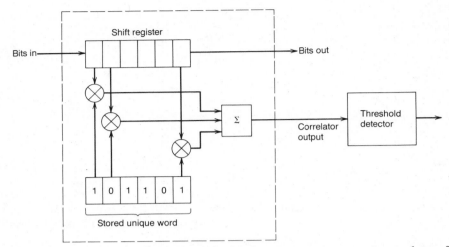

Figure 6.11 A unique word correlator. The Intelsat system uses separate correlators for the bit streams received in the P and Q QPSK channels. The correlator output should activate the threshold detector only when the bits in the shift register match the stored unique word. (*Source:* Robert M. Gagliardi, *Satellite Communications*, Lifetime Learning Publications, Belmont, CA, 1983, p. 248. Copyright © 1984 Wadsworth, Inc. Reprinted by permission of Van Nostrand Rheinhold Co., Inc.)

Intelsat uses a fourth unique word, UW3, only in traffic bursts. It is given by

$$\text{UW3:} \quad P \quad 0111 \quad 1000 \quad 1001 \quad 1000 \quad 0111 \quad 0110$$
$$Q \quad 0111 \quad 1000 \quad 1001 \quad 1000 \quad 0111 \quad 0110$$

At the receiving end of a link, incoming bits are clocked into a shift register, where they are compared with a stored version of the expected UW. The circuit that does this is called a *UW correlator*, illustrated in Figure 6.11 [24]. When the bits in the shift register match those stored in the memory, the correlator produces its maximum output voltage pulse. Each pair of bits that fails to match reduces the output amplitude. Depending on the framing error probability that can be tolerated, a TDMA system will accept as correct a received UW that differs from the expected one by not more than a specified number of bits. This means that when the output pulse from the UW correlator exceeds some minimum value, this signifies that the UW has been received and that a new frame is in progress. Thus the UW correlator provides an accurate time reference for a receiving terminal.

A TDMA terminal may fail to recognize the UW, or it may interpret some other bit sequence as the UW. The first error is called a *miss* and the second is called a *false alarm*; both could result from bit errors that make received sequences different from what was transmitted. In addition a false alarm could occur if the bits in the transmitted information happen to match the stored UW closely enough. An excellent derivation and discussion of UW false alarm and miss probabilities appears in Chapter 8 of reference 21; our representation and notation follow that closely.

Let E be the number of bits by which a received N-bit sequence can differ from the stored UW and still be counted as correct. In an acceptable sequence there may be I errors and $N - I$ correct bits so long as $I \leq E$. For a bit error probability p, the probability that an N-bit UW will be received correctly is the probability P of all combinations of I errors and $N - I$ nonerrors in N consecutive received bits. This is given by

$$P = \sum_{I=0}^{E} \binom{N}{I} p^I (1 - p)^{N-I} \tag{6.19}$$

where $\binom{N}{I}$ is the binomial coefficient

$$\binom{N}{I} = N!/[I!(N - I)!] \tag{6.20}$$

The probability of a miss is Q, where

$$Q = 1 - P = \sum_{I=E+1}^{N} \binom{N}{I} p^I (1 - p)^{N-I} \tag{6.21}$$

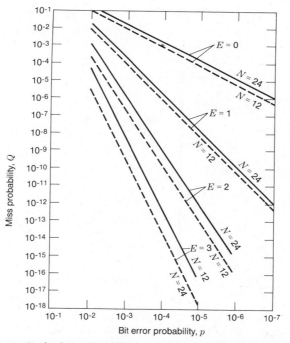

Figure 6.12 Unique word miss probability as a function of bit error probability (p), unique word length (N), and number of errors (E) in unique word correlation threshold. (*Source:* K. Feher, *Digital Communications: Satellite/Earth Station Engineering*, copyright © 1983, p. 378. Reprinted by permission of Prentice-Hall, Inc., Englewood Cliffs, NJ.)

Figure 6.12 [21] is a plot of Q versus bit error rate p for selected UW lengths N and thresholds E. For an E of zero, Q is about two orders of magnitude larger than p, but if two or three errors can be tolerated (E values of 2 or 3), then the miss probability can be made vanishingly small. Intelsat requires that the unique word miss probability be smaller than 10^{-8} at an E_b/N_0 of 7 dB [17].

A false alarm will occur when a sequence of N received bits matches the expected UW to within I positions, $I \le E$. If the incoming bit stream is random, then the probability that N successive bits will match the UW is 2^{-N}. The total number of combinations within which I errors can occur is

$$\sum_{I=0}^{E} \binom{N}{I}$$

and hence the probability of a false alarm is

$$F = 2^{-N} \sum_{I=0}^{E} \binom{N}{I} \tag{6.22}$$

Figure 6.13 illustrates the dependence of F on word length and bit error probability.

Once frame sync has been established, a TDMA terminal can operate so that it examines the UW correlator only during some time window in which the UW

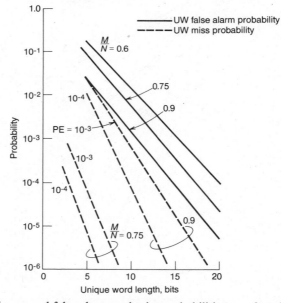

Figure 6.13 Unique word false alarm and miss probabilities as a function of word length for selected error rates. The number of bits in the unique word is N; M is the number of bits in a received sequence that must match the unique word stored in the correlator. (*Source:* Robert M. Gagliardi, *Satellite Communications*, Lifetime Learning Publications, Belmont, CA, 1983, p. 249. Copyright © 1984 Wadsworth, Inc. Reprinted by permission of Van Nostrand Rheinhold Co., Inc.)

is expected to occur. This *windowing* can significantly reduce the probability of a false alarm. See reference 21 for a detailed discussion.

Reception of the complement of the UW will cause the UW correlator to put out a large negative pulse instead of the usual positive pulse. This may be useful in identifying the start of a superframe (see the earlier material on TDM) and in maintaining tighter network synchronization. In the Inmarsat system, for example, every sixth reference burst carries the complement of the UW [22].

The four unique words UW0, UW1, UW2, and UW3 of the Intelsat TDMA system were cleverly chosen to permit use of a single unique word correlator. The 12-bit pattern

$$0111 \quad 1000 \quad 1001$$

appears in both P and Q channels for the first 12 symbols of all the unique words. We will call it the *reference sequence*. Upon two successive receptions (within a specified detection threshold) of the reference sequence, the unique word is declared received.

Which of the four possible unique words was received is decided by logic, which looks at the correlator output for the second 12-bit pattern. It compares the bits in the P channel and in the Q channel with the reference sequence and calculates the number of bits d_{P2} and the number of bits d_{Q2} that the patterns received by the P and Q channels differ from the reference sequence. It then defines two logical variables XP and XQ such that XP is 1 if d_{P2} is more than six; otherwise XP is zero. Likewise XQ is 1 if d_{Q2} is more than six; otherwise it is zero. Which unique word was received is determined by XP and XQ as follows [17].

Phase Ambiguity Removal

Most carrier recovery schemes for QPSK are subject to a *phase ambiguity error*; they may establish the phase of the reference carrier with a plus or minus 90° or with a 180° uncertainty [18]. This may cause the demodulated P or Q channel

Table 6.2
How XP and XQ Are Used to Determine Which Unique Word Was Transmitted

	$XQ = 0$	$XQ = 1$
$XP = 0$	UW0	UW2
$XP = 1$	UW1	UW3

Source: Intelsat TDMA/DSI System Specification (TDMA/DSI Traffic Terminals) (BG-42-65E Rev. 2.), Intelsat, Washington, D.C., June 23, 1983. (Reprinted with permission of Intelsat, Washington, D.C. 20008.)

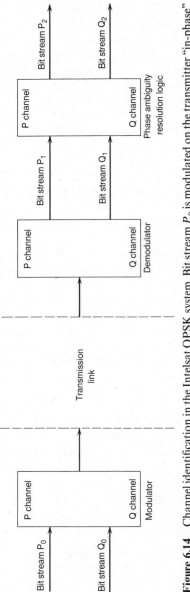

Figure 6.14 Channel identification in the Intelsat QPSK system. Bit stream P_0 is modulated on the transmitter "in-phase" carrier and bit stream Q_0 is modulated on the transmitter "in-quadrature" carrier. At the receiver bit stream P_1 is recovered in-phase with the recovered carrier and bit stream Q_1 is recovered in-quadrature with the recovered carrier. If the carrier recovery loop made a phase error, P_1 may be either P_0 or Q_0 or their complements. The phase ambiguity resolution logic removes the phase error and aligns P_2 with P_0 and Q_2 with Q_0. (Reprinted with the permission of the International Telecommunications Satellite Organization from *Intelsat TDMA/DSI System Specification (TDMA/DSI Traffic Terminals)* (BG-42-65E Rev. 2), Intelsat, Washington, DC, June 23, 1983.)

or both to be the complement of what was transmitted. In addition, the P and Q channels may be interchanged.

Any phase ambiguity error introduces errors into the recovered unique words on both P and Q channels, and these characteristic errors may be used to remove the ambiguity. In the Intelsat system [17] this is done by looking at the numbers of bits d_{P1} and d_{Q1} in which what are assumed to be the P and Q channels at the receiver differ from the reference sequence. In Intelsat notation (Figure 6.14) the transmitted QPSK channels are called $P0$ and $Q0$; the received channels with a possible phase ambiguity (i.e., at the demodulator output) are called $P1$ and $Q1$, and the channels sent on to the terrestrial interface after the phase ambiguity has been resolved are called $P2$ and $Q2$. The way in which $P2$ and $Q2$ are connected to $P1$ and $Q1$ resolves the ambiguity. Table 6.3 [17] illustrates how d_{P1} and d_{Q1} are used to resolve the phase ambiguity.

Station Identification Code

The station identification code (SIC) identifies the transmitting station. It plays no role in frame or slot synchronization beyond identifying the station that transmitted the burst. Operational code lengths range upward from six bits [24]. Intelsat uses an 8-bit code that it calls the terminal number. This is used in the housekeeping channels, but unlike an SIC it is not identified as part of the reference burst.

Traffic Bursts

Traffic bursts begin like reference bursts, and in most TDMA systems a reference burst is simply a traffic burst without any traffic. The first part is called the preamble, and it normally includes a CR/BTR sequence, a unique word, and housekeeping information. The second part of a traffic burst is called the traffic data burst, or, in the Intelsat systems, a DSI (digital speech interpolation) or a

Table 6.3
Phase Ambiguity Resolution

Condition	$d_{Q1} \leq 6$	$d_{Q1} > 6$
$d_{P1} \leq 6$	$P2 = P1$	$P2 = Q1$
	$Q2 = Q1$	$Q2 = P1$
$d_{P1} > 6$	$P2 = Q1$	$P2 = P1$
	$Q2 = P1$	$Q2 = Q1$

Source: Intelsat TDMA/DSI System Specification (TDMA/DSI Traffic Terminals), (BG-42-65E Rev. 2.), Intelsat, Washington, D.C., June 23, 1983. (Reprinted by permission of Intelsat, Washington, D.C. 20008.)

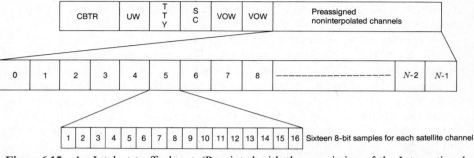

Figure 6.15 An Intelsat traffic burst. (Reprinted with the permission of the International Telecommunications Satellite Organization from *Intelsat TDMA/DSI System Specification (TDMA/DSI Traffic Terminals)* (BG-42-65E Rev. 2), Intelsat, Washington, DC, June 23, 1983.)

DNI (digital noninterpolation) sub-burst, depending on whether or not the satellite link involved is using demand access. We discuss DSI and DSI sub-bursts in Sections 6.4 and 6.5; here we present the DNI sub-burst as a representative format for transmitting preassigned TDMA traffic.

Figure 6.15 illustrates the format of an Intelsat traffic burst and the DNI sub-burst that it includes. The DNI sub-burst consists of 128 time slots called satellite channels (SCs), numbered 0 through 127. Each SC contains 128 bits, which, for telephone transmission, are the 15 consecutive 8-bit samples of one voice channel that must be transmitted in each 2 ms Intelsat TDMA frame. Thus the total traffic data burst is 16,384 bits or 8192 symbols long. Each satellite channel accomodates a bit rate of 64 kbps; data channels at higher bit rates are transmitted using multiple SCs. The maximum data rate that the Intelsat TDMA system can carry is 128 times 64 kbps or 8.192 Mbps.

Guard Times

Guard times are empty time slots that separate TDMA bursts. They prevent overlap and make it easier for receiving stations to separate incoming bursts, and they play the same role as guard bands in FDM/FM/FDMA systems. Guard times add to system overhead and represent transponder time that does not earn revenue; hence there is some motivation for making them as short as possible. On the other hand, shorter guard times require more precise and hence more expensive synchronization techniques (described below), and this provides economic motivation for long guard times. The guard times selected represent a compromise between these two factors.

The required guard time may be estimated [21] by calculating the maximum uncertainty that could be observed between bursts and then doubling it. The results and the system specifications are commonly expressed in terms of symbols. Thus the Intelsat TDMA guard time is specified as 60 symbols, corresponding to 993 ns or almost 1 μs.

TDMA Synchronization and Timing

Orbital Effects

Each station in a TDMA network has a different range to the satellite, and all these ranges vary with time (see Chapter 2). This complicates timing and synchronization in TDMA systems. Our presentation of the basic problems involved is patterned after that in [18].

Let the range from the satellite to the Nth earth station at time t be $r_N(t)$. Assume that the satellite is transmitting reference bursts that begin at times $t = 0$, $T_{FS}, 2T_{FS}, 3T_{FS}, \ldots$ where T_{FS} is the frame time (i.e., the length of the frame expressed in convenient time units) at the satellite. Let there be M frames in a superframe, so that the superframe duration at the satellite is MT_{FS}.

The reference burst that left the satellite at $t = 0$ will reach the Nth earth station at time t_A, which satisfies the equation

$$t_A = \frac{r_N(t_A)}{c} \tag{6.23}$$

where c is the free space velocity of light. For convenience we will assume that during any one frame the range may be approximated by its value at the time that the reference burst for that frame left the satellite. (In other words, we will approximate the true behavior of the range by letting it make step-function changes at the beginning of each frame at the satellite.) Then the reference bursts will arrive at the earth station at times $r_N(0)/c$, $r_N(T_{FS})/c$, $r_N(2T_{FS})/c) \ldots$ If we represent the arrival time of the jth frame by t_{Nj}, then

$$t_{Nj} = \frac{r_N(jT_{FS})}{c}, \qquad j = 0, 1, 2, 3 \ldots \tag{6.24}$$

The frame time at the Nth station, T_{NF}, is the time interval between the arrival times of successive frames. The frame rate R_{NF} is the reciprocal of the frame time.

$$T_{NF} = t_{Nj} - t_{Nj-1} = \frac{r_N[jT_{FS}] - r_N[(j-1)T_{FS}]}{c} \tag{6.25}$$

$$R_{NF} = \frac{1}{T_{NF}} \tag{6.26}$$

Thus each member of a TDMA network will perceive a different frame rate, and this frame rate will vary with time as the range to the satellite changes. Since the bit rate should be an integer multiple of the frame rate, each station will transmit with its own bit rate. This requires TDMA stations to use bit stuffing techniques (see Chapter 5) to multiplex bit streams running at slightly different rates.

To enter a TDMA network and establish and maintain synchronization, an earth station must know its range to the satellite. The propagation time required for a burst to leave a transmitting earth station, pass through the transponder, and arrive at a receiving earth station is the minimum time required for an adjustment in transmission time to be detected by the rest of the network. By the equations developed in Chapter 2, the maximum range from an earth station to a geosynchronous satellite is about 41,756 km. This corresponds to a round-trip time delay of 278 ms. This is far too long to use as a frame time since a sample of each voice channel must be transmitted at an average rate of once per 0.125 ms.

Satellite Loop-Back Synchronization

In satellite loop-back or self-locking synchronization, each earth station receives its own transmissions through the satellite and adjusts its timing so that its traffic bursts fall within the appropriate window. Each station adjusts its clock to a reference time called the *start of transmit frame* or *SOTF* defined so that, if it transmitted at the SOTF, this reference burst would leave the transponder at the same time as the control station's reference bursts [18]. This scheme permits all stations in the network to synchronize their clocks with respect to the satellite. Any station could become the reference station by transmitting whenever its own clock indicated SOTF and the overall network synchronization would be unchanged.

A station that wishes to enter the network must be able to transmit some initial bursts and adjust their timing without disrupting the traffic already in progress. It can do this by using low-power bursts (say 25 dB below normal TDMA operating power) that are either unmodulated or are modulated by a short pseudonoise (*PN*) sequence (see Section 6.3). These have no effect on the ongoing traffic but they can be separated from it by specialized detectors. The station can either step its transmissions through the transmit frames until it hits the correct slot or use a calculated burst time based on the estimated range of the satellite as a starting point [21].

Cooperative Synchronization [17, 21]

Satellite loop-back synchronization requires that an earth station be able to hear its own transmissions as they are repeated by the transponder. This may not be possible in multibeam or satellite-switched TDMA systems, and furthermore it does not provide for the sort of synchronized network control that can monitor the synchronization of all stations and automatically shut down any network members who are out of sync and unaware of the problems they are causing. For this reason the Intelsat TDMA system uses *cooperative synchronization*; a control station listens to the bursts and sends timing instructions back to each transmitting station.

The key ingredient in these instructions is the time delay, D_N, that must elapse between the start of a receive multiframe (SORMF) and the start of a transmit multiframe (SOTMF). Knowing the time of SORMF, a station may count clock pulses to determine the start of each transmit frame and the time it should transmit its traffic burst in that transmit frame. In the Intelsat TDMA system, the control stations transmit D_N values to all network members once every 32 multiframes. Once a station has acquired sync, it compensates for changes in satellite range automatically as it receives new delays. When entering the network, a station uses the D_N supplied to it by the network as a starting point for positioning its SOTMF and its initial acquisition bursts. The station can be shut down quickly if the control station determines that it is disrupting the network and sends it a "don't transmit" (DNTX) code.

6.3 CODE DIVISION MULTIPLE ACCESS

Code Division Multiple Access (CDMA) is a scheme in which, on the average, a number of users occupy all of a transponder bandwidth all of the time. Their signals are encoded so that information from an individual transmitter can be detected and recovered only by a properly synchronized receiving station that knows the code being used. This provides a decentralized satellite network, as only the pairs of stations that are communicating need to coordinate their transmissions. Subject to transponder power limitations and the practical constraints of the codes in use, stations having traffic can access a transponder on demand without coordinating their frequency (as in FDMA) or their time slot (as in TDMA) with any central authority. Each receiving station has its own code, called its "address" [25], and a transmitting station simply modulates its transmission with the address of the intended receiver whenever it wishes to send a message to that receiver.

While the purposes (secrecy and jam resistance) of *spread-spectrum* (*SS*) communications systems are quite different from CDMA, the two techniques are identical. Since much about SS is classified, this has severely restricted discussion of CDMA in the open literature and its adoption for commerical satellite systems. In addition, CDMA is more suited for a military tactical communications environment where many small groups of mobile stations communicate briefly at irregular intervals than to a commercial environment where large volumes of traffic pass continuously between a small number of fixed locations. For these reasons, to our knowledge, there has been no adoption of CDMA for commercial satellite systems, and our discussion of CDMA will be brief. For more information about SS and CDMA the reader should consult the May 1983 issue of *IEEE Transactions on Communications* and particularly references 26 and 27. An excellent discussion of the potential application of CDMA appears in reference 24.

Spread-Spectrum Transmission and Reception

The techniques used for spread-spectrum transmission are called *direct-sequence pseudonoise* (PN), *frequency hopping* (FH), *time hopping* (TH) and *chirp* [28]. We will discuss only the first two.

In a PN system [29] (Figure 6.16), a conventional digital modulator generates a sequence of bits at rate R_b. If transmitted by BPSK, this bit sequence would require a bandwidth w_b approximately equal to R_b. Each bit is divided into pieces called chips. Each chip corresponds to a member of a *PN sequence*, and a chip is transmitted as a $+1$ or a -1 depending on whether its corresponding element in the sequence is a $+1$ or a -1. Expanding on an example from [29], if the PN sequence were $+1, +1, +1, -1, +1, -1, -1$ (leftmost member transmitted first)

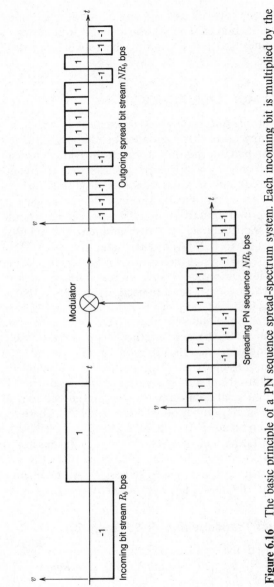

Figure 6.16 The basic principle of a PN sequence spread-spectrum system. Each incoming bit is multiplied by the same PN sequence. In this example the incoming bit stream is $-1\,1$ and the PN sequence is $1\,1\,1\,-1\,1\,-1\,-1\,-1$.

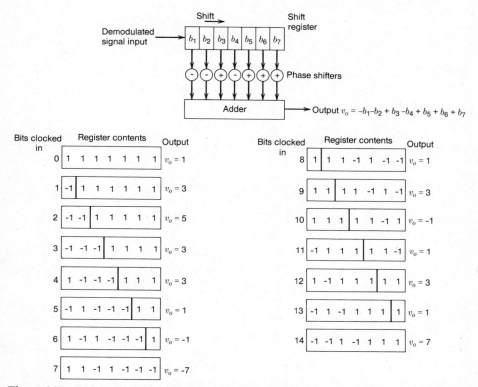

Figure 6.17 PN despreading using a matched filter. The incoming bits are clocked into a shift register, passed through phase shifters, and added. The signs of the phase shifters correspond to the modulating PN sequence. The figure sketches the shift register contents and the output voltage value as each chip in the sequence from Figure 6.16 is clocked in. We have assumed that the process began with all ones in the shift register.

as illustrated in Figure 6.16, a $+1$ in the original bit stream would be transmitted by the chip sequence $+1, +1, +1, -1, +1, -1, -1$ and a -1 in the original bit stream would be transmitted by the chips $-1, -1, -1, +1, -1, +1, +1$. In other words, each incoming bit is multiplied by the same PN sequence. If the PN sequence has N members, then the radio channel occupies a bandwidth $W = NR_b = Nw_b$ and transmits symbols (i.e., the chips) at a rate of $R_c = NR_b$. Thus the spectrum of the original bit stream is spread over a wide bandwidth.

At the receiver the original bits may be recovered by demodulating the chips and putting them into a matched filter, which consists of a delay line, phase shifters whose values correspond to the PN sequence at the transmitter, and an adder. Figures 6.17 and 6.18 illustrate this for the sequence of Figure 6.16. Note that the adder output will be $+7$ when the original bit was a $+1$ and -7 when the original bit was a -1.

The receiving process is equivalent to multiplying the incoming signal by the same PN sequence that was used to spread the bit stream at the transmitter. This

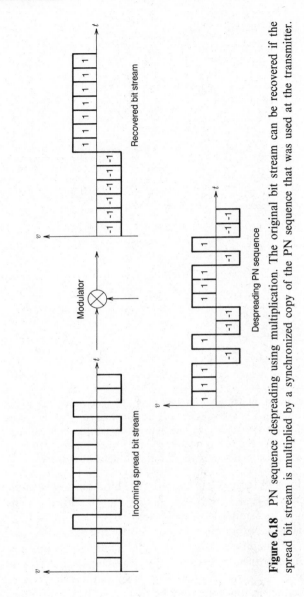

Figure 6.18 PN sequence despreading using multiplication. The original bit stream can be recovered if the spread bit stream is multiplied by a synchronized copy of the PN sequence that was used at the transmitter.

"collapses" the spectrum of the wanted signal back to the original bandwidth w_b, the value it originally had at baseband. Suppose an interferer transmits a single carrier at some frequency in the bandwidth W and the receiver receives P_I W of power from the interferer, measured at the antenna. When the interfering carrier is multiplied in the matched filter by the PN sequence of the wanted transmitter, its power P_I will be spread over the bandwidth W and the power density of the interfering signal will be proportionally reduced. This reduction in the power density of a CW interfering signal is called the *processing gain* G_p of the spread-spectrum system. As a ratio it is numerically equal to the number of chips per bit [29]. An interfering signal that was spread with some other PN sequence will not be despread when multiplied by the PN sequence used at the receiver.

The PN sequence used in this example is what is called a *maximal length sequence*; these are easily generated with a shift register and some modulo-2 adders [30]. A modulo-2 adder is simply an exclusive-or circuit; the rules for modulo-2 addition are[3]

$$0 \oplus 1 = 1$$

$$0 \oplus 0 = 0$$

$$1 \oplus 1 = 0$$

We will have more to say about PN sequences in the next chapter. Figure 6.19 illustrates the generator for this particular sequence and shows how the sequence is generated.

A generator incorporating a shift register with k stages can generate n PN sequences with minimum correlation with each other, where [30]

$$n = 2^k - 1 \tag{6.27}$$

Note that n is equal to N, the number of chips used to spread each bit from the original incoming data stream. When the matched filter of Figure 6.18 receives the particular PN sequence for which it was designed, its output will be $+N$ or $-N$. When it receives any other member of the sequence its output will be $+1$ or -1.

For the example we have been using, all seven members of the set of PN sequences are

(1)	$+1$,	$+1$,	$+1$,	-1,	$+1$,	-1,	-1
(2)	$+1$,	$+1$,	-1,	$+1$,	-1,	-1,	$+1$
(3)	$+1$,	-1,	$+1$,	-1,	-1,	$+1$,	$+1$
(4)	-1,	$+1$,	-1,	-1,	$+1$,	$+1$,	$+1$
(5)	$+1$,	-1,	-1,	$+1$,	$+1$,	$+1$,	-1
(6)	-1,	-1,	$+1$,	$+1$,	$+1$,	-1,	$+1$
(7)	-1,	$+1$,	$+1$,	$+1$,	-1,	$+1$,	-1

[3] When talking about logical operations like modulo-2 addition we will use 0 and 1 to represent the two binary digits. As discussed in the previous chapter, these are modulated and detected as -1 and $+1$, and we will represent them in this way when discussing PN sequence detection, for example.

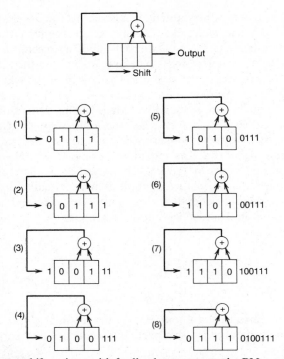

Figure 6.19 Using a shift register with feedback to generate the PN sequence 1 1 1 0 0 1 0. The top sketch shows the shift register configuration. The modulo-2 sum of the rightmost 2 bits is shifted into the register at every clock pulse and the rightmost bit is shifted out. The register begins with 1 1 1, generates the wanted sequence, and ends with 1 1 1. Steps (1) through (8) illustrate the process, showing the bit that came out on the last clock pulse and the bit that will go in on the next clock pulse.

This group illustrates several characteristic properties of maximal length PN sequences: (1) the number of $+1$s and the number of -1s differ by 1; (2) all sequences from the same generator are cyclic permutations of each other; (3) no sequences from the same generator are complements of each other; (4) there are no repeated groups of bits of length $> k$ (3 in this case).

In a frequency hopping (FH) spread-spectrum system, the instantaneous frequency of the transmitter is stepped over some bandwidth W that is much larger than that required to carry the information being transmitted. The sequence of frequencies used (called the hop sequence) is generated by a PN sequence. If the PN sequences of multiple FH transmitters are uncorrelated, then all can share the same bandwidth with only a small probability that two or more will use the same instantaneous frequency at the same time. In analogy to a direct sequence system the processing gain of an FH system is numerically equal to the ratio of the hopped bandwidth to the baseband information bandwidth [21].

Applicability of CDMA to Commercial Systems [27]

While FDMA, TDMA, and CDMA all have their proponents, a claim can be made on theoretical grounds that in a perfect system each will support the same number of users in a given bandwidth. Differences in capacity between them arise from practical considerations like intermodulation distortion, synchronization, and the like. The appeal of CDMA lies in its potential for uncoordinated access, but this application requires a large processing gain. A common rule of thumb is that the number of uncoordinated users that a CDMA system can support is $G_p/10$. This puts severe bandwidth or PN sequence length constraints on SS systems and severely limits their attractiveness for multiple access. In fact CDMA will probably be adopted only for those systems that must use SS for security reasons or interference rejection.

6.4 ESTIMATING CHANNEL REQUIREMENTS

Thus far in this chapter and in the previous one we have talked about the number of voice channels that can be carried by different modulation and multiple access schemes without specifying how the number of channels itself is to be determined. This is a problem that has concerned telephone engineers for almost one hundred years, but it will probably be unfamiliar to many readers. We will give an abbreviated summary here for which we are indebted to reference 31; for an excellent introduction to the general field of telecommunication traffic engineering, see that reference.

Measuring Traffic

The terminology that describes the measurement of telecommunications traffic originated in telephone engineering and retains the name "call" to describe the use of a channel to link two points that are exchanging information. We will follow the same convention, recognizing that a call could be a telephone call, a link between computers, a television transmission, and so forth.

At its simplest, the *quantity* of traffic carried by a link in some time interval T is the sum of the durations of each individual call that the link carried during T. If the link carried one call and it lasted 100 s, then the quantity of traffic carried was 100 call-seconds or 1 ccs, where ccs is the abbreviation for "hundred call-seconds"[4] [32]. An equal traffic quantity of 1 ccs would be represented by two 50-s calls, four 25-s calls, ten 10-s calls, and so on. Another name for the quantity of traffic is the total holding time, where "holding" means using a channel

---

[4] The abbreviation ccs is used in the United States both to represent one hundred call-seconds (a measurement of the quantity of traffic) and to represent one hundred call-seconds per hour (a measurement of the traffic intensity). This inconsistent use of ccs to abbreviate the units of two different quantities is unfortunate, but apparently it is firmly embedded in U.S. telephone engineering practice. It leads to the seemingly contradictory statement that $1E = 36$ ccs. See p. 36–10 of reference 32 for a discussion of this question.

(i.e., holding or occupying a telephone line). Mathematically [31] if a link can carry N channels and t_x represents the total time that x out of the N channels are in use during time interval T, then

$$T = \sum_{x=0}^{N} t_x \qquad (6.28)$$

The quantity of traffic (total holding time) h is given by

$$h = \sum_{x=0}^{N} x t_x \qquad (6.29)$$

Traffic intensity (also called traffic flow and traffic load) is measured in *erlangs* (E), where $1E$ is one call-second per second. A single channel occupied for a total time t out of a period T carries $t/T\,E$. The maximum load for a single channel is $1E$, and the number of erlangs carried by a link cannot be greater than the number of channels on the link [31]. In general the traffic intensity A in erlangs is given by

$$A = \frac{h}{T} = \left(\frac{1}{T}\right) \sum_{x=0}^{N} x t_x = \sum_{x=0}^{N} x \left(\frac{t_x}{T}\right) \qquad (6.30)$$

On the right-hand side of Eq. (6.30) x is the number of channels occupied simultaneously and (t_x/T) is the proportion of the time T for which x channels are occupied simultaneously. Thus A is the average number of channels occupied simultaneously during time T, and we may write Eq. (6.30) as

$$A = nh \qquad (6.31)$$

where n is the average number of calls carried per unit time and h is the average duration of a call.

The Basic Traffic Equation

Assume that the traffic between two points is a stationary process in the statistical sense, and let $p(x, t) = p(x)$ be the probability that x channels will be busy at time t. Let dt be a time increment sufficiently short that it can contain only one event, where an event is either the initiation of a new call (i.e., the occupancy of a new channel) or the termination of a call (i.e., vacating a previously occupied channel). Assume that the probability that any one call will not terminate in time interval dt is e^{-dt}. (This assumption says that the length of calls is distributed exponentially and that the probability that a call will end is not influenced by its previous history—i.e., by how long the call has already lasted.) The probability that *all* x independent calls in progress at time t will not terminate in time increment dt is

$$(e^{-dt})^x = e^{-x\,dt} \qquad (6.32)$$

It is certain that either (1) no call out of the x in process will terminate in dt, or else (2) that at least one call will terminate in dt. The probabilities of these two things happening must add to unity. Hence the probability that at least one call will terminate in dt is given by

$$1 - e^{-x\,dt} = 1 - (1 - x\,dt + \text{higher-order terms}) \simeq x\,dt \qquad (6.33)$$

If A is the traffic intensity carried by the network, then the probability that a new call will be initiated in time dt is $A\,dt$.

Assume that the link under consideration is in steady-state (equilibrium) operation and that the number of busy channels x is less than full capacity. Steady-state operation means that the average value of x is constant over any finite time interval and this requires that, over any infinitesimal time interval dt, the probability that a call is made (the probability that the value of x increases by 1) is equal to the probability that a call ends (the probability that x decreases by 1). The probability that x increases by 1 is the probability that x channels are in use and that a new call is initiated. If these events are independent, then the probability of both happening simultaneously is the product of the individual probabilities, and the probability that x will increase by 1 in time dt when the traffic intensity is A is given by $p(x)A\,dt$. The probability that one call will terminate during dt when $x + 1$ calls were in progress at the beginning of the interval is given by $p(x + 1)(x + 1)\,dt$.

Equating the probability that a call will end in dt to the probability that a new call will be made in dt yields

$$p(x)A\,dt = (x + 1)p(x + 1)\,dt \qquad (6.34)$$

Eliminating dt and rewriting A as nh from Eq. (6.31) we have

$$np(x) = \left(\frac{1}{h}\right)(x + 1)p(x + 1) \qquad (6.35)$$

which is called the *basic traffic equation*.

The two solutions to Eq. (6.35) in common use with permanently assigned links (i.e., the original telephone problem) differ in their definition of x. In the *Erlang B model*, x is the number of busy channels, $0 \le x \le C$, where C is the total number of channels available. The probability that the xth channel is busy is $p(x)$, and the probabilities are related by the normalization condition.

$$\sum_{x=0}^{C} p(x) = 1 \qquad (6.36)$$

For traffic intensity AE, this leads to a probability $p(C)$, frequently written $B(C, A)$, that the last available channel (the Cth channel) is busy, given by

$$p(C) = B(C, A) = \frac{A^C}{\left[C! \sum_{i=0}^{C} \left(\dfrac{A^i}{i!}\right)\right]} \qquad (6.37)$$

In the *Poisson model* x is the number of busy channels and waiting calls, and thus x can range from zero to infinity since the number of calls that can wait is not limited to the number of available channels. Thus the normalization condition (6.36) becomes

$$\sum_{x=0}^{\infty} p(x) = 1 \qquad (6.38)$$

The probability that x will exceed C when the traffic intensity is A is given by $P(C, A)$ or $P(x > C)$ where

$$P(C, A) = P(x > C) = \sum_{x=C+1}^{\infty} e^{-A} \frac{A^x}{x!} \qquad (6.39)$$

Some authors write $P(x \geq C)$ instead of $P(x > C)$ for $P(C, A)$; the difference is that the summation then begins with $x = C$.

The Erlang B distribution incorporates the assumption that blocked calls are lost (i.e., that a caller who receives a busy signal will not wait but will seek alternate means of communication [33]), and this is more realistic than the Poisson assumption that blocked callers will wait. For this reason, the Erlang B distribution is preferred in practice for estimating channel requirements. But for small numbers of channels the two models Eqs. (6.37) and (6.39) give similar results. For example, to carry an average traffic intensity A of 13 E, the Erlang B model predicts that 25 channels are needed to make the probability of a blocked call 0.001 and 18 channels are needed to make the probability of a blocked call 0.05. The Poisson model requires 26 and 19 channels for these two cases. In these examples the 1 channel difference between the models is larger than their mathematical difference because the predictions must be rounded up to the next integer number of channels.

For a convenient algorithm to calculate channel requirements according to the Erlang B model, the reader should consult reference 34. Reference [35] provides some interesting examples.

Channel Requirements for Fixed Assignment and Demand Assignment Schemes

Equation (6.37) specifies the number of channels required between two locations to carry AE of traffic with a stated grade of service (blocking probability). If these locations are two of many earth stations communicating through a transponder, then the total number of channels that the multiple access scheme must provide will vary widely, depending on whether channel assignments are fixed or are made on demand in response to the traffic. For example, with fixed assignment the total number of channels required by any one earth station is the sum of the channels required by Eq. (6.37) to connect that earth station with each of the other stations with which it might exchange traffic. But if, on the other hand, any channel can be allocated to any path on demand, than the number needed by a station is simply that calculated by Eq. (6.37) as required to carry the total traffic handled by the station.

As an illustration based on a blocking probability of 0.001, a station handling 20E of traffic intensity will need 35 channels if any channel can be assigned on demand to any destination. If the 20E are divided equally between 5 destinations, then with fixed assignments the station must maintain sufficient channels to carry 4E to each of 5 destinations. This requires 12 channels per destination or 60 in all. Thus in this particular example, demand assignment requires only about 58 percent as many channels as fixed assignment to move the same amount of traffic between the same number of destinations. For other examples see pp. 375ff of reference 2.

Speech Interpolation and Prediction

Our discussion thus far has focused on how the number of channels needed to provide a given grade of service is calculated and how this number may be reduced by demand assignment. The number of demand assignment channels needed for voice communications may be reduced even further through the use of speech interpolation or speech prediction schemes that take advantage of the intermittent nature of voice communication. Originally developed for analog signals on submarine cable systems, the technique of reassigning channels during speech pauses is frequently called *TASI* for *time-assigned speech interpolation*. Digital implementations are also called *digital speech interpolation* or *DSI*.

The average telephone user listens as much as he or she talks. This "talk–listen effect" means that each half of an occupied full-duplex telephone channel transmits speech for about half the time. Furthermore, a speaker must pause between words. In telephone terminology, the words between the pauses are called *speech spurts*. The pauses occupy 25 to 35 percent of the time that a person is speaking; expressed another way, a continuous speaker actually uses a channel for 0.65 to 0.75 of the time. Multiplying these numbers by 0.5 to account for the talk–listen effect, we find that each side of a telephone conversation needs to occupy a transmission channel for about 0.33 to 0.38 of the time. This fraction of time is called the *telephone load activity factor*, τ_L. If other factors such as the time that a caller is waiting for the other party to answer are included, the average value of τ_L may drop as low as 0.25. This is the number recommended by CCITT; while many designs are based on it, τ_L values of 0.35 and 0.4 are also used. [7, 36]

When a talker pauses between words or to listen, a speech interpolation system takes away the channel that had been used and assigns it to someone else. When the talker again speaks, the system must give him or her a new channel. The delay between the onset of sound and the completion of the channel assignment and connection is called a *clip*, because the first part of the talker's speech has been cut off or clipped. The time lapse required for the equipment to recognize the onset of sound and make a connection is called a *connect clip*, while the time that the system must spend waiting for a channel to become available (i.e., the time that the system must wait for someone to pause and free a channel) is called a *competitive clip*. The fraction of speech lost to competitive clips is called the *freeze-out fraction*. Speech interpolation systems typically are designed to limit the freeze-out fraction to 0.5 percent and to hold the percentage of clips longer than 50 ms to less than 2 percent [36].

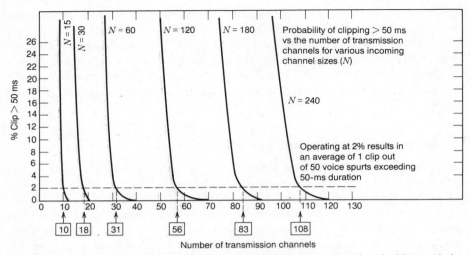

Figure 6.20 Competitive clipping performance of a DSI system. (Reprinted with permission from S. J. Campanella, "Digital Speech Interpolation," *Comsat Technical Review*, **6**, 127–158 (Spring 1976).)

The probability that a speech spurt will be competitively clipped for longer than t s is given by

$$P(n, N, t) = \sum_{x=n}^{N} \frac{N!\Theta^x(1 - \Theta)^{N-x}}{x!(n - x)!} \tag{6.40}$$

where

$$\Theta = \tau_L e^{-t/L} \tag{6.41}$$

Here $L = 1.5$ s is the mean time of a speech spurt, τ_L is the activity factor, n is the number of transmission channels available, and N is the number of incoming voice (telephone) channels [36]. For a given N, n may be found from Eq. (6.40) by trial-and-error solution for the n that gives the required value of P. The results are displayed in Figure 6.20. The Intelsat system is designed more conservatively than the figure would require. It provides 127 DSI channels for each 240 incoming terrestrial channels [17], while the figure indicates that 108 channels would be satisfactory.

Channel requirements for speech transmission can be reduced still further by predictive encoding techniques. In these systems, digitized voice samples from each channel are compared to predictions (based on past history) of what the samples should be. Those samples that are predietable are not transmitted; a prediction circuit at the receiver regenerates them. One such system, called SPEC for *speech predictive encoded communications*, achieves a 25 percent reduction in the number of transmitted samples for 25 percent of the time [36].

6.5 PRACTICAL DEMAND ACCESS SYSTEMS

In this section we will present several demand access (DA) schemes that are or soon will be in commercial service. These range from the Intelsat TDMA system, which incorporates both DA and PA in the same format, to the pure DA schemes of SPADE and the Inmarsat system.

Demand Access in the Intelsat TDMA System [17]

The Intelsat TDMA system is designed to accommodate both DA and PA digital traffic in each traffic burst. Intelsat calls a transmission that might contain DSI a *DSI burst*, but the channels involved may not all carry DSI voice.

The problem of describing the operation of the Intelsat DA system is complicated by the use of three different names for channels. A satellite channel (SC) is a numbered time slot in each TDMA traffic burst. A terrestrial channel (TC) is an incoming or outgoing voice line (analog or digital) to be carried by the system. Each terrestrial channel is numbered according to a system used by the destination earth station. At the point where a TC connects to the TDMA/DSI equipment, it becomes an international channel (IC). The ICs are numbered from 1 to 240, and a DSI equipment module can accomodate 240 terrestrial channels. A "TC to IC Table" gives the relationship between TC channel numbers and IC channel numbers for each station. See reference 21 for further discussion of this point.

Each DSI burst carries 127 satellite channels plus an assignment channel; see Figure 6.21 for the details of the burst structure. Each satellite channel contains 16 8-bit samples, so the traffic occupies 16,256 bits, and an additional 128 bits are allocated to the assignment channel. Thus the DSI burst is 16,384 bits or 8192 symbols long and lasts for 135.593220 μs. This is the same as for the DNI sub-burst of Section 6.2. Under normal operating conditions the satellite channels are numbered from 1 to 127, beginning with the SC immediately following the assignment channel. Noninterpolated (i.e., PA) terrestrial channels carried by the system are assigned to the highest numbered satellite channels beginning with 127, whereas DSI terrestrial channels are assigned to the lowest numbered satellite channels beginning with 1.

Under overload conditions (more than 127 terrestrial channels requiring transmission at the same time), the system can add up to 16 extra DA DSI channels by robbing satellite channels 1 through 112 of one bit per sample. The *overload satellite channels* are numbered 255 through 240; 255 is filled first, then 254, then 253, and so on. Overload channel 255 is formed by robbing the least significant bit from satellite channels 1 through 7. The slot formerly occupied by the least significant bit of satellite channel 1 carries the most significant bit of 16 successive samples of the terrestrial channel carried by overload channel 255.

Information about SC assignments is carried by the *assignment channel* (*AC*). The assignment channel of each DSI burst carries three 16-bit assignment messages. The first 8 bits are the number of the SC being assigned and the second 8 bits are the number of the IC that the transmitting earth station is assigning to that SC. The 48 bits of the three assignment messages are Golay encoded (see

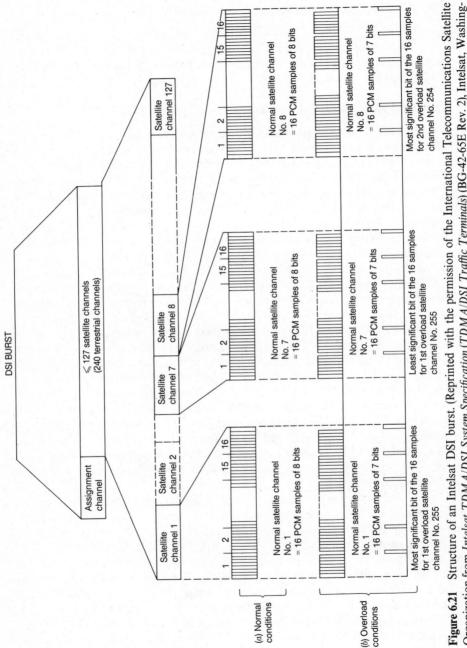

Figure 6.21 Structure of an Intelsat DSI burst. (Reprinted with the permission of the International Telecommunications Satellite Organization from *Intelsat TDMA/DSI System Specification (TDMA/DSI Traffic Terminals) (BG-42-65E Rev. 2)*, Intelsat, Washington, DC, June 23, 1983.)

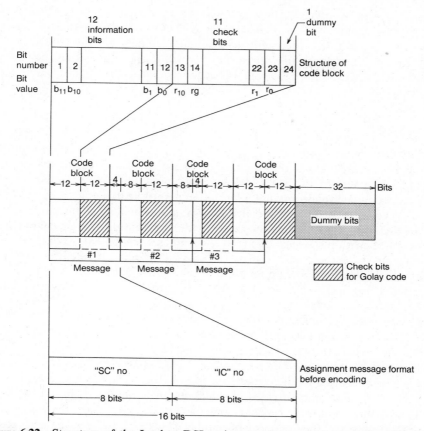

Figure 6.22 Structure of the Intelsat DSI assignment message channel. (Reprinted with the permission of the International Telecommunications Satellite Organization from *Intelsat TDMA/DSI System Specification* (*TDMA/DSI Traffic Terminals*) (BG-42-65E Rev. 2), Intelsat, Washington, DC, June 23, 1983.)

Chapter 7) to occupy a 96-bit slot in the assignment channel as illustrated by Figure 6.22. The remaining 32 bits of the 128-bit assignment channel are dummies.

Assignment messages have two forms and three functions. A DSI terminal can transmit only three assignment messages in one frame. In case of conflict, priority goes in decreasing order to new assignments, reassignments, disconnects, and refreshment messages. The first transmission of an SC number, IC number pair (where the IC number is nonzero) indicates the demand assignment of a time slot. Transmission of an SC number followed by a zero is a command to release (i.e., to free) that SC. Repeated transmissions of the same message are called *refreshment messages* and do not change the allocation of SCs. Assignment messages transmitted on one TDMA frame must be executed by the time the next frame is received, 2 ms later. Once an SC assignment is made, it is not changed unless the

SC is inactive and another TC has become active and needs an SC. In other words, satellite channels are not reassigned at the end of every speech spurt unless the traffic load demands it. Assignments to overload channels are made only when all of the normal channels are full. Terrestrial channels that had been assigned to overload channels are reassigned to normal channels as soon as the end of a speech spurt permits.

SPADE

SPADE brings together in one system almost all of the ideas about multiplexing and multiple access that we have discussed thus far. It carries traffic using demand assignment SCPC FDMA with QPSK modulation of the individual carriers. It disseminates coordinating information about the dynamic channel assignments over a TDMA subsystem called the common signaling channel (*CSC*). The name SPADE is a somewhat forced acronym representing "Single-channel-per-carrier *PCM Multiple Access Demand Assignment Equipment*." While SPADE was developed for the obsolescent INTELSAT IV family of satellites, Intelsat's published specifications foresee its use through the INTELSAT V series [12]. As the initial investment in SPADE equipment is written off, SPADE will probably be replaced by the Intelsat TDMA system and the Intelsat SCPC system [37]. There are some slight differences between the frequency plan used for SPADE on INTELSAT IV-A satellites and those used on INTELSAT IV and V; we will confine the discussion that follows to INTELSAT V.

Figure 6.23 shows the SPADE frequency plan at IF. The 70-MHz center frequency is upconverted to the center of a transponder. The two halves of the 36-MHz bandwidth are each divided into 800 channels numbered in two ways. In one system, the center frequency of channel 400 is 22.5 kHz below the 70-MHz IF center frequency, and the frequency of channel 401 is 22.5 kHz above the IF center frequency. Thus channel 400 is centered at 69.9775 MHz and channel 401 is centered at 70.0225 MHz. Center frequencies of the other are placed at 45-kHz increments from these frequencies so that the IF center frequency of channel N is given by

$$f_N = 0.045\,N + 51.9775 \text{ MHz} \qquad (6.42)$$

The center of the transponder band (70 MHz) is marked by a pilot carrier provided by the network's reference station. To avoid interference with the pilot, the channels immediately above and below it (401 and 400) are not used, and these portions of the spectrum are unoccupied. In addition, channels 1 and 2 are left vacant to accommodate the common signaling channel (CSC) that provides network control information.

The two halves of a telephone conversion are carried by a pair of channels above and below the pilot. A total of 397 such pairs are available, and this is the number of simultaneous duplex channels that a 36-MHz transponder using SPADE can provide. For convenient reference the channels above the pilot are given a second set of numbers identified by primes whereby channels K and K'

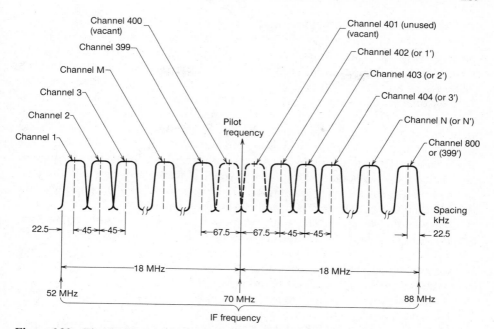

Figure 6.23 The SPADE frequency plan at IF. (Reprinted with permission of the International Telecommunications Satellite Organization from *Standard A Performance Characteristics of Earth Stations in the INTELSAT IV, IVA, and V Systems having a G/T of 40.7 dB/K* (BG-28-72E Rev. 1), Intelsat, Washington, DC, December 15, 1982.)

carry the two sides of the same telephone conversation. In the primed system channel N' corresponds to channel (401 + N') in the unprimed system. The center frequencies of two paired channels, say M and M', differ by 18.045 MHz [12]. The reader should note that this primed numbering and pairing scheme that is now the Intelsat standard differs from an earlier one in which primed and unprimed channels were exactly 18 MHz apart. The earlier numbering system is widely referenced in the literature; see for example reference 38 or reference 3. The advantage of using a system in which the center frequencies of the two halves of a telephone conversation always differ by the same amount is that a single synthesizer can generate the center frequencies of both channels.

In the SPADE system, a channel pair is automatically allocated to an incoming call by a demand assignment signaling and switching (DASS) unit [39]. This is basically a special-purpose computer that maintains in its memory a file of empty channels and randomly assigns one of these on request to an incoming call. Random assignment minimizes the probability of two stations wanting the same channel at the same time. The station to whom an outgoing call comes is called the *originating earth station*, and its DASS makes the assignment. Say it assigns channel pair M. When this is done, the originating earth station broadcasts a message on the common signaling channel (CSC) notifying the destination earth station that it has a call for it on channel pair M. Channel pair M is allocated to the first earth station whose request for it is broadcast by the satellite. Thus, if the originating earth station hears a request for pair M from someone else before it hears the satellite broadcast its own request, the originating earth station's DASS issues a new channel pair and the request process begins again. Likewise if the destination earth station hears some other station request pair M before it hears the originating station's request, the destination earth station adds channel M to its list of busy channels and ignores the originating station's request. Since the originating earth station will also find out that the channel is busy, the destination earth station simply waits for another request. When the destination earth station receives a request for a free channel, it acknowledges it over the CSC and the assignment process is complete. The destination and originating earth stations then test the channel pair and the call can proceed.

The voice traffic on the telephone channels is carried by conventional 64 kbps QPSK occupying an IF bandwidth of about 38 kHz. The nominal bit error rate is 1×10^{-6} and the threshold bit error rate is 1×10^{-4} [12]. SPADE channels are not reassigned between speech spurts, but the carriers are turned off during long pauses and listening periods to save transponder power.

The SPADE CSC operates in a TDMA mode using a center frequency 18.045 MHz below the pilot and occupying a total bandwidth of 160 kHz [12]. The modulation is two-phase PSK at 128 kbps and the frame length is 50 ms. Each station transmits one 1 ms burst per frame, and the SPADE system can accommodate one reference station plus 49 other stations [39]. Details of the structure of the SPADE reference burst and data burst appear in Figure 6.24.

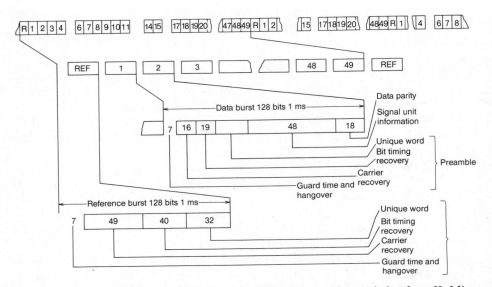

Figure 6.24 SPADE reference and data bursts. (Reprinted with permission from K. Miya, *Satellite Communications Engineering*, p. 126. Copyright © Lattice Publishing Co., Tokyo, Japan 1975.)

The Inmarsat System

Our discussion thus far has concentrated on using demand access for allocating channels between a fixed number of earth stations in response to varying traffic conditions. In maritime satellite communications the problem is different; a large number of mobile earth stations need to be able to send and receive messages at irregular intervals. Stations come and go as ships sail and dock, and the network must be able to accommodate whatever stations are active at a particular time. Demand access is a necessity for such a system.

The Inmarsat system uses a variety of modulation types and modes of operation. The access procedure is slightly different for each type; in this section we will restrict our discussion to demand access analog telephony between ship earth stations (SES) and coastal earth stations (CES). Unless otherwise indicated, our technical data come from reference 22, and we have based our operational information on that reference and on reference 40.

Any Inmarsat station may request a channel at any time by transmitting a request burst on the request channel. This is a random access channel; we will discuss random access in Section 6.6. A CES may start the process by assigning an available channel to a particular ship; a ship may start the process by asking a particular CES for a channel. Figure 6.25 illustrates the request burst that the SES transmits. It is 172 bits long and is sent at 4800 bps using coherent BPSK on either of Inmarsat's two request channels, 1638.600 and 1642.950 MHz. Inmarsat stations use these two frequencies alternately. A request burst lasts for 35.83 ms.

There are 339 Inmarsat telephone channels. These consist of frequency pairs spaced 25 kHz apart beginning at 1535.025 MHz for SES reception and 1636.525 MHz for SES transmission. For any channel pair, the SES transmitting and receiving frequencies differ by 101.500 MHz and the SES receiving frequency is lower than the transmitting frequency. The channels are numbered from 1 for the lowest frequency to 339 for the highest.

Figure 6.26 illustrates the format of the 348-bit assignment message that is sent by the CES in response to a ship station's request for a channel. It is transmitted via BPSK at 1200 bps with a frame length of 0.29 s.

Once a SES has received a channel assignment, further signaling required to set up the call is carried out on the channel itself.

Notes to Figure 6.25:

1. The first bit transmitted is written to the left. This corresponds to least significant bit.

2. Error detection shall be Bose-Chaudhuri-Hocquenghem (BCH) 63, 39.

Notes to Figure 6.26:

Number of bits

1. The first bit transmitted is written to the left (see the telegraph character "*A*" above. In the signaling channel this corresponds to the least significant bit.

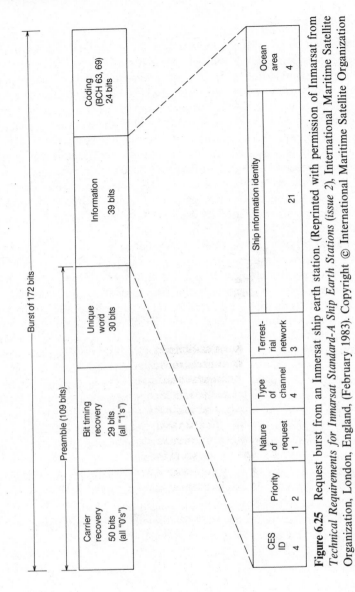

Figure 6.25 Request burst from an Inmarsat ship earth station. (Reprinted with permission of Inmarsat from *Technical Requirements for Inmarsat Standard-A Ship Earth Stations (issue 2)*, International Maritime Satellite Organization, London, England, (February 1983). Copyright © International Maritime Satellite Organization (INMARSAT) 1983.

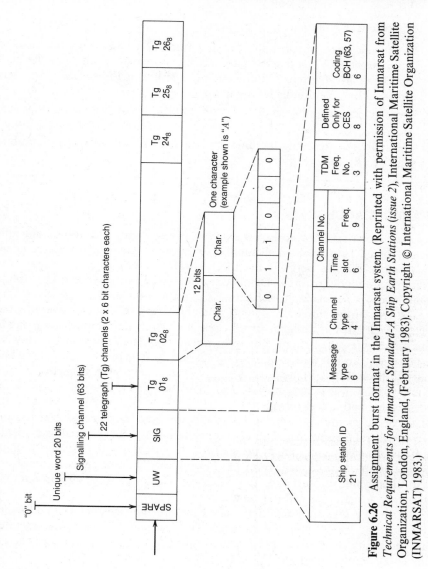

Figure 6.26 Assignment burst format in the Inmarsat system. (Reprinted with permission of Inmarsat from *Technical Requirements for Inmarsat Standard-A Ship Earth Stations (issue 2)*, International Maritime Satellite Organization, London, England, (February 1983). Copyright © International Maritime Satellite Organization (INMARSAT) 1983.)

Notes to Figure 6.26 (*continued*):

2. In the telegraph channel the first bit transmitted indicates the type of character field. When the first bit is a 0, the subsequent 5-bit character field represents an ITA No. 2 character; when it is a 1, the subsequent 5 bits represent line conditions for signaling.

3. Error detection coding shall be Bose-Chaudhuri-Hocquenghem (BCH) 63, 57.

4. The spare bit preceding the unique word shall be a 0. All other spare bits shall be 1s.

6.6 RANDOM ACCESS

Preassigned access systems work best when relatively constant amounts of traffic are being passed between a small number of stations. Demand access becomes more attractive when a network must carry traffic whose volume, origins, and destinations are highly variable. Demand access requires an overhead investment in system control (for example, the network coordination station and the assignment messages in the Inmarsat system), and this overhead reduces overall transponder capacity. If the traffic involved consists of short data bursts occurring at random times, then the system control overhead may be eliminated by allowing each station to transmit at will. Some transmissions will be lost to interference and will have to be repeated, but the overhead involved in acknowledging correct messages and repeating lost ones may be less than that which would have to be invested in a demand access scheme.

The classic example of appropriate traffic for random access is that passing between a central computer and an interactive terminal. The terminal user spends most of his or her time looking at the screen and then typing in data, but nothing is transmitted until the carriage return key is pressed. The user expects to wait for the central computer to respond, and occasional delays in transmission required for repeating a lost burst will not be noticed.

Systems like this are commonly called *packet radio* or *packet broadcasting* systems since packets of data are involved. The first packet system was operated by the University of Hawaii, and for that reason this type of a random access scheme is frequently called an *ALOHA channel.*

The basic analysis [41] of an ALOHA system assumes that packets are generated by a Poisson process in which the probability of a packet originating in time interval ΔT is proportional to ΔT. The packets all have the same length, T_S, and the rate at which packets are transmitted is λ packets/second. Some packets "collide" and are lost; the rate at which packets are successfully received is λ' packets per second, and $\lambda' < \lambda$. The normalized channel utilization G, which Martin [19], commenting on Abramson's paper, defines as the fraction of the time that the channel is used for sending original (i.e., not repeated) packets is given by

$$G = \lambda T \tag{6.43}$$

The two rates are related by

$$\lambda = \lambda' e^{-2\lambda' T} \tag{6.44}$$

Martin shows that the mean number of times that a packet will have to be retransmitted to overcome collisions is given by

$$N = e^{2\lambda' T} \tag{6.45}$$

and that G may be calculated from

$$G = \frac{1}{2N} \log_e(N) \tag{6.46}$$

The maximum value of G is 1/2e or 0.184; this corresponds to an N of 2.7. Hence under conditions of maximum throughput, an ALOHA channel is used 18.4 percent of the time for sending original packets, and the average packet must be repeated 2.7 times before it is received without interference.

In a satellite system, each earth station can hear its own packets repeated by the satellite and thus can determine whether a retransmission is necessary without waiting for an acknowledgment from the addressee. The likelihood of a collision during a retransmission can be reduced by having each station use a random number generator to determine the waiting period before retransmission. This decreases the probability that the same two packets will collide a second time without forcing any one station always to wait longer than the others before retransmitting.

The throughput of an ALOHA system can be increased by introducing some amount of network control. The simplest system for doing this is called *slotted ALOHA* because it allows participating stations to transmit packets beginning only at specified times. Thus colliding packets will interfere totally and the situation where the beginning of one packet interferes with the end of another is avoided. This doubles the utilization to 38.8 percent [19].

While packet networks are important for data transmission, packet broadcasting has not yet caught on as a practical multiple-access technique. Most satellite packet networks that have been proposed or put into operation (see reference 23) appear to be demand-access systems rather than pure random-access ALOHA systems. See reference 42 for a description of a demonstration system that was scheduled for operation in late 1984.

6.7 MULTIPLE ACCESS WITH ON-BOARD PROCESSING

In all of the multiple-access systems discussed so far, the role of the satellite is simply to amplify and retransmit the uplink signals. This will change with the availability of larger launch vehicles and more sophisticated digital logic, and future communications satellites will perform a significant amount of on-board processing.

One intriguing possibility, described in a recent FCC filing by Hughes Communications Galaxy, Inc. [43], is a satellite system that would use FDMA on the uplink and TDMA on the downlink. The Hughes system envisions 96 uplink carriers, each SQPSK modulated at 1.544 Mbps. These would be individually demodulated at the satellite and time division multiplexed onto a single 150-Mbps QPSK downlink. The uplink would retain the simplicity of FDMA; earth stations simply transmit on the correct frequency without any need for uplink synchronization. Demodulation and remodulation at the satellite avoids any re-transmission of uplink noise, and the use of a single digitally modulated carrier on the downlink eliminates the need for backoff. Synchronization is provided by the satellite itself; receiving earth stations simply track the slow changes in reference burst times that result from orbital motion. Hughes claims a 6-dB (C/N) advantage for this system over conventional FDMA—5 dB from the absence of a backoff and 1 dB from the elimination of retransmitted uplink noise.

As described earlier in this chapter, most TDMA technology and operating procedures evolved under the implicit assumption that all earth stations in the network would be able to receive all transmissions as they were relayed by the satellite. This provides complete interconnectivity; any station can transmit to any other station at any time. But the spot beams that will be required for frequency sharing and to provide the necessary fade margins (see Chapter 8) and large E_b/N_0 values for high data rates at frequencies above 10 GHz will destroy this interconnectivity. Stations will be able to hear only those transmissions that come down in their own local spot beam. The satellite will have to incorporate the necessary equipment to interconnect different uplink and downlink spot beams in response to changing traffic demands. This is called *satellite-switched* TDMA (SS-TDMA). See reference 44 for a general discussion.

The number of spot beams that a spacecraft can provide is limited. An alternative to leaving spot beams on fixed locations is to have a single scanning beam that continuously scans a satellite's service area. This requires that the spacecraft have the capability to receive a traffic burst from one location, store it, and re-transmit it when the scanning beam is on the intended recipient. Such systems have been proposed for NASA's ACTS (Advanced Communications Technology Satellite) spacecraft and for INTELSAT VI [45].

6.8 SUMMARY

Multiple access is the process by which a large number of earth stations inter-connect their links through a satellite. In frequency division multiple access (FDMA), stations are separated by frequency, while in time division multiple access (TDMA), they are separated in time. In code division multiple access (CDMA), stations use spread-spectrum transmissions with orthogonal codes to share a transponder without interference. Multiple access may be preassigned or demand, depending on whether or not it responds to changing traffic loads.

Frequency division multiple access with frequency division multiplexing and FM modulation (FDM/FM/FDMA) is the oldest and currently (1984) the most widely used multiple access scheme. In it each earth station is assigned frequency bands for its uplink transmissions. Because of the TWT backoff required to reduce intermodulation distortion, the spectral efficiency (i.e., the number of channels that can be carried per MHz of bandwidth) degrades with the number of stations that access a transponder. Companded single sideband (CSSB) is an FDMA scheme in which each voice channel requires only 4 kHz at RF for transmission. Besides high spectral efficiency, its main advantage over FDM/FM/FDMA is that CSSB performance depends only on the number of channels carried and not on the number of accesses. CSSB is more sensitive to interference than FDM/FM/FDMA, and this may alter the relative advantages of the two analog FDMA schemes in practical applications.

In time division multiple access (TDMA), earth stations transmit in turn. Since only one carrier is present at a time, no TWT backoff is required and thus full transponder EIRP is available. TDMA performance does not degrade with the number of accesses. TDMA transmissions are organized into frames; a frame contains one or two reference bursts that synchronize the network and identify the frame and a series of traffic bursts. Each participating station transmits one traffic burst per frame. Frames are grouped into larger structures called superframes. Frames, superframes, and individual traffic bursts are identified by standardized bit sequences called unique words. One of the major technical problems in implementing TDMA is synchronization. Once synchronization is acquired, it must be maintained dynamically to compensate for orbital motion of the spacecraft.

In code division multiple access (CDMA) stations transmit at the same time and in the same frequency bands using spread-spectrum (SS) techniques. CDMA avoids the centralized network control required for synchronization in TDMA, but it is difficult to implement in practice. Its use will probably be limited to those systems (chiefly military) that must employ SS for reasons other than multiple access.

The number of channels required to carry an anticipated volume of traffic between earth stations may be calculated from the Erlang B model. In demand assignment systems, these channels may be drawn from a pool available to all members of a network. This requires fewer total channels than if each station is permanently assigned the number of channels needed to carry the anticipated volume of traffic between it and all of the other members of the network. The number of voice channels required can be further reduced if the network reassigns channels during the quiet periods between speech spurts. This technique is called speech interpolation.

In applications like interactive computing that are characterized by burst transmissions separated by long quiet periods, random access may be used. Each station transmits at will, waits for an acknowledgment, and transmits again if none is received. This eliminates the requirement for centralized control of a demand-access system. Such schemes are called ALOHA systems after the first successful implementation.

REFERENCES

1. John G. Puente, William G. Schmidt, and Andrew M. Werth, "Multiple-Access Techniques for Commercial Satellites," *Proceedings of the IEEE*, **59**, 218–229 (February 1971). (Reprinted in [46], pp. 28–39.)
2. H. L. VanTrees, E. V. Hoverstein, and T. P. McGarty, "Communications Satellites: Looking to the 1980s," *IEEE Spectrum*, **14**, 42–51 (December 1977). (Reprinted in [47], pp. 16–28).

3. K. Miya, Ed., *Satellite Communications Engineering*, Lattice Publishing Co., Tokyo, Japan, 1975.

4. Roger L. Freeman, *Telecommunication Transmission Handbook* (2nd ed.), John Wiley & Sons, New York, 1981.

5. J. L. Dicks, P. H. Schultze, and C. H. Schmitt, "Systems Planning," *Comsat Technical Review*, **2**, 439–475, (Fall 1972). (Combined with several other papers and reprinted in [47], pp. 141–197.)

6. K. Miya, Ed., *Satellite Communications Technology*, KDD Engineering and Consulting, Tokyo, Japan, 1981.

7. *Transmission Systems for Communications* (5th ed.), Bell Telephone Laboratories, Holmdel, NJ, 1982.

8. Jack L. Dicks and Martin P. Brown, Jr., "Frequency Division Multiple Access (FDMA) for Satellite Communication Systems," *IEEE Eascon Conference Record*, 1974, pp. 167–178. (Reprinted in [46], pp. 84–95.)

9. N. K. M. Chitre and J. C. Fuenzalida, "Baseband Distortion Caused by Intermodulation in Multicarrier FM Systems," *Comsat Technical Review*, **2**, 147–171 (1972). (Reprinted in [46], pp. 125–137.)

10. J. C. Fuenzalida, O. Shimbo, and W. L. Cook, "Time-Domain Analysis of Intermodulation Effects Caused by Nonlinear Amplifiers," *Comsat Technical Review*, **3**, 89–141 (1973). (Reprinted in [46], pp. 138–164.)

11. K. Jonnalagadda and L. Schiff, "Improvements in Capacity of Analog Voice Multiplex Systems Carried by Satellite," *Proceedings of the IEEE*, **72**, 1537–1547, (November 1984).

12. Wilbur L. Pritchard, "Satellite Communications—An Overview of the Problems and Programs," *Proceedings of the IEEE*, **65**, 294–307 (March 1977). (Reprinted in [47], pp. 2–15.)

13. *Standard A Performance Characteristics of Earth Stations in the INTELSAT IV, IVA, and V Systems Having a G/T of 40.7 dB/K* (BG-28-72E Rev. 1), Intelsat, Washington, DC, December 15, 1982.

14. K. Jonnalagadda, "Single Sideband, Amplitude Modulated, Satellite Voice Communication System Having 6000 Channels per Transponder," *RCA Review*, **43**, 464–488 (September 1982).

15. E. Laborde and P. J. Freedenberg, "Analytical Comparisons of CSSB and TDMA/DSI Satellite Transmission and Techniques," *Proceedings of the IEEE*, **72**, 1548–1555 (November 1984).

16. R. J. Brown, Jr., M. L. Guha, R. A. Hedinger, and M. L. Hoover, "Companded Single Sideband (CSSB) Implementation on COMSTAR Satellites and Potential Application to INTELSAT V Satellites," *Proceedings of the IEEE International Conference on Communications*, E2.1.1–E2.1.6 (1983).

17. *Intelsat TDMA/DSI System Specification (TDMA/DSI Traffic Terminals)* (BG-42-65E Rev. 2), Intelsat, Washington, DC, June 23, 1983.

18. J. J. Spilker, Jr., *Digital Communications by Satellite*, Prentice-Hall, Englewood Cliffs, NJ, 1977.

19. James Martin, *Communications Satellite Systems*, Prentice-Hall, Englewood Cliffs, NJ, 1978.

20. O. Gene Gabbard and Pradman Kaul, "Time-Division Multiple Access," *IEEE Eascon Conference Record*, 1974, pp. 170–184. (Reprinted in [46], pp. 96–101.)

21. K. Feher, *Digital Communications: Satellite/Earth Station Engineering*, Prentice-Hall, Englewood Cliffs, NJ, 1983.

22. *Technical Requirements for Inmarsat Standard-A Ship Earth Stations* (Issue 2), International Maritime Satellite Organization, London, England (February 1983).

23. V. K. Bhargava, D. Haccoun, R. Matyas, and P. Nuspl, *Digital Communications by Satellite*, John Wiley & Sons, New York, 1981.

24. Robert M. Gagliardi, *Satellite Communications*, Lifetime Learning Publications, Belmont, CA, 1984.

25. J. W. Schwartz, J. M. Aein, and J. Kaiser, "Modulation Techniques for Multiple Access to a Hard-Limiting Satellite Repeater," *Proceedings of the IEEE*, **54**, 763–777 (May 1966). (Reprinted in [47], pp. 308–322.)

26. R. A. Scholtz, "The Origins of Spread-Spectrum Communications," *IEEE Transactions on Communications*, *COM-30*, 822–854 (May 1982).

27. R. L. Pickholtz, D. L. Schilling, and L. B. Milstein, "Theory of Spread-Spectrum Communications-A Tutortial," *IEEE Transactions on Communications*, *COM-30*, 855–884 (May 1982).

28. Marlin P. Ristenbatt and James L. Daws, Jr., "Performance Criteria for Spread Spectrum Communications," *IEEE Transactions on Communications*, *COM-25*, 756–761 (August 1977). (Reprinted in [2], pp. 360–366.)

29. Charles E. Cook and Howard S. Marsh, "An Introduction to Spread Spectrum," *IEEE Communications Magazine*, **21** (2), 8–16 (March 1983).

30. Marlin P. Ristenbatt, "Alternatives in Digital Communications," *Proceedings of the IEEE*, **61**, 703–721 (June 1973). (Reprinted in [2], pp. 212–230.)

31. D. Bear, *Principles of Telecommunication Traffic Engineering* (paperback edition), Peter Peregrinus, Ltd., Stevenage, England, 1980.

32. *Reference Data for Radio Engineers*, (6th ed.) Howard W. Sams and Co., Indianapolis, 1975.

33. Burwell Goode, "SBS TDMA-DA System with VAC and DAC," *Proceedings of the IEEE*, **72**, 1594–1610 (November 1984).

34. T. N. Shimi and Y. L. Park, "Efficient Approaches to Erlang Loss Function Computations," *Comsat Technical Review*, **13**, 143–155 (Spring 1983).

35. S. Shimura, Ed., *International Submarine Cable Systems*, KDD Engineering and Consulting, Tokyo, Japan, 1984.

36. S. J. Campanella, "Digital Speech Interpolation," *Comsat Technical Review*, **6**, 127–158 (Spring 1976). (Reprinted in [47], pp. 108–138.)

37. *SCPC/PSK (4ϕ) and SCPC/PCM/PSK (4ϕ) System Specification* (BG-9-21E Rev. 3), Intelsat, Washington, DC, March 31, 1982.

38. J. G. Puente and A. M. Werth, "Demand-Assigned Service for the Intelsat Global Network," *IEEE Spectrum*, **8**, 59–69 (January 1971). (Reprinted in [47], pp. 382–292.)

39. Burton I. Edelson and Andrew M. Werth, "SPADE System Progress and

Application," *Comsat Technical Review*, **2**, 221–241 (1972). (Reprinted in [46], pp. 63–73.)

40.   K. Miya, Ed., *Satellite Communications Technology*, KDD Engineering and Consulting, Tokyo, Japan, 1981.

41.   Norman Abramson, "The Throughput of Packet Broadcasting Channels," *IEEE Transactions on Communications, COM-25*, 117–128 (January 1977). (Reprinted in [47], pp. 393–404.)

42.   Christian Bertin, "PACKSATNET—An Alternative to Data Networks," *Proceedings of the IEEE*, **72**, 1532–1536 (November 1984).

43.   *Application of Hughes Communications Galaxy, Inc., for a Ka Band Domestic Communications Satellite System.* Submitted by Hughes Communications Galaxy, Inc., to the Federal Communications Commission, Washington, DC, on December 15, 1983.

44.   Takuro Muratani, "Satellite-Switched Time-Domain Multiple Access," *IEEE Eascon Conference Record*, 1974, pp. 189–196. (Reprinted in [47], pp. 413–420.)

45.   S. J. Campanella and J. V. Harrington, "Satellite Communications Networks," *Proceedings of the IEEE*, **72**, 1506–1519 (November 1984).

46.   Ivan Kadar, Ed., *Satellite Communications (AIAA Selected Reprint Series*, Vol. 18), American Institute of Aeronautics and Astronautics, New York, 1976.

47.   H. L. VanTrees, Ed., *Satellite Communications*, IEEE Press, New York, 1979.

PROBLEMS

1. You are designing an FDM/FM/FDMA analog link that will occupy 36 MHz of an INTELSAT V transponder. The uplink and downlink center frequencies of your occupied band are 5985.5 MHz and 3760.5 MHz. The distance from the satellite to your earth station is 40,000 km. The saturation uplink flux density for your uplink is -75 dBw/m^2 and the satellite's G/T is $= -11.6$ dBK^{-1}. At saturation the transponder EIRP for your downlink is 29 dBw and the earth station's G/T is 41 dBK^{-1}. The transponder is linear in that the EIRP in dBw is *BO* dB below the saturation value when the uplink flux density is backed off *BO* dB below saturation. The intermodulation carrier to noise ratio, $(C/N)_I$ in dB, is related to the backoff *BO* in dB by

$$(C/N)_I = 7.86 + 0.714 BO$$

In other words, at saturation the value of $(C/N)_I$ is 7.86 dB. Find the maximum overall carrier-to-noise ratio $(C/N)_o$ in dB that this link can achieve. What backoff must be used to achieve it? (When you need a frequency in your calculations, use the uplink or downlink center frequency as appropriate.) Make your calculations for beam center.

Problems 2 through 6 all involve a satellite and earth stations with the same specifications.

Five earth stations share one transponder of a 6/4 GHz satellite. The satellite and earth station characteristics are given below:

Satellite	Transponder BW = 36 MHz
	Transponder gain = 90 dB (max)
	Input noise temp. = 550 K
	O/P Power = 6.3 W (max)
	4 GHz antenna gain = 20 dB
	6 GHz antenna gain = 22 dB
Earth station	4 GHz antenna gain = 60.0 dB
	6 GHz antenna gain = 61.3 dB
	Rx System Temp. = 100 K
Path loss	At 4 GHz = 196 dB
	At 6 GHz = 200 dB

2. The stations all operate in a TDMA mode. The original speech signals are sampled at 8 kHz, using 8 bits/sample. The sampled signals (PCM) are then multiplexed into 40 Mbps streams at each station, using QPSK.

a. Find the bit rate for each PCM signal.

b. The number of speech signals (as PCM) that can be sent by each earth station, as a single access.

c. The frame time for TDMA without storage.

3. Assume that the TDMA system uses a 125-μs frame time. Find the number of channels/earth station when:

a. No time is lost in overheads, preambles, and the like.

b. A 5-μs preamble is added to the beginning of each earth station's transmission.

c. A 5-μs preamble is added to each frame and a 2-μs guardband is allowed between each transmission.

4. Assume that storage for the speech signals is available at each earth station. If a 750-μs frame time is used instead of 125-μs, find the new channel capacities of the earth stations for the cases in Problem 3 above.

5. Find the earth station transmitter power and received (C/N) when the system is operated:

a. In TDMA with the transponder saturated by each earth station in turn.

b. In FDMA with 5-dB input and output backoff.

6. Assume that the transponder is linear with 90 dB gain up to saturation. Ignore antenna beam shape losses.

a. The FDMA system employs frequency modulation using baseband bandwidths of 252 kHz to contain 60 channels of telephone data. If the rms frequency deviation of each 252 kHz baseband is 500 kHz, how many channels can each earth station send, assuming equal sharing of the 36-MHz RF bandwidth in the transponder?

    b.   Calculate the (S/N) for a 0 dBm test tone in the telephone channel at the top of the baseband, including psophometric weighting but excluding preemphasis improvement.

    c.   How does the capacity of the system compare for FDMA and TDMA?

    7.   One SCPC voice channel of the SPADE system carries QPSK at a symbol rate of 32,000 symbols per second in a bandwidth of 38 kHz. It achieves a bit error rate of 1×10^{-4}; without coding this requires an E_b/N_0 of 9.4 dB. The single carrier for this SPADE channel is radiated by an INTELSAT V transponder with an EIRP (for this one carrier) of 0 dBW to an earth station 40,000 km away. The channel center frequency is 4095 MHz. Answer questions (a) through (c) below about this channel and its receiver.

    a.   Tabulated EIRP's for INTELSAT V transponders are about 29 dBW. Why does this single SPADE carrier have a 0-dBW EIRP?

    b.   What value of (C/N) in decibels must be present at the input to the earth station receiver? Ignore any uplink or intermodulation effects.

    c.   What is the minimum G/T value in decibels that the earth station must have to achieve the specified bit error rate *with a 3-dB margin*?

    8.   A digital communication system uses a satellite transponder with a bandwidth of 50 MHz. Several earth stations share the transponder using QPSK modulation. Standard data rates used in the system are 80 kbps and 2 Mbps. The RF bandwidth required to transmit the QPSK signals is 0.75 of the bit rate. The TDMA frame is 125 μs in length, and a 1 μs guard time is required between each access. A preamble of 48 bits must be sent by each earth station at the start of each transmitted data burst.

    a.   What is the symbol rate of each earth station's transmitted data burst?

    b.   Calculate the number of earth stations that can be served by the transponder if each station sends data at 80 kbps.

    c.   Calculate the number of earth stations that can be served if each earth station sends data at 2 Mbps.

    d.   Repeat the calculations in (b) and (c) for a frame that is 1000 μs in length.

    9.   The capacity of the system described in Problem 8 can be increased substantially by using satellite switched TDMA. Assume that each earth station sends some data to every other earth station in every frame, and that it takes one microsecond to reposition the satellite antenna beam from one earth station to another. Only the downlink antenna beam is switched; the uplink uses a common zone beam. The frame length to be used is 1000 μs, and the extra antenna gain at the satellite is traded for an increase in the data rate by using 16-phase PSK. In this scheme, the 50-MHz bandwidth transponder can carry a symbol rate of 28.75 Msps corresponding to a data rate of 115 Mbps. Other parameters of the system are unchanged.

    a.   Find the number of earth stations that can share the transponder when each earth station sends data at 2 Mbps.

    b.   Find the total data throughput of the transponder after all preamble bits have been removed.

7

ENCODING AND FORWARD ERROR CORRECTION FOR DIGITAL SATELLITE LINKS

7.1 ERROR DETECTION AND CORRECTION

The transmission of information over a satellite communication system always results in some degradation in the quality of the information. In analog links the degradation takes the form of a decrease in signal-to-noise ratio. We saw in Chapter 5 that by using wideband FM we can trade bandwidth for power and achieve a good baseband signal-to-noise ratio with a low carrier-to-noise ratio in the RF signal. In digital links we measure degradation of the information content of a signal in terms of the bit error rate. By using phase shift keying, usually coherent QPSK, we can again trade bandwidth for signal power and achieve good bit error rates with low carrier-to-noise ratios.

A fundamental difference between analog and digital signals is that we can improve the bit error rate of a digital signal by the use of error correction techniques. No such technique is available for analog signals since once the information is contaminated by noise, it is extremely difficult to remove the noise, as we cannot in general distinguish between the signal and the noise electronically. (There are techniques that attempt to distinguish between signal and noise in television pictures, by using the correlation properties of the picture. They have been used successfully to enhance the quality of images of the moon and other planets obtained by the *Voyager* and similar space probes. However, the time taken to process the picture and the computer power needed make such techniques impractical for regular TV and voice transmissions.) In a digital system, we can add extra *redundant bits* to our data stream, which can tell us when an error occurs in the data and can also point to the particular bit or bits that have been corrupted. Systems that can only detect errors use *error detecting codes*. Systems that can detect and correct errors use *forward error correction* (FEC).

Some confusion exists in the literature over the term *coding*, since it is applied to several different processes, not all of which are concerned with error detection and correction. In the popular sense, *coding* is used to describe the rearrangement of information to prevent unauthorized use. This process is known technically as *encryption*. It is widely used on both analog and digital signals that are sent by cable and radio links. Digital signals are much more amenable to encryption, which can be achieved by convolving the data bits with a long pseudonoise (PN) sequence to destroy the intelligibility of the baseband data. To recover the information, the PN sequence used in the encryption process must be known to the recipient; this information is contained in the *key* to the code, which must be changed at frequent intervals to maintain good security. We will not be concerned any further, in this chapter, with encryption. It is, however, an important aspect of communications for commercial and military users who are concerned about interception and improper use of their transmitted data.

Coding is also applied to many processes that change data from one form to another. For example, pulse code modulation (PCM) changes analog data into binary words for transmission over a digital link. It is fundamental to the transmission of voice by digital techniques, and uses a device commonly called a *codec*, short for coder–decoder. The term *coding* is also applied to devices that scramble a digital data stream to prevent the occurrence of long strings of 1s or 0s.

Throughout this chapter we shall use the term *coding* to refer to error detection or error correction. This implies that additional (redundant) bits are added to the data stream to form an error-detecting or error-correcting code. It is possible, in theory, to generate codes that can detect or correct every error in a given data stream. In practice, there is a trade-off between the number of redundant bits added to the information data bits and the rate at which information is sent over the link. The *efficiency* of a coding scheme is a measure of the number of redundant bits that must be added to detect or correct a given number of errors. In some FEC systems the number of redundant bits is equal to the number of data bits, resulting in a halving of the data rate for a given channel transmission rate. (This is called a *rate one-half* scheme.) The loss of communication capacity is traded for a guaranteed low error rate; for example, commercial users transmitting financial data are willing to pay the extra cost in return for guaranteed accuracy in their accounting data, and military users are willing to lower data rates in order to obtain reliable communication in the presence of enemy jamming.

In this chapter we first discuss the techniques of error detection, and how they can be implemented in a satellite communication link, and then consider forward error correction. FEC in some form can be used on digital satellite links to improve bit error rates under low (C/N) conditions. Propagation disturbances, particularly attenuation by rain on links operating at 14/11 GHz and higher frequencies, result in reduced (C/N) for a small percentage of the time. Rather than designing the link with a very high (C/N) in clear weather, it is economic to add FEC to the link and allow the (C/N) to degrade during periods of rain attenuation. In clear air, the FEC system will correct the few errors that do occur; when rain attenuation is present, it maintains the bit error rate at an acceptable level with the degraded (C/N).

Finally, we examine some techniques that can be used to correct errors by requesting retransmission of corrupted data blocks. These are called *ARQ* (automatic repeat request) systems. They are widely used for packet transmission systems, where data are sent in blocks with variable delays, rather than being sent in real time. ARQ systems require a return channel; they cannot be used in one-direction transmission of data, and FEC must be used in such cases.

The operator of a digital satellite communication link has an option of providing FEC as part of the link, or of providing only a basic transmission channel. At 6/4 GHz and on many 14/11 GHz links, the channel is provided without error correction or detection. A minimum BER is guaranteed by the operator for a specified percentage of time, based on the link design and projected performance. The user is then free to add error detection or FEC to the data sent to and received from the link. If digital speech is sent, error detection or correction is rarely applied. With digital data, some measures must be taken to guard against error, and the user will normally provide the necessary equipment.

Links operating at frequencies above 10 GHz are subject to increases in BER during propagation disturbances. The link will be designed with a margin of a few decibels so that the BER falls below an acceptable level, typically 10^{-6}, for only a small percentage of any month or year. The total time for which the margin is exceeded by propagation effects will be less than 0.5 percent of any month in a well designed system. During the remaining 99.5 percent of the month, the E_b/N_0 of the received signal will be well above threshold, and very low BER will result. There may, in fact, be no errors for long periods of time and billions of bits can be transmitted with complete accuracy. Under these conditions, FEC and error-detection systems do nothing for the communication system. However, unless we can detect a falling E_b/N_0, coding may have to be applied all the time to be certain it is available when E_b/N_0 approaches threshold. To that extent, coding is an insurance against the possibility of bit errors; for most of the time it is unnecessary, but when it is needed, it proves invaluable.

Common carriers, who supply communication links to users on a dial-up or leased basis, do not generally apply FEC to their links, nor do they define the *protocols* to be used. (A protocol defines an operating procedure in a link.) These are user-supplied services and must be defined by the user for the data to be sent. In such cases, the error detection and correction equipment will be located at the customers' premises, whereas the earth station may be a long distance away and accessed via terrestrial data links.

The situation may be very different in a single-user network such as a military communication system. The earth station may be located at the data source, and coding and protocol form an integral part of the network operation. A similar situation can arise in carefully controlled systems such as Intelsat's, where the link operator specifies the user's earth station and operating parameters in detail.

Coordination between users is essential in an international system that interconnects the communication systems of different countries. In such cases, all protocols, codes, and FEC techniques must be very carefully defined and standardized. Because of the large number of earth stations in the Intelsat system and their wide geographic distribution, some paths between certain earth stations may have lower than usual E_b/N_0 ratios. It is feasible for those earth stations to employ FEC on the lower quality routes, without FEC being needed by all users.

7.2 CHANNEL CAPACITY

In any communication system operating with a noisy channel, there is an upper limit on the information capacity of the channel. Shannon [1] examined channel capacity in mathematical terms, and his work led to significant developments in information theory and coding.

For an additive white Gaussian noise channel, the capacity H is given by

$$H = B \log_2 \left(1 + \frac{P}{N_0 B} \right) \text{bps} \tag{7.1}$$

where B is the channel bandwidth in hertz
P is the received power in watts,
N_0 is the single sided noise power spectral density in watts per hertz

Equation (7.1) is commonly known as the Shannon-Hartley law. We can rewrite Eq. (7.1) specifically for a digital communication link by putting $H = 1/T_b$, where T_b is the bit duration in seconds. The energy per bit is E_b, giving

$$E_b = PT_b = \frac{P}{H} \tag{7.2}$$

Then substituting $E_b/N_0 = P/HN_0$ in Eq. (7.1) yields

$$\frac{H}{B} = \log_2 \left(1 + \frac{E_b}{N_0} \frac{H}{B} \right) \tag{7.3}$$

The ratio H/B is the *spectral efficiency* of the communication link, the ratio of bit rate to the bandwidth of the channel. Figure 7.1 shows the ratio $\log_2(H/B)$ plotted against E_b/N_0 in dB for the case when $H < B$ and the link operates at a bit rate H bps. Regardless of the bandwidth used, the E_b/N_0 cannot go below -1.6 dB (ln 2) if we are to operate at capacity. This is known as the Shannon bound. It sets a lower theoretical limit on the E_b/N_0 we can use in any communication link, regardless of the modulation or coding schemes. A link operating with $H < B$ is said to be *power limited* because it does not use its bandwidth efficiently.

Figure 7.2 shows the case for a link with $H > B$, operating at capacity. In this case, we can increase the capacity for a given bandwidth without limit, but only by providing very large E_b/N_0 ratios, implying very high transmitter power. As we saw in Chapter 5, satellite links must operate with E_b/N_0 in the range of 5 to 25 dB, which limits the theoretical spectral efficiency to below 4 bits/Hz. When $H > B$, the link is said to be *bandwidth limited*, implying that we could increase

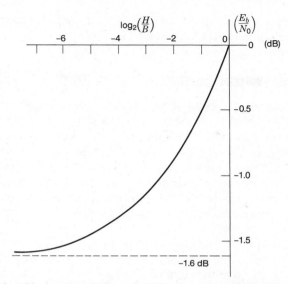

Figure 7.1   Relationship between H/B and E_b/N_0 for power limited case and low E_b/N_0 ratio.

capacity by using the available transmitter power in a wider bandwidth. Practical links using PSK do not achieve capacities anywhere near the Shannon theoretical capacity H. Shannon's theory assumes essentially zero bit errors; to achieve a bit error rate of 10^{-10} in a QPSK link requires a theoretical E_b/N_0 of 13 dB with a spectral efficiency of 2 bits/Hz. Equation (7.3) predicts $E_b/N_0 = 1.77$ dB for this case. What coding, in particular FEC, can do for us is to improve the link performance under conditions of low E_b/N_0, such as during periods of rain attenuation, so that the BER of the link does not rise excessively. This takes us closer to

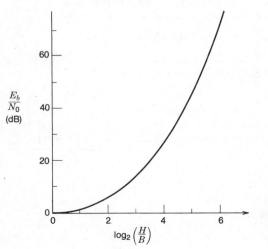

Figure 7.2   Relationship between H/B and E_b/N_0 for bandwidth limited case and high E_b/N_0 ratio.

the Shannon capacity in the region of low E_b/N_0 while not increasing excessively the bandwidth required for transmission.

7.3 ERROR-DETECTION CODING

Error detection coding is a technique for adding redundant bits to a data stream in such a way that an error in the data stream can be detected. Unless one redundant bit is added for every data bit, the exact position of a single bit error cannot be determined. Usually, one redundant bit is added for every N data bits; this allows a single error within that *block* of N bits to be detected. A simple example of an error detecting code system that has been in use for many years for transmission of information is the 8-bit ASCII code [2]. The ASCII code is widely used for transmission of computer and teleprinter data over telephone lines and radio links.

The 8-bit ASCII code consists of 128 characters, each having seven data bits plus a single parity bit. The seven data bits have an internationally agreed-upon interpretation and represent the alphabet in upper- and lowercase letters, numerals 0 to 9, and a set of useful commands, symbols, and punctuation marks. The eighth bit, the parity bit, is used for detection of error in the seven data bits of the character. For example, in a system using *even parity*, the parity bit is 0 when the sum of the data bits is even, and 1 when the sum is odd. Thus the sum of the data bits plus the parity bit is always made even, or 0 in modulo-2 arithmetic. Figure 7.3 shows an example of even and odd parity coding. In *odd parity*, the sum of the data bits plus the parity bit is always odd, or 1 in modulo-2. Single errors in the seven data bits, or the parity bit, are detected at the receiving end of the link by checking the eight received bits of each character for conformity with the

	Data Bits	Parity Bit	Sum (Modulo-2)
Even parity	0101101	0	0
Odd parity	0101101	1	1

Figure 7.3a Example of even and odd parity for a 7-bit ASCII word.

Even Parity Example

	Received Codeword	Sum of Bits	Error Detected?
One error	01010010	1	Yes
Two errors	01010110	0	No
Three errors	11010110	1	Yes

Figure 7.3b Example of error detection in a 7-bit ASCII word with even parity. Error bits are underlined.

parity rule. In modulo-2 arithmetic, $0 \oplus 0 = 0, 0 \oplus 1 = 1, 1 \oplus 0 = 1$, and $1 \oplus 1 = 0$. Similarly, $0 \otimes 0 = 0$, $0 \otimes 1$ or $1 \otimes 0 = 0$, and $1 \otimes 1 = 1$. All the codes that we will be considering are binary, and modulo-2 arithmetic will be used throughout this chapter.

Suppose we have a system using even parity, which transmits the character A in ASCII code, as illustrated in Figure 7.3. At the receiving end of the link we check the eight bits by modulo-2 addition. If the sum is 0, we conclude that the character is correct. If the sum is 1, we detect an error. Should two bits of the character be corrupted, the modulo-2 sum is 0, so we cannot detect this condition. We can easily calculate the improvement in error rate (assuming that we discard corrupted characters) that results with parity in a seven-bit character.

For example, let the probability of a single-bit error occurring in the link be p, and let us suppose that p is not greater than 10^{-1}.

The probability $P_e(k)$ of k bits being in error in a block of n bits is given by the binomial probability function

$$P_e(k) = \binom{n}{k} p^k (1 - p)^{n-k} \tag{7.4}$$

where p is the probability of a single-bit error occurring, and

$$\binom{n}{k} = \frac{n!}{k!(n-k)!} \tag{7.5}$$

For example, in the ASCII *codeword* of 8 bits, we have one parity bit, which allows us to detect one error, and seven data bits. Two errors cannot be detected, although three can. The probability that there are two errors is much higher than the probability of four errors, so when parity is used, the probability of an undetected error occurring in an ASCII word is P_{wc} where

$$P_{wc} \simeq P_e(2) = \binom{8}{2} p_c^2 (1 - p_c)^6 \tag{7.6}$$

where p_c is the single-bit error probability for the 8-bit word. When p_c is small, $(1 - p_c)^6 \simeq 1$ so

$$P_{wc} \simeq \binom{8}{2} p_c^2 = 28 p_c^2 \tag{7.7}$$

If we had not used parity, the probability P_{wu} of a single error in the 7-bit word with bit error probability p_u is

$$P_{wu} \simeq \binom{7}{1} p_u (1 - p_u)^7 = 7 p_u \tag{7.8}$$

Thus the improvement in error rate for the ASCII words is approximately $4p$, provided $p_c \simeq p_u$.

For example, if we have $p_c = p_u = 10^{-3}$, the probability of error for uncoded 7-bit words is 7×10^{-3} and for words with a single parity bit added it is 2.8×10^{-5}.

Codeword		Message
011	011	011
110	110	110
Message bits	Parity bits	

Example of (6, 3) systematic code-words. The parity bits are at the right of the message bits.

Codeword		Message
001	0111	0111
101	1100	1100
Parity bits	Message bits	

Example of (7, 4) systematic code-words. The parity bits are at the left of the message bits.

Figure 7.4   Examples of (6, 3) and (7, 4) systematic codewords.

With a transmission bit error rate of 10^{-6}, the word error rate with parity is approximately 2×10^{-11}. This corresponds to one character error in the transmission of the text of this book 20,000 times over a link with a BER of 10^{-6}.

Some caution is needed in making the assumption that the bit error rate is the same for uncoded and coded transmissions. When we added the single parity bit to a 7-bit ASCII character, the transmission bit rate went up from 7 bits per character to 8 bits per character. The increase in bit rate will result in an increase in BER if the channel remains unchanged, so not all the expected decrease in character error rate will be achieved in practice. This is particularly true when we add several parity check bits to our data bits, to obtain error-correction codes.

The example given above for parity error detection is one case of *block error detection*. We have transmitted our data as *blocks*, in this case eight–bit blocks consisting of seven data bits and one redundant parity bit. In general, we can transmit n bits in a block, made up of k message bits plus r check bits. The n-bit block is called a *codeword* and coding schemes in which the message bits appear at the beginning of the codeword, followed by the check bits, are called *systematic codes*. Figure 7.4 shows an example of some systematic codes.

Linear Block Codes

Linear block codes are codes in which there are 2^k possible message blocks of k bits, to which are added $(n - k)$ redundant check bits generated from the

k message bits by a predetermined rule. In a *systematic linear block code* the first k bits of the codeword are the message and the remaining $(n - k)$ bits are the check bits. A codeword with n bits of which k bits are data is written as (n, k). The (6, 3) and (7, 4) codewords in Figure 7.4 are examples of systematic linear block codes.

The general form of a linear block codeword C is

$$C = DG \tag{7.9}$$

where D is the k-bit message and G is a generator matrix that creates the check bits from the data bits.

Multiplication of matrices can be performed only when the number of columns in the first matrix equals the number of rows in the second matrix. Thus D is a matrix of k columns and one row, and G must be a matrix with k rows and n columns. Multiplication is carried out in modulo-2 arithmetic, which is just like regular multiplication but without a carry. To multiply matrix D by matrix G we first multiply each element in the first column of G by the corresponding elements in D, and then modulo-2 sum the results. Repeating the procedure n times yields the matrix C, a codeword of n bits. The following very simple example for a (3, 2) code illustrates the process.

Let D be a 2-bit data word 10 and G be a generator matrix with

$$G = \begin{bmatrix} 1 & 0 & 1 \\ 0 & 1 & 0 \end{bmatrix}$$

Then the codeword $C = DG$ is

$$C = \begin{bmatrix} 1 & 0 \end{bmatrix} \begin{bmatrix} 1 & 0 & 1 \\ 0 & 1 & 1 \end{bmatrix}$$

$$= [(1 \otimes 1) \oplus (0 \otimes 0)][(1 \otimes 0) \oplus (0 \otimes 1)][(1 \otimes 1) \oplus (0 \otimes 1)]$$

$$= (1 \oplus 0)(0 \oplus 0)(1 \oplus 0)$$

$$= 1 \ 0 \ 1$$

Thus C is a (3, 2) codeword having two data bits and one parity bit, with even parity. The (3, 2) code in this example could detect one error in the 3-bit codeword.

The reader may wish to verify that the other code words in this set are

$$D = 00, \qquad C = 000$$

$$D = 01, \qquad C = 011$$

$$D = 11, \qquad C = 110$$

The general form of G is

$$G = [I_k \quad P_k]_{k \times n} \tag{7.10}$$

where I_k is the identity matrix of order k and P is an arbitrary k by $n - k$ matrix. No procedures exist for designing the P matrix, but some P matrices lead to codes with desirable properties such as high rate efficiency, ability to detect or correct more than one error, and so on. The subject of code design is involved, and the interested reader is referred to Shamnugam [3] for a review of coding for communication purposes, or to Lin and Costello [4] for a more detailed treatment of the subject. We restrict our survey here to the basic ideas behind the application of coding in communication links.

An example of a generator matrix for a (6, 3) block code is G where

$$G = \begin{bmatrix} 100 & 011 \\ 010 & 101 \\ 001 & 110 \end{bmatrix} \tag{7.11}$$

The message length is 3, the codeword length is 6. The matrix G has the form $[I:P]$ as in Eq. (7.10). There are eight possible codewords (000) through (111). The codewords are generated by Eq. (7.9) as

$$C = DG = [D] \begin{bmatrix} 100 & 011 \\ 010 & 101 \\ 001 & 110 \end{bmatrix} \tag{7.12}$$

For example, if the message word D is 001 the corresponding codeword $C(001)$ is

$$C(001) = [001] \begin{bmatrix} 100 & 011 \\ 010 & 101 \\ 001 & 110 \end{bmatrix} = 001 \quad 110 \tag{7.13}$$

The remaining codewords for this (6, 3) code are given in Table 7.1. The reader should verify these words using Eq. (7.12). Note that the check bits are not unique to a particular message; in fact, they cycle in a symmetrical pattern.

The codewords C are sometimes called *code vectors*. We now need a technique to detect errors in the codewords, which is accomplished by use of a *parity check matrix H*, defined as

$$H = [P^T:I_{n-k}]_{(n-k) \times n} \tag{7.14}$$

Table 7.1
Codewords of the Example
(6, 3) Code

Message bits	Codewords
000	000 000
001	001 110
010	010 101
011	011 011
100	100 011
101	101 101
110	110 110
111	111 000

where P^T is the transpose of P_k in Eq. (7.10). P^T is obtained by interchanging the rows and columns of the matrix P. I_{n-k} is an identity matrix for the check bits. Error detection is achieved by multiplying a received codeword R by the transpose of the parity check matrix, H^T. If the codeword R is correct, then $R = C$ and H^T is defined by

$$C \times H^T = 0 \qquad (7.15)$$

If the codeword R is in error, such that

$$R = C + E \qquad (7.16)$$

where E is an *error vector*, the error will be detectable if E is not zero. We check for errors by finding the *error syndrome S*. The syndrome is a single word of length $n - k$, where $n - k$ is the number of parity check bits in the codeword, formed by multiplication of the received codeword with the parity check matrix. It is always zero if the received codeword has no errors.

$$S = RH^T$$
$$= [C + E]H^T = CH^T + EH^T \qquad (7.17)$$

Since CH^T is zero by definition in Eq. (7.15)

$$S = EH^T \qquad (7.18)$$

Thus the syndrome S is zero if R is a valid codeword. (It may also be zero for some combinations of multiple errors.) If S is nonzero, we know that an error has occurred in transmission of our codeword.

Example 7.3.1

An example of a code that can detect two errors in a message of length 4 bits is a (7, 4) *Hamming code*. This code has 4 message bits and 3 check bits in a 7-bit codeword.

The generator matrix G for this code is

$$G = \begin{bmatrix} 1000 & 111 \\ 0100 & 110 \\ 0010 & 101 \\ 0001 & 011 \end{bmatrix}$$
$$\quad\; I_4 \qquad P_3$$

The 16 possible codewords of this (7, 4) code are shown in Table 7.2. Using Eq. (7.14) we can form the parity check matrix H.

$$H = \begin{bmatrix} 1110 & 100 \\ 1101 & 010 \\ 1011 & 001 \end{bmatrix}$$
$$\quad\; P^T \qquad I_3$$

Table 7.2
Codewords and Weights for the (7, 4) Hamming Code Used in Examples 7.3.1 and 7.3.2

Data, D	Codeword, C	Weight
0000	0000000	—
0001	0001011	3
0010	0010101	3
0011	0011110	4
0100	0100110	3
0101	0101101	4
0110	0110011	4
0111	0111000	3
1000	1000111	4
1001	1001100	3
1010	1010010	3
1011	1011001	4
1100	1100001	3
1101	1101010	4
1110	1110100	4
1111	1111111	7

Now consider a message $D = (1010)$, one of 16 possible messages generated from four message bits. The corresponding codeword is C, which we generate from

$$C = DG = 1010 \quad 010$$

The syndrome $S = CH^T$ is zero. If we now insert an error into the codeword so that the received code vector R is corrupted in the second bit (underlined)

$$R = 1\underline{1}10010$$

Then the syndrome $S = RH^T$ is

$$S = [1110010] \begin{bmatrix} 1 & 1 & 1 \\ 1 & 1 & 0 \\ 1 & 0 & 1 \\ 0 & 1 & 1 \\ 1 & 0 & 0 \\ 0 & 1 & 0 \\ 0 & 0 & 1 \end{bmatrix} = 110$$

Since S is not zero, we have detected an error.

Example 7.3.2
Let us suppose that 2 bits and then 3 bits in our (7, 4) Hamming codeword are corrupted. Let the received codeword be $R = 101\underline{1}01\underline{1}$ when the message D

is 1010 and the transmitted codeword $C = 1010010$. The syndrome $S = RH^T$ is

$$S = [1011011] \begin{bmatrix} 111 \\ 110 \\ 101 \\ 011 \\ 100 \\ 010 \\ 001 \end{bmatrix} = 010$$

which indicates that the received codeword is incorrect.

If we had three errors, and received, for example, $R = 1110100$, the syndrome is $S = 000$. The errors are not detected, because this is a legitimate codeword in the (7, 4) set. The (7, 4) Hamming code can always detect two errors, but cannot reliably detect three errors. We use a general formula to determine the error-detection capabilities of a linear block code in Section (7.4).

Error Correction with Linear Block Codes

The linear block codes described in the previous section have the capability of locating errors in a received codeword. Generally, a larger number of errors can be detected than can be located, but when an error can be located within a codeword, it can then be corrected. This is a powerful feature of coding and one reason for the widespread use of digital transmission techniques, especially by satellite.

As an example of the error-correction properties of linear block codes, consider again the (7, 4) Hamming code used in Example 7.3.1. The syndrome was generated by using Eq. (7.17), $S = RH^T$. In Example 7.3.1, the received codeword $R = 1110010$ was a corruption of the transmitted codeword $C = 1010010$, with the second bit changed from 0 to 1. The syndrome S was 110, indicating the presence of an error in the received codeword. The syndrome $S = 110$ is identical to the second row of the H^T matrix, pointing to the second bit as the error bit. For single errors, the syndrome is always the row of the H^T matrix corresponding to the bit that is in error. If the error occurs in a message bit, we can correct the message by inverting the appropriate bit in the message part of the received codeword R. If the error occurs in a parity check bit, we may not need to take any action, if only the message bits are transmitted to the receiver output.

7.4 ERROR DETECTION AND CORRECTION CAPABILITIES OF LINEAR BLOCK CODES

Some linear block codes are better for error detection or correction than others. There are some basic theorems that define the capabilities of linear block codes in terms of the *weight, distance*, and *minimum distance*. The weight (or Hamming weight), w, of a code vector C is the number of nonzero components of C. The distance (or Hamming distance), d, between two code vectors C_1 and C_2 is the number of components by which they differ. The minimum distance of a block code is the smallest distance between any pairs of codewords in the entire code.

These definitions lead to some useful rules that tell us how many errors a given code can detect or correct. First, the minimum distance of a linear block code is the minimum weight of any nonzero codeword. The minimum distance of the (6, 3) code in Table 7.1 is 3. For the (7, 4) Hamming code considered in Examples 7.3.1 and 7.3.2 and tabulated in Table 7.2, the minimum distance is also 3. Second, the number of errors that can be detected in a code with minimum distance d_{min} is $(d_{min} - 1)$. The number of errors that can be corrected is $\frac{1}{2}(d_{min} - 1)$, rounded to the next lowest integer if the number is fractional.

Example 7.4.1

Consider the (7, 4) Hamming code shown in Table 7.2. The minimum distance of this code is 3, so up to $(3 - 1) = 2$ errors can be detected. The number of errors that can be corrected is $(3 - 1) \times \frac{1}{2} = 1$.

To correct two errors, we must have a code with a minimum distance of at least 5, so that $\frac{1}{2}(5 - 1) = 2$. When the code can correct only one error, the syndrome of the corrupted codeword points to the position in the codeword at which the error occurred. When more than one error exists in a codeword of a code capable of multiple error correction, the syndrome will be nonzero. However, finding the two (or more) bits that are in error becomes more complex.

It is necessary to examine the received codeword to determine which, out of all the codewords that could have been transmitted, the received codeword is closest to. The syndrome is used to indicate the most likely correct codeword. When multiple errors have to be corrected, the procedure becomes extremely cumbersome: as a result, this general approach to error correction is not used in practice.

A subset of linear block codes called *binary cyclic codes* has been developed for which implementation of error-correction logic is much simpler. The codes can be generated using shift registers, and error detection and correction can also be achieved with shift registers and some additional logic gates. A large number of binary cyclic codes have been found, many of which have been named after the people who first proposed them. The best known are the Bose-Chaudhuri-Hocquenghem codes (*BCH codes*), which were independently proposed by three workers at about the same time in 1959–60 [5, 6]. In the following section we look briefly at the techniques used to generate cyclic codes and to correct errors in received codewords.

7.5 BINARY CYCLIC CODES

Algebraic Structure of Cyclic Codes [3]

A binary cyclic code C is made up of (n, k) codewords, V, of the form

$$V = (v_0, v_1, v_2, v_3 \ldots v_{n-1}) \tag{7.19}$$

If the codeword V is shifted to the right by i bits to obtain a new word $V^{(i)}$

$$V^{(i)} = (v_{n-i}, v_{n-i+1}, \ldots v_0, v_1 \ldots v_{n-i-1}) \tag{7.20}$$

then $V^{(i)}$ is also a codeword. Thus there are n possible codewords in an (n, k) binary cyclic code.

The codeword V can be represented by a code polynomial $V(x)$ of the form

$$V(x) = v_0 + v_1 x + v_2 x^2 + \ldots v_{n-1} x^{n-1} \tag{7.21}$$

The coefficients v_0, v_1, and so on, are 0s or 1s obtained by modulo-2 addition and multiplication. The variable x has an index p corresponding to the position of the pth bit in the codeword. In general, we can obtain the code polynomial $V^{(i)}(x)$ for the code vector $V^{(i)}$ as

$$V^{(i)}x = v_{n-i} + v_{n-i+1}x + v_{n-i+2}x^2 + \ldots v_0 x^i + v_1 x^{i+1} + \ldots v_{n-i-1}x^{n-1} \tag{7.22}$$

$V^{(i)}(x)$ is the remainder resulting from dividing $x^i V(x)$ by $x^n + 1$, so

$$x^i V(x) = q(x)(x^n + 1) + V^{(i)}(x) \tag{7.23}$$

where $q(x)$ is a polynomial called the quotient.

We can generate codewords V in two ways. We can form *nonsystematic* codewords by using a *generator polynomial* $g(x)$ such that we form a *code polynomial* $V(x)$ from a *data polynomial* $D(x)$

$$V(x) = D(x) g(x) \tag{7.24}$$

where

$$D(x) = d_0 + d_1 x + d_2 x^2 + \ldots d_{k-1} x^{k-1}$$

and

$$g(x) = g_0 + g_1 x + g_2 x^2 + g_3 x^3 + \ldots g_{n-1-k} x^{n-1-k}$$

The coefficients d and g are all 0s or 1s. The codeword V will not necessarily contain the original message D when the codeword is formed using Eq. (7.24).

An alternative method of forming the codeword V is to use the generator polynomial $g(x)$ to make a codeword having the form

$$\begin{gathered} V = (r_0, r_1, r_2 \ldots r_{n-k-1}, d_0, d_1, d_2 \ldots d_{k-1}) \\ n - k \text{ parity check bits} \quad k \text{ message bits} \end{gathered} \tag{7.25}$$

where

$$r(x) = r_0 + r_1 x + r_2 x^2 + \ldots r_{n-k-1} x^{n-k-1}$$

The polynomial $r(x)$ is the *parity check polynomial* of the code and is the remainder obtained when $x^{n-k} D(x)$ is divided by $g(x)$.

$$x^{n-k} D(x) = q(x) g(x) + r(x) \tag{7.26}$$

where $q(x)$ is the quotient.

The code polynomial $V(x)$ is formed by

$$V(x) = r(x) + x^{n-k}D(x) \tag{7.27}$$

An example of the generation of a cyclic code using these two methods will help to illustrate the process [3].

Example 7.5.1

The generator polynomial $g(x) = 1 + x + x^3$ can be used to generate a $(7, 4)$ cyclic code. Let a data word be $D = 1010$. The corresponding message polynomial is $D(x) = 1 + x^2$

From Eq. (7.24), the codeword polynomial is

$$V(x) = D(x)\,g(x)$$

or

$$
\begin{aligned}
V(x) &= (1 + x^2)(1 + x + x^3) \\
&= 1 + x + x^3 + x^2 + x^3 + x^5 \\
&= 1 + x + x^2 + x^5
\end{aligned}
$$

since $x^3 + x^3 = (1 + 1)x^3 = 0x^3 = 0$ in modulo-2 arithmetic.

The codeword V is

$$V = 1 + 1x + 1x^2 + 0x^3 + 0x^4 + 1x^5 + 0x^6 = 1110010$$

Note that the message $D = 1010$ is not evident within this codeword—it does not have the structure given in Eq. (7.25). The systematic codeword form of Eq. (7.25) must be generated using Eqs. (7.26) and (7.27). The last four bits of the required codeword are the data bits $D = 1010$. The first three bits are parity check bits (r_1, r_2, r_3) obtained from

$$\frac{x^{n-k}D(x)}{g(x)} = q(x) + r(x)$$

Using the $g(x)$ generator polynomial and $D(x)$ message polynomial given above, we can find $r(x)$ as the remainder of dividing $x^3(1 + x^2)$ by $1 + x + x^3$. Using algebraic division we obtain

$$
\require{enclose}
\begin{array}{r}
x^2 \\
x^3 + x + 1 \enclose{longdiv}{x^5 + x^3 } \\
\underline{x^5 + x^3 + x^2} \\
x^2
\end{array}
$$

Note that there are no minus signs in modulo-2 arithmetic. Subtraction is the same as addition.

The remainder $r(x) = x^2$, so $r = 001$. Hence the codeword is

$$V = 0011010$$

Table 7.3

Systematic and Nonsystematic Codewords of (7, 4) Cyclic Code with Generator Polynomial $g(x) = 1 + x + x^3$

Message Codeword D	Nonsystematic Codeword $V(x) = D(x)\,g(x)$	Systematic Codeword V
0000	0000000	0000000
0001	0001101	1010001
0010	0011010	1110010
0011	0010111	0100011
0100	0110100	0110100
0101	0111001	1100101
0110	0101110	1000110
0111	0100011	0010111
1000	1101000	1101000
1001	1100101	0111001
1010	1110010	0011010
1011	1111111	1001011
1100	1011100	1011100
1101	1010001	0001101
1110	1000110	0101110
1111	1001011	1111111

Source: Reprinted from K. S. Shamnugam, *Digital and Analog Communication Systems*, John Wiley & Sons, New York, 1979, Table 9.3, p. 466, with permission.

Table 7.3 shows the full 16 codewords for systematic and nonsystematic (7, 4) cyclic codes generated from

$$g(x) = 1 + x + x^3$$

Generation of Cyclic Codes

An encoder to generate the (7, 4) systematic code in Table 7.3 is shown in Figure 7.5. The message input D forms the first 4 bits of the 7-bit codeword and is transmitted when the output switch is set to position 1 (transmitting the most significant bit first). As the four data bits are transmitted, the gate is open allowing the data bits to be shifted into the registers r_0, r_1, r_2. The feedback connection between r_0 and r_1 generates the appropriate $g(x)$ function and is equivalent to a division operation. When the four data bits have been transmitted and fed to the shift register, the switch is set to position 2 and the contents of the register are transmitted. The feedback connections around the shift register correspond to the coefficients of the generator polynomial $g(x)$, and the contents of the register after k data bits have passed though the gate is the remainder, r, which forms the parity check part of the codeword.

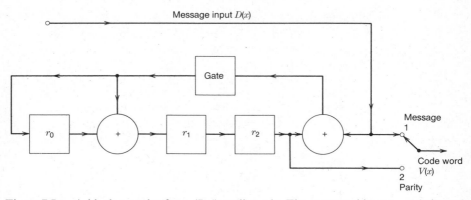

Figure 7.5a A block encoder for a (7, 4) cyclic code. The message bits are read through the switch in position 1 to form the first four bits of the codeword. The message bits are also read into the registers r_0, r_1, and r_2 via the feedback logic to generate three parity bits that are added to the four message bits by setting the switch to position 2.

Codeword

1	1	0	1	0	0	0

Parity   Message
bits   bits

Figure 7.5b Example of a codeword of a (7, 4) systematic code. See Table 7.3 for a full listing of this code.

For a detailed discussion of cyclic code generation and error correction, the reader should refer to references 4 and 7.

Error Detection and Correction with Cyclic Codes

The detection of errors in a cyclic code is achieved by calculating the syndrome of the received codeword R; if the syndrome is zero, we have a valid codeword. If it is nonzero, one or more errors have occured. The syndrome of a cyclic code is calculated as the remainder obtained when the polynomial of the received codeword, $R(x)$ is divided by the generator polynomial $g(x)$

$$\frac{R(x)}{g(x)} = P(x) + \frac{S(x)}{g(x)} \qquad (7.28)$$

where $P(x)$ is the quotient.

The syndrome has order $n - k - 1$ or less. If $E(x)$ is the error that has been added to the transmitted codeword $V(x)$ to get $R(x)$, then

$$R(x) = V(x) + E(x) \qquad (7.29)$$

and

$$\frac{R(x)}{g(x)} = \frac{V(x)}{g(x)} + \frac{E(x)}{g(x)} \qquad (7.30)$$

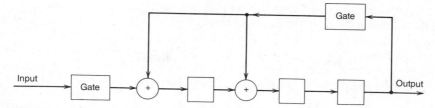

Figure 7.6 Syndrome circuit for the $(7, 4)$ cyclic code generated by $g(x) = 1 + x + x^3$. (*Source:* S. Lin and D. J. Costello, Jr., *Error Control Coding: Fundamentals and Applications,* copyright © 1983, p. 100. Reprinted by permission of Prentice-Hall, Inc., Englewood Cliffs, NJ.)

If we generate $V(x)$ from $V(x) = D(x)\,g(x)$ we find

$$E(x) = [P(x) + D(x)]\,g(x) + S(x) \qquad (7.31)$$

The error pattern $E(x)$ tells us how to correct the received codeword to get back to the original codeword V.

The strength of cyclic codes is the ease with which the syndrome of Eq. (7.28) may be calculated and the error correction of Eq. (7.31) applied, using digital logic circuits. A circuit for decoding and correcting single errors in the $(7, 4)$ code generated by the circuit of Figure 7.5 is shown in Figure 7.6.

The buffer register holds the received codeword R, which is 7 bits long. As each codeword is read into the buffer, it is entered into the syndrome calculation circuit. The three input AND gate acts as an error detector; if the output of the gate is 0, the current bit in the received codeword is correct. If the gate output is 1, the current codeword bit is incorrect and must be inverted. As the codeword is shifted out of the buffer register, the output gate inverts any bit that the error-detection gate indicates is incorrect.

BCH and Burst Error Correction Codes

The BCH codes form the most powerful group of cyclic error-correction codes yet devised and are readily implemented with shift register and logic circuits. For this reason, they are widely used in systems employing error correction. They have the following properties:

1. Length of shift register: m stages.
2. Block length: $n = 2^m - 1$ $(m \geq 3)$.
3. Number of errors that can be corrected: t $(t < 2^{m-1})$.
4. Number of parity check bits: $n - k \leq mt$.
5. Minimum distance: $d_{\min} \geq 2t + 1$.

For example, a $(1023, 923)$ BCH code has $t = 10$ and $d_{\min} \geq 21$. This code requires the transmission of 100 parity check bits for every 923 data bits, giving a code rate of 923/1023.

The codes discussed thus far have the property of correcting random errors, such as those caused by the peaks of thermal noise. Under some conditions, such as transient interference or cochannel interference caused by multipath, errors may occur in bursts. Special codes have been developed that can correct *burst errors*, that is, errors that occur in adjacent bits. The number of parity check bits needed to correct q sequential bit errors has a lower bound given by

$$n - k \geq 2q \tag{7.32}$$

A code capable of correcting burst errors of length q is called a q *burst-error correcting-code*. Table 7.4 lists some burst-error-correcting cyclic codes.

An alternative technique for burst error correction is to use interleaving of bits from a series of codewords that have good short burst-error-correction properties [4]. The individual bits of a given codeword are spaced l bits apart, so that a burst error of length l corrupts only one bit in each of l codewords. This single error can be corrected at the receiver, rendering the transmission invulnerable to longer error bursts. This technique is more valuable for terrestrial microwave links and troposcatter circuits that suffer short, deep fades due to multipath than for satellite links where the major causes of bit error are rain attenuation and depolarization.

Rain effects change slowly and cause outages lasting for seconds or minutes, so no amount of coding can reasonably correct the resulting loss of thousands or millions of bits.

Table 7.4
Examples of Burst-Error-Correcting Codes

Code (n, k)	No. of Parity Check Bits	Burst Error Correcting Ability q	Code Rate
7,3	4	2	3/7
15,9	6	3	9/15
511,499	12	4	499/511
1023,1010	13	4	1010/1023
131,119	12	5	119/131
290,277	13	5	277/290
34,22	12	6	22/34
169,155	14	6	155/169
103,88	15	7	88/103
96,79	17	8	79/96
56,38	18	9	38/56
59,39	20	10	39/59

Source: Reprinted from K. S. Shamnugam, *Digital and Analog Communication Systems*, John Wiley & Sons, New York, 1979, Table 9.5, p. 475, with permission.

Golay Codes

The Golay code and the single error-correcting Hamming codes are examples of *perfect codes*, in which all possible patterns of a given number of errors are corrected. The Golay code is a (23, 12) cyclic code that corrects all patterns of three or fewer bit errors. It has a minimum distance of 7. A closely related form of the Golay code has one overall parity check bit added to form a (24, 12) code with a minimum distance of 8. The (24, 12) code will detect all patterns of 7-bit errors and correct all patterns of 3-bit errors. It also has the advantage of a coding rate of one-half; rate one-half coding is easier to implement than other rates because the 2:1 ratio between message bits and code bits simplifies clock synchronization between the input data stream and the output coded stream.

7.6 PERFORMANCE OF BLOCK ERROR CORRECTION CODES

Calculation of the improvement of bit error rate with block encoding requires a comparison of the uncoded error rate to that obtained after correction of blocks of encoded data. For the perfect codes given in Section 7.5, the improvement can be determined exactly; when other codes are used, only an upper and lower bound on the error rate after coding can be established.

For the perfect codes, for which t errors can be corrected, the word error probability is

$$P_{we} = \sum_{i=t+1}^{n} \binom{n}{i} q_c^i (1 - q_c)^{n-i} \qquad (7.33)$$

where q_c is the bit error rate on the transmission link for the coded data, and n is the codeword length in bits.

If the data were sent uncoded, as k data bits, the word error probability would be

$$P_{wu} = 1 - (1 - q_u)^k \qquad (7.34)$$

where q_u is the bit error rate when uncoded data are transmitted on the same link. Because the bit rate for the shorter words of k bits is always lower than for the longer codewords of n bits, q_u is always lower than q_c on a given link. Equation 7.33 is an upper bound for the codeword error probability with other than perfect codes [7].

Figure 7.7 shows a comparison of the performance of a number of cyclic codes when implemented in a coherent PSK link [8]. The curves are for an ideal link with no modem implementation margin and show symbol error probability as a function of E_b/N_0 at the demodulator input.

The (7, 4) code is a single-error-correcting Hamming code. The remainder are BCH codes; the (127, 64) code corrects 10 errors, the (1023, 688) code corrects 36 errors. The (127, 113) BCH code is a double-error-correcting code that has been specified for use in 120 Mbps transmission using the INTELSAT V TDMA system [9].

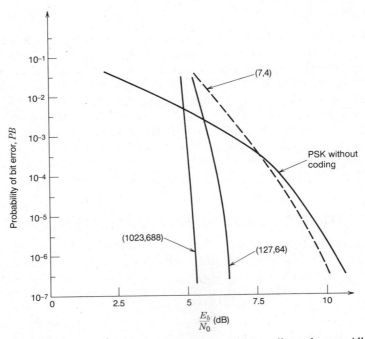

Figure 7.7 Probability of error in a PSK link for various coding schemes. All codes are approximately rate one-half. Note that below 5 dB E_b/N_0, coding does not improve the error rate. (*Source:* K. Feher, *Digital Communications: Satellite/Earth Station Engineering*, copyright © 1983, p. 285. Reprinted by permission of Prentice-Hall, Inc., Englewood Cliffs, NJ.)

Note that by the use of FEC, we can reduce the E_b/N_0 needed to achieve a 10^{-6} error rate from a theoretical 10.5 dB without coding to 7 dB with (127, 64) coding. The symbol error rate now increases very rapidly as the 7 dB E_b/N_0 value is approached; the degradation in error rate is very rapid, falling by a factor of 10^5 for less than 2 dB change in received signal power. The threshold is even more abrupt with (1023, 688) coding.

7.7 CONVOLUTIONAL CODES

Convolutional codes are generated by a tapped shift register and two or more modulo-2 adders wired in a feedback network. The name is given because the output is the convolution of the incoming bit stream and the bit sequence that represents the impulse response of the shift register and its feedback network [10]. Figure 7.8 illustrates an example of a convolutional encoder, which has been widely described in the literature [10, 11, 12, 13]. As each incoming information bit propagates through the shift register, it influences several outgoing bits, spreading the information content of each data bit among several adjacent bits. An error in any

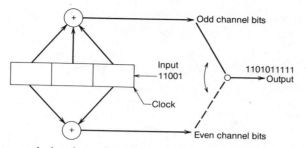

Figure 7.8 The convolutional encoder used in the text example. (Reprinted with permission from Marlin P. Ristenbatt, "Alternatives in Digital Communication," *Proceedings of the IEEE*, **61**, 703–721 (June 1973). Copyright © 1973 IEEE.)

one output bit can be overcome at the receiver without any information being lost. The process is somewhat like forming an image from a hologram, where information is distributed more or less uniformly over a two-dimensional field. The image can be reconstructed from only a portion of the field, although resolution is lost if a significant part of the hologram is discarded.

The *state* of a convolution encoder is defined by the shift register contents that will remain after the next input bits are clocked in. If the shift register is K bits long and input bits enter in groups of k, then the encoder has $2^{(K-k)}$ states. Putting in a group of k input bits causes the encoder to change states; a change of state is called a *state transition*. From a given state, a convolutional encoder can go to only 2^k other states (although one of these 2^k options may be to remain in the starting state). Each state transition is associated with a unique sequence of input bits and is accompanied by a unique sequence of output bits. The quantity K, which measures the length of the shift register, is called the *span* or the *constraint length* of the encoder. If v output bits are transmitted for every k input bits, then the encoding rate is k/v [10].

A decoder keeps track of the encoder's state transitions and reconstructs the input bit stream. Transmission errors are detected because they correspond to a sequence of transitions that could not have been transmitted. When an error is detected, the decoder begins to construct and keep track of all the possible tracks (sequences of state transitions) that the encoder might be transmitting. At some point, which depends on its speed and memory, the decoder selects the most probable track and puts out the input bit sequence corresponding to that track. The other tracks that it had been carrying are discarded. The algorithm used for decoding is named for A. J. Viterbi (see reference 12), and for this reason convolutional codes are sometimes called Viterbi codes.

An Illustrative Example

Consider the encoder of Figure 7.8. We will define the shift register's state by the values of its leftmost two bits before the next input bit is clocked in. After

Table 7.5
State Transitions in the Example Convolutional Encoder

Starting State	Input	Register Contents After Input	Output	Ending State
a:(0, 0)	0	000	00	**a**:(0, 0)
	1	100	11	**b**:(1, 0)
b:(1, 0)	0	010	10	**c**:(0, 1)
	1	110	01	**d**:(1, 1)
c:(0, 1)	0	001	11	**a**:(0, 0)
	1	101	00	**b**:(1, 0)
d:(1, 1)	0	011	01	**c**:(0, 1)
	1	111	10	**d**:(1, 1)

the input bit enters, the two bits that define the state are found in the rightmost two locations. We will call the states **a**, **b**, **c**, and **d**, defined in terms of bits as follows

$$\mathbf{a}:(0, 0) \qquad \mathbf{c}:(0, 1)$$

$$\mathbf{b}:(1, 0) \qquad \mathbf{d}:(1, 1)$$

Suppose the register is in state **a**, (0, 0), and a 0 enters. Immediately afterwards the register contents are (0, 0, 0) and the output symbol is $0 \oplus 0 \oplus 0 = 0$ and $0 \oplus 0 = 0$ or (0, 0). Thus when the register is in state **a** and a 0 enters, the output is (0, 0) and the encoder remains in state **a**. Now suppose that the register is again in state **a**, (0, 0), and this time a 1 enters. The register contents become (1, 0, 0) and the output bits are $1 \oplus 0 \oplus 0 = 1$ and $1 \oplus 0 = 1$ or (1, 1). Since the register contents are now (1, 0, 0), the encoder is in state **b**. Thus when the register is in state **a** and a 1 enters, the encoder changes to state **b** and emits the symbol (1, 1) in the process. Table 7.5 illustrates this process and all of the other allowed transitions for this encoder.

Another way to display the operation of this encoder is with the state transition diagram of Figure 7.9. In it a solid line indicates the trajectory that the detector follows when the input bit is a 0 and a dashed line indicates the trajectory that corresponds to a 1. The vertices are labeled with their corresponding states **a**, **b**, **c**, and **d**. The bit pairs at the middle of each trajectory indicate the output corresponding to that transition. Note that from each state it is possible to go to no more than two other states.

To illustrate a process by which the output of a convolutional encoder can be decoded, it is convenient to work with a *trellis diagram* like the one shown in Figure 7.10. Here the states of the encoder are plotted vertically and time is plotted horizontally. The transitions are shown as arrows joining the states occupied by

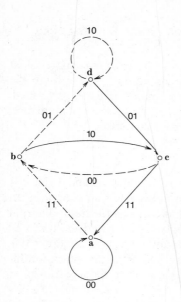

Figure 7.9 State transition diagram for the example convolution encoder. Solid lines indicate trajectories that follow input of a 0 and dashed lines indicate transitions corresponding to 1 inputs. Vertices are the stable states **a**, **b**, **c**, and **d** of the encoder. Output bits are written near the middle of the corresponding transitions.

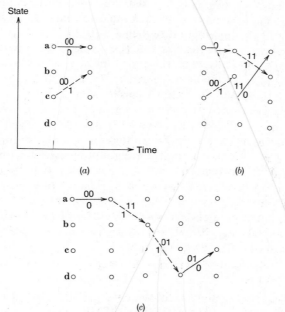

Figure 7.10 Trellis diagrams used in decoding the output of a convolutional encoder. The letters **a**, **b**, **c**, **d**, represent the four stable states of the encoder used in the text example, and the arrows represent transitions. A solid arrow is a transition resulting from an input 0 and a dashed arrow is a transition resulting from an input 1. The input that causes each transition is written below the arrow and the output that is generated by the transition is written above the arrow. (a) Two choices for the first transition. (b) Alternative choices for the first and second transitions. The top one is correct because the other arrows do not form a continuous path. (c) Complete path corresponding to the received bit sequence 00 11 01 01.

the encoder at successive times. If the bit sequence to be decoded was received correctly, then only one set of arrows will form a continuous path through the trellis, and that path traces the sequence of operations that the encoder performed. The corresponding input bits are the decoded version of the received sequence.

As an example, assume that we received an error-free message

$$00 \quad 11 \quad 01 \quad 01$$

which we know was encoded on the encoder of Figures 7.8, and 7.9 and Table 7.5. The 00 could represent an input 0 and a transition from state **a** to state **a** or an input 1 and a transition from state **c** to state **b**. We draw both on the trellis diagram as in Figure 7.10a. The 11 could represent a transition from **a** to **b** and an input 1 or a transition from **c** to **a** and an input 0. When we draw both as in Figure 7.10b, the true path is obviously **a–a–b** since no other combination of arrows is continuous. Using the correct path to continue decoding, we note that the only transition beginning at **b** and putting out 01 ends at **d** and represents an input 1. Likewise the only transition beginning at **d** that puts out an 01 results from an input 0 and ends at **c**. We now have established the complete path as shown in Figure 7.10c and decoded the incoming bit sequence as 0110.

In this example we were fortunate in being able to identify the starting state (**a**) of the encoder after only two transitions. In general it may be necessary to follow the first $4K$ transitions of an encoder of span K before the starting point and the subsequent path of the encoder can be identified.

A transmission error is detected when a group of bits is received that does not correspond to an allowed transition. The error may be corrected by following all of the paths that might have been transmitted for a period of time and retaining the path whose bits have the largest correlation coefficient with the bit stream actually received. The correlation coefficient between two bit sequences is found by replacing each 0 by a -1, multiplying corresponding bits in each sequence, and summing the result.

As a simple illustration of the error-detection and -correction process, assume that the bit stream 00 11 01 01 was transmitted but because of a transmission error we received

$$00 \quad 11 \quad \underline{1}1 \quad 01$$

where the incorrect bit is underlined. We carry out the steps of Figure 7.11a and 7.11b and arrive at state **b** when the $\underline{1}$ is received. There is no transition from state **b** that produces a 11 output, so we know that an error was made. What we received as 11 must have been sent as a 01, representing a transition to **d**, or as a 10, representing a transition to **c**. We must draw and keep track of both alternative paths as in Figure 7.11b. If the encoder was at state **d** when the last bit pair came in, the 01 represents a transition to **c**. On the other hand, if it started at state **c**, then the 01 represents another error because the only transitions from **c** are to **a** with a 11 output and to **b** with a 00 output. This means that there are two possible paths leaving **c** and we must draw both of them as in Figure 7.11c.

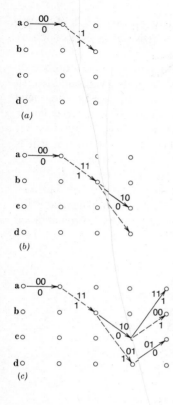

Figure 7.11   The use of alternative paths to decode a bit sequence containing transmission errors. See Figure 7.10. (*a*) Encoder path prior to transmission error. (*b*) Possible encoder paths corresponding to received bit sequence 00 11 11. (*c*) Possible encoder paths corresponding to received bit sequence 00 11 11 01.

At this point we must choose between candidate paths **b–c–a**, **b–c–b**, and **b–d–c**. To make the choice we calculate the correlation coefficient of the bit sequence corresponding to each path with the received sequence 11   01. Here are the calculations.

b–c–a	1	−1	1	1
Received	1	1	−1	1
Product	1	−1	−1	1

Sum of products = 0

b–c–b	1	−1	−1	−1
Received	1	1	−1	1
Product	1	−1	1	−1

Sum of products = 0

b–d–c	−1	1	−1	1
Received	1	1	−1	1
Product	−1	1	1	1

Sum of products = 2

The path with the largest correlation coefficient (sum of products) is **b–d–c**. We choose it as correct and it is. Thus the received sequence is correctly decoded as 0110.

The number of bit errors that can be corrected depends on the number of bits transmitted, the sequence in which the errors occur, and the number and length of the candidate paths that the decoder carries before discarding all but the one(s) with the highest correlation coefficient(s). As an illustration for the particular decoder and bit stream used here, the reader may wish to verify that two bit errors may be corrected if the received stream is 01 11 11 01 but not if it is 00 10 11 01, where again incorrectly received bits are underlined.

In general a Viterbi decoder must be able to retain information on $2^{k(K-1)}$ paths for satisfactory performance [14]. There is of course a trade-off between cost, BER, k, and K in practical convolution codes. Figure 7.12 [14] illustrates the performance of one such code; the reader should consult that reference for a detailed discussion of Viterbi decoder performance.

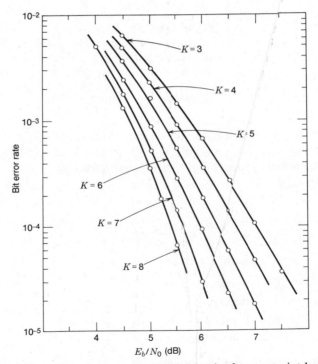

Figure 7.12 Performance of a rate one-half Viterbi codes for constraint length (K) values ranging from 3 to 8 and 32-bit paths. (Reprinted with permission from Jerrold A. Heller and Irwin Mark Jacobs, "Viterbi Decoding for Satellite and Space Communications," *IEEE Transactions on Communications*, **COM-19**, 835–848 (October 1981). Copyright © 1981 IEEE.)

7.8 IMPLEMENTATION OF ERROR DETECTION ON SATELLITE LINKS

Error detection is invariably a user-defined service, forming part of the operating protocol of a communication system in which the earth–satellite–earth segment may only be a part. It allows the user to send and receive data with a greatly reduced probability of error, and a very high probability that uncorrected errors are identified and located within a block of data, so that the existence of an error is known even if the exact bit or word in error cannot be determined. The penalty for the user is a reduced transmission rate, just as in FEC. Implementation of error correction by use of error detection and retransmission requires the use of protocols. A protocol is an agreed-upon set of actions that define how each end of the data link proceeds so that data are transmitted in an accurate and ordered fashion through the link.

Error detection is readily accomplished using the coding techniques described in the previous sections. In most communications systems it is not sufficient simply to detect an error; it must be corrected. When an error-detection code is used, a retransmission of the block of data containing the error must be made so that the correct data are acquired at the receiving terminal. The usual technique for obtaining a retransmission is for the receiving terminal to send an *acknowledge* (*ACK*) signal to the transmitting end when it receives an error-free block of data. If an error is detected in the block, a *not acknowledge* (*NAK*) signal is sent, which triggers a retransmission of the erroneous block of data. This is called an automatic repeat request (ARQ) system.

Such systems work well on terrestrial data links with relatively low data rates and short time delays. Their implementation on satellite links is more difficult due to the long transmission delay; as a result FEC is generally preferred for satellite paths.

There are three basic techniques for retransmission requests, depending on the type of link used. In a one-way, *simplex link*, the ACK or NAK signal must travel on the same path as the data, so the transmitter must stop after each block and wait for the receiver to send back a NAK or ACK before it retransmits the last data block or sends the next one. With a one-way delay of 240 ms on a satellite link, the data rate of such a system will be very low and is suitable only for links in which data are generated slowly, as when someone is typing on a terminal keyboard. Satellite links usually establish two-way communication (duplex channel) by the use of FDM, SCPC, or TDM, as discussed in Chapter 6. The ACK and NAK signals can be sent on the return channel while data are sent on the go channel. However, if the data rate is high, the acknowledgment will arrive long after the block to which it relates was transmitted.

In a *stop-and-wait* system, the transmitting end sends a block of data and waits for the acknowledgment to arrive on the return channel. The delay is the same as in the simplex case, but implementation is simpler. Figure 7.13 shows an example of a stop-and-wait sequence.

In a *continuous transmission* system using the *go-back-N* technique, data are sent in blocks continuously and held in a buffer at the receive end of the link.

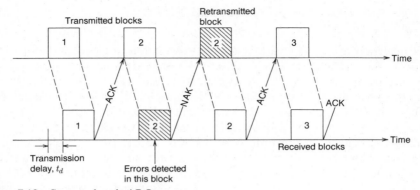

Figure 7.13 Stop-and-wait ARQ system.

Each data block is checked for errors as it arrives, and the appropriate ACK or NAK is sent back to the transmitting end, with the block number appended. When a NAK(N) is received, the transmit end goes back to block N and retransmits all subsequent blocks, as illustrated in Figure 7.14. This requires the transmitter on a satellite link to hold at least 480 ms of data, to allow time for the data to reach the receive end and be checked for errors and for the acknowledgment to be sent back to the transmit end. Since there is a delay in transmission only when a NAK is received, the *throughput* on this system is much greater than with the stop-and-wait method. Throughput is the ratio of the number of bits sent in a given time to the member that could theoretically be sent over an ideal link.

If sufficient buffering is provided at both ends of the link, only the corrupted block need be retransmitted. This system is called *selective repeat* ARQ. In Figure 7.14, block 3 is corrupted, and blocks 4, 5, 6, and 7 are transmitted before the NAK message is received. At this point, we could transmit block 3 only if blocks 4, 5, 6, and 7 are stored at the receive end. On receiving a correct version of

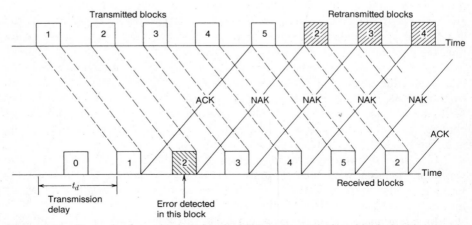

Figure 7.14 Example of a go-back-N-blocks ARQ system. N is three blocks in this example.

block 3, the receive buffer substitutes it for the corrupted version and releases the data for retransmission. In systems handling data rates of megabits per second, the buffer requirements for continuous transmission systems become very large.

Reference 15 contains a good survey of error-detection techniques for use in satellite communication systems and of the various ARQ systems that can be implemented. Some hybrid ARQ systems are described that combine FEC with retransmission of blocks when uncorrected errors are detected. This combines the error-correction properties of the FEC code for a limited number of errors with the error-detection properties of the same code when too many errors are present for all of them to be corrected.

Figures 7.15 and 7.16 [15] show how throughput can be increased as the BER increases for a number of hybrid ARQ systems. In Figure 7.15, curves 1 through 6 are for selective repeat ARQ systems. The improvement in performance over the go-back-N system is clearly shown. Curve 1 is an ideal selective repeat system with infinite buffering, and curves 2 and 3 show the effect of finite buffer size (512 and 1024 bits) when selective repeat ARQ is used. Curves 4, 5, and 6 are for selective repeat ARQ systems with finite buffer (512 bits) and FEC of 3, 5, and 10 bits per block [16]. In each case, the block length n is 1024 bits.

Figure 7.16 shows the effect of using convolutional FEC codes with selective repeat ARQ. The block length is 1024 bits with 1000 data bits. There is a marked

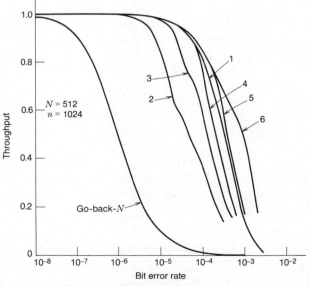

Figure 7.15 Throughput of various ARQ schemes with buffer size $N = 512$ (curves 2, 4, 5, 6), $N = 1024$ (curve 3), and infinite buffer size (curve 1). Curves 1, 2, 3 are all selective repeat ARQ without error correction; curves 4, 5, and 6 are for error correction parameter $t = 3, 5$, and 10. (Reprinted with permission from S. Lin, D. J. Costello, Jr., and M. J. Miller, "Automatic-Repeat-Request Error-Control Schemes," *IEEE Communications Magazine*, **20**, p. 13 (December 1984). Copyright © 1984 IEEE.)

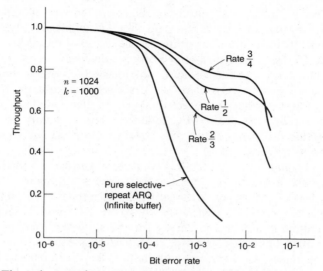

Figure 7.16 Throughput performance of three hybrid ARQ schemes using convolutional codes. (Reprinted with permission from S. Lin, D. J. Costello, Jr., and M. J. Miller, "Automatic-Repeat-Request Error-Control Schemes," *IEEE Communications Magazine*, **20**, p. 14 (December 1984). Copyright © 1984 IEEE.)

improvement over the simpler FEC codes when the BER falls below 10^{-4}, and throughput is maintained above 0.75 for BERs down to 10^{-2} with the rate 3/4 code. A complex decoder is needed with convolutional codes, as discussed in Section 7.7.

Example 7.8.1
Calculate the frequency of retransmission, throughput, and buffer requirements of a satellite link capable of carry data at rates of

(a) 24 kbps   (b) 1 Mbps

when a block length of 127 bits is used and the one-way path delay is 240 ms, for a bit error rate of 10^{-4} and a double-error-detecting code (127, 120) using the following ARQ schemes.

1. Stop and wait.
2. Continuous transmission with transmit buffer only (go-back-N).
3. Continuous transmission with buffers at both ends of the link (selective repeat).

For a 127-bit code block, the probability of one or two errors is given by Eq. (7.4)

$$P_e(k) = \binom{n}{k} p^k (1 - p)^{n-k}$$

where $k = 2$, $n = 127$, and p is the probability of a single bit error, which is 10^{-4} in this example.

Thus the probability of an error being detected in the block of 127 bits is

$$P_e(1 \text{ or } 2 \text{ errors}) = 127 \times 0.0974 \times 10^{-4} + \frac{127 \times 126}{2} \times 0.095 \times 10^{-10}$$

$$= 1.245 \times 10^{-3}$$

This means that, on average, one in every 803 received blocks has one or two detectable errors.

1. In a stop-and-wait system we must send 127 bits and wait for an acknowledgment. Transmission of 127 bits takes approximately 5 ms at 24 kbps and 127 μs at 1 Mbps. Waiting for the acknowledgment takes 480 ms, so in both systems the transmission rate is approximately two blocks per second, or 254 bps. A block error would be detected, on average, after transmission of 803 blocks, which takes 400 s. Thus the throughput of the system is dominated by the path delay and waiting time when no transmission takes place. The inefficiency of stop and wait methods on satellite links rules out this technique completely for continuous data transmission.

2. Go-back-N system.
   a. The time to send 803 blocks at 24 kbps is 4.25 s. Thus every 4.25 s (on average) we must stop and wait 480 ms for a retransmission of 94 blocks of data. This slows the throughput of the system by about 12 percent. We need a buffer for $94 \times 127 = 11,938$ bits at each end of the link.
   b. If the data rate is 1 Mbps, we will detect block errors every 100 ms; since it takes 480 ms to call for a retransmission, the system spends most of the time retransmitting data, giving a throughput of about 172 kbps, on average. This is well below the potential capacity of the 1 Mbps link. We need a buffer for about 480k bits at the transmitter.

3. Selective repeat system. The only time lost in a selective repeat system is in the retransmission of blocks in which an error occurred. For an error in every 803 blocks, the efficiency is $803/804 = 99.87$ percent.
   a. In the 24 kbps system, the average data rate is 23.97 kbps. Transmit and receive buffers must hold 11,520 bits.
   b. In the 1 Mbps system, the average data rate is 998.7 kbps. Transmit and receive buffers must hold 480k bits.

7.9 SUMMARY

The transmission of data over a satellite communication link is likely to result in some errors occurring in the received data, for at least a small percentage of time, because of noise added by the transmission system. Many links guarantee only 10^{-6} bit error rate and may not achieve this accuracy during periods of rain or other propagation disturbances. Bit errors contribute to the baseband (S/N) ratio when digital speech is sent, but it is rarely necessary to correct bit errors in speech; the listener can make such corrections because there is a lot of redundancy in speech. When data are sent over a link, the receiving terminal does not know in advance what form the data take and can only detect or correct errors if extra, redundant bits are added to the transmitted data.

Coding of data provides a means of detecting errors at the receiving terminal. Error-detecting codes allow the presence of one or more errors in a block of data bits to be detected. Error-correcting codes allow the receiving terminal equipment to locate and correct a limited number of errors in a block of data. When error detection is employed, some form of retransmission scheme is needed so that the data block can be sent again when it is found to be in error. Retransmission schemes use ARQ (automatic repeat request) techniques and are easiest to apply in packet switched data networks, where data are not transmitted in real time. The long round-trip delay (480 ms) in a satellite link makes simple stop-and-wait systems unattractive for real-time data transmission. Throughput can be increased by providing data storage at both ends of the link and using continuous transmission in which corrupted data blocks are retransmitted by interleaving them with subsequent data block transmissions.

Forward error correction (FEC) provides a means of both detecting and correcting errors at the receiving terminal without retransmission of data. FEC codes add redundant parity check bits to the data bits in a way that allows errors to be located within a codeword. In general, twice as many errors can be detected by an FEC code as can be corrected. FEC has the advantage over error detection that a single unit at each end of the link (a codec) can insert and remove the FEC code and make corrections as required. Because more errors can be detected than can be corrected, hybrid schemes using both FEC and ARQ have been proposed. ARQ schemes can achieve virtually error-free transmission when the link error rate is not excessive. Provided that all errors introduced by the link are detected, retransmission of the corrupted data block can be repeated until the data are received correctly. In an FEC system, there is always a finite possibility that an error will not be corrected, resulting in incorrect data being received.

Linear block codes are a class of error-detecting and -correcting codes that are easy to implement when the codewords are short. Hamming codes fall into this group. Binary cyclic codes are popular linear block codes for long codewords or data blocks. They can be generated and detected using shift registers and logic gates. The BCH and Golay codes are examples of binary cyclic codes.

Convolutional encoding is a more powerful technique than linear block encoding. The information contained in any one data bit is spread through several bits of the codeword. However, decoding convolutional codes is a complex process that requires decisions on the most likely transmitted data sequence when a codeword is received in error. Convolutional encoding achieves the best improvement in error rate, especially when the link BER is high. It provides good resistance to interference and deliberate jamming, making it popular in military communication systems.

Interference tends to cause burst errors in which many sequential bits are corrupted. Special burst-error correction codes are available with the capability of correcting errors in a number of adjacent bits. Scrambling and interleaving of data bits are other ways in which the effect of burst errors can be reduced.

REFERENCES

1. C. E. Shannon, "A Mathematic Theory of Communications," *Bell Syst. Tech. J.*, Part I, 379–423; Part II, 623–656 (1948).
2. L. W. Couch, *Digital and Analog Communication Systems*, Macmillan, New York, 1983.
3. K. S. Shamnugam, *Digital and Analog Communication Systems*, John Wiley & Sons, New York, 1979.
4. S. Lin, and D. J. Costello, Jr., *Error Control Coding: Fundamentals and Applications*, Prentice-Hall, Englewood Cliffs, NJ, 1983.
5. A. Hocquenghem, "Codes Corecteurs d'Erreurs," *Chiffres*, **2**, 147–156 (1959).
6. R. C. Bose and D. K. Ray-Chaudhuri, "On a Class of Error Correcting Binary Group Codes," *Info. Control*, **3**, 68–79 (March 1960).
7. W. W. Peterson and E. J. Weldon, Jr., *Error-Correcting Codes*, 2nd ed., MIT Press, Cambridge, MA, 1970.
8. K. Feher, *Digital Communications: Satellite/Earth Station Engineering*, Prentice-Hall, Englewood Cliffs, NJ, 1983.
9. T. Maratani, H. Saithoh, K. Koga, Y. Mizuno, and Y. J. S. Snyder, "Application of FEC Coding to the Intelsat TDMA Systems," *Proc. 4th Intl. Conf. on Digital Satellite Comm.*, Montreal (October 1978).
10. Marlin P. Ristenbatt, "Alternatives in Digital Communication," *Proceedings of the IEEE*, **61**, 703–721 (June 1973). (Reprinted in [11], pp. 212–230.)
11. Harry L. VanTrees, Ed., *Satellite Communications*, IEEE Press, New York, 1979.
12. J. J. Spilker, Jr., *Digital Communications by Satellite*, Prentice-Hall, Englewood Cliffs, NJ, 1977.
13. G. David Forney, Jr., "The Viterbi Algorithm," *Proceedings of the IEEE*, **61**, 268–278 (March, 1973). (Reprinted in [11], pp. 286–296.)
14. Jerrold A. Heller and Irwin Mark Jacobs, "Viterbi Decoding for Satellite and Space Communications," *IEEE Transactions on Communications* COM-19, pp. 835–848, October 1971. (Reprinted in [11], pp. 273–286.)
15. S. Lin, D. J. Costello, Jr., M. J. Miller, "Automatic-Repeat-Request Error-Control Schemes," *IEEE Communications Magazine*, **20**, 5–7 (December 1984).
16. S. Lin and P. S. Yu, "A Hybrid ARQ System with Parity Retransmission for Error Control of Satellite Channels," *IEEE Transactions on Communications*, *COM-30*, pp. 689–694, July 1982.

PROBLEMS

1. Alphanumeric characters are transmitted as 7-bit ASCII words, with a single parity bit added, over a link with a transmission rate of 9.6 kbps.
a. How many characters are transmitted each second?
b. If a typical page of text contains 500 words with an average of five characters per word and a space between words, how long does it take to transmit a page?
c. If the bit error rate on the link is 10^{-5}, how many characters per page are detected as having errors?
How many undetected errors are there?
d. On average how many pages can be transmitted before
  (i) a detected error occurs
  (ii) an undetected error occurs?
e. If the BER increases to 10^{-3}, how many detected and undetected errors are there in a page of the text?

2. A (6, 3) block code is formed from a generator matrix

$$G = \begin{bmatrix} 100 & 110 \\ 010 & 101 \\ 001 & 011 \end{bmatrix}$$

a. Draw up a table of the eight codewords in this code.
b. A message 111 000 100 001 110 011 is to be sent using this code. Convert the message to its coded form.
c. What is the minimum distance of this code? How many errors can be detected in a codeword? How many errors can be corrected in a codeword?

3. The following series of codewords is received on a link using the (7, 4) Hamming code given in Table 7.2.

   1100001   1001100   1110011   1110100   0011101

a. Are there any errors in the received codewords? (Check against Table 7.2.) Which words contain errors?
b. Calculate the syndrome for any code words that have detected errors.
c. Find the correct codewords using the parity check matrix in Example 7.1. Check your own answer by comparison with Table 7.2.

4. Find the weight and minimum distance of the systematic (7, 4) cyclic code in Table 7.3.
a. How many errors can this code detect?
b. How many errors can it correct?

5. Analysis of a 56-kbps data link shows that it suffers burst errors that corrupt several adjacent bits. The statistics for burst errors on this link are given in Table P.5.

Table P.5
Statistics for Burst Errors on a Link in Problem 5

No. Adjacent Bits Corrupted	Probability of Occurrence
2	4×10^{-2}
3	2×10^{-3}
4	3×10^{-4}
5	1×10^{-6}
6	2×10^{-9}
7	5×10^{-11}
8	1×10^{-12}
9	3×10^{-14}
10	2×10^{-17}

a. Using Table 7.4, select a burst error correcting code that will reduce the probability of an uncorrected burst error below 10^{-10}.
b. Calculate the data rate for messages sent over the link using the code you selected.
c. Estimate the average bit error rate for the coded transmission.

6. A rate $\frac{1}{2}$ convolutional encoder is connnected as shown in Figure P7.6

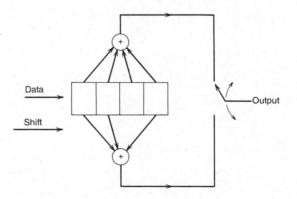

Figure P7.6

a. Determine and draw the state-diagram representation of the encoder.
b. Determine how the encoder would encode the binary sequence 10110. Assume that the left-most bit is transmitted first.
c. Assuming that the encoder started with 0000 in the shift register prior to accepting the first data bit, find the most probable decoding of a received message 00 11 10 10 00. Assume that the left-most bit was received first. How many bit errors does the received encoded message contain?

7. A satellite link carries packet data at a rate of 256 kbps. The data are sent in 255-bit blocks using a (255, 247) code that can detect three errors. The

probability of a single bit error, p_b, varies from 10^{-6} under good conditions to 10^{-3} under poor conditions. The one way link delay is 250 ms.

    a.   If no error detection is used, what is the message data rate for the link?

    b.   For a link BER of 10^{-6}, find the probability of detecting an error in a block of 255 bits. Hence find how often an error is detected.

    c.   Estimate the probability that a block of 255 bits contains an undetected error when the link BER is 10^{-3}.

    d.   Find the message data throughput when the link BER is 10^{-6} and a stop-and-wait ARQ system is used, assuming one retransmission always corrects the block.

    8.   Repeat Problem 7 using a block length of 1024 bits and a (1024, 923) code that can detect 22 errors in a block.

The (1024, 923) code can correct 10 errors. Find the average number of blocks that can be transmitted before an uncorrected error occurs when the BER is 10^{-3}. Repeat the analysis for a BER of 10^{-2}. *Note:* The probability of an unlikely event (11 or more errors in a block of 1024 data bits with $p_b = 10^{-3}$ in this case) can be calculated from the Poisson distribution more easily than from the binomial distribution. The Poisson distribution is given by

$$P(x = k) = \frac{\lambda^k e^{-\lambda}}{k!}$$

where $\lambda = Np_b$. N is the block length, and k is the number of bits in error.

8

PROPAGATION ON SATELLITE–EARTH PATHS AND ITS INFLUENCE ON LINK DESIGN

In Chapter 4 we introduced the concept of a link power budget. The key equation in that development was Eq. (4.11), repeated here as Eq. (8.1) and by this time presumably familiar to the reader.

$$P_r = \text{EIRP} + G_r - L_p - L_a - L_{ra} \text{ dBW} \tag{8.1}$$

This equation indicates how the received power in dBW depends on the transmitter EIRP, the receiving antenna gain, and the various losses in the system. Everything on the right-hand side is independent of time except the atmospheric loss, L_a.

Usually L_a is written as the sum of a constant term called the *atmospheric absorption* and a variable term called the *attenuation*, A. At most frequencies of commercial interest, the atmospheric absorption is relatively unimportant (a few tenths of a decibel), and it disappears into the general uncertainty about the other terms in Eq. (8.1). The attenuation is zero in clear weather, but it can increase to large values during unfavorable propagation conditions. Rapid fluctuations in attenuation are called *scintillations*, while longer term increases are called *fades*. Rain fades are a particular problem above 10 GHz.

Attenuation can affect all types of satellite links; those that employ orthogonal polarizations to transmit two different channels on a common or overlapping frequency band are also degraded by *depolarization*. This is a conversion of some of the energy in a transmitted signal from one polarization to another. Absent under ideal propagation conditions, it can cause co-channel interference and crosstalk in dual-polarized satellite links. Rain is a primary cause of depolarization.

Both attenuation and depolarization come from interactions between the propagating electromagnetic waves on a satellite link and whatever is in the atmosphere at the time. The atmospheric constituents may include free electrons, ions, neutral atoms, molecules, and hydrometeors (a term that seems pedantic but that conveniently describes raindrops, snowflakes, sleet, hail, free-floating ice crystals, etc.), and most of these come in a variety of sizes. Their interactions with radio waves depend strongly on frequency, and effects that dominate 30-GHz propagation, for example, may be negligible at 4 GHz. The converse is also true. With one major exception (Faraday rotation, Section 8.2), almost all propagation effects become more severe as frequency increases. A convenient dividing line is 10 GHz, and most authorities discuss propagation below 10 GHz separately from propagation above 10 GHz. But this division is arbitrary, and 10 GHz is not a sharp cutoff frequency for any atmosphere propagation phenomenon.

The literature of satellite-path propagation is extensive, and we can only cover some of the more important points here. References 1 and 2 summarize almost all aspects of the subject and provide extensive bibliographies. For an excellent general treatment of the subject of microwave propagation, see reference 3.

8.1 QUANTIFYING ATTENUATION AND DEPOLARIZATION

Attenuation is the decibel difference between the power received at a given time t and the power received under ideal propagation ("clear weather") conditions.

$$A(t) = P_{r_{\text{clear-weather}}} \, \text{dBW} - P_r(t) \, \text{dBW} \tag{8.2}$$

Depolarization is more difficult to quantify and its description requires some background information.

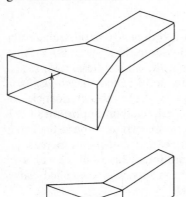

Figure 8.1 Orthogonally polarized waveguide horn antennas. The top horn transmits and receives vertically polarized waves. Turned on its side as in the lower sketch, it transmits and receives horizontally polarized waves. The arrow indicates the origin of the electric field vector.

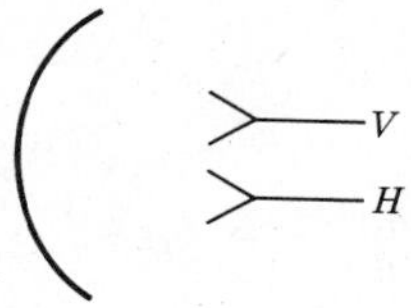

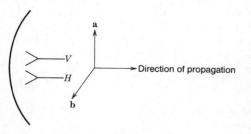

Figure 8.2 Schematic representation of a dual-polarized antenna. A pair of orthogonally polarized horns illuminates a common parabolic reflector.

Orthogonal polarization frequency sharing works because any transmitting antenna radiates waves with a polarization that is characteristic of that antenna and, when receiving, the same antenna is "blind" (i.e., cannot receive) waves whose polarization is orthogonal to the characteristic polarization [4]. Consider, for example, the pair of horn antennas shown in Figure 8.1. The top horn transmits vertical polarization and is blind to horizontal, and the bottom horn transmits horizontal polarization and is blind to vertical. (There are infinitely many orthogonal polarization pairs, of course; horizontal and vertical are just easy to visualize.) If the two horns were used to illuminate a single reflector, the result would be a dual-polarized antenna, drawn schematically as in Figure 8.2. Two such antennas (one for transmitting and one for receiving) would allow two channels (one with vertical polarization and another with horizontal polarization) to share the same frequency. If the antennas were ideal, there would be no interference between the channels.

To illustrate the process by which depolarization is measured [5, 6], imagine a dual-polarized antenna transmitting orthogonally polarized signals. We will call the two polarizations V and H for vertical and horizontal, but these could be any orthogonal pair. Let the complex phasor amplitudes of the transmitted electric field vectors with polarization V and H be **a** and **b**, respectively, as shown in Figure 8.3. The antenna is excited so that **a** and **b** are equal.

If the transmission medium between the antennas were clear air, at the receiving antenna **a** would give rise to a V polarization wave of amplitude $\mathbf{a}_c$ and **b** would cause an H polarization wave of amplitude $\mathbf{b}_c$. The subscript c stands for co-polarized; these fields have the same polarization as their transmitted counterparts. See Figure 8.4.

If rain or ice or some other depolarizing medium is in the propagation path, some of the energy in **a** will couple into a small H polarized field component whose amplitude at the receiving antenna is $\mathbf{a}_x$, and **b** will give rise to a small V polarized component $\mathbf{b}_x$. The unwanted $\mathbf{b}_x$ will interfere with the wanted $\mathbf{a}_c$ and the unwanted $\mathbf{a}_x$ will interfere with the wanted $\mathbf{b}_c$. This interference will cause

Figure 8.3 Fields excited by a dual-polarized antenna. The field radiated by the V horn has the vertically polarized electric field vector indicated by **a** and the field radiated by the H horn has the electric field vector indicated by **b**.

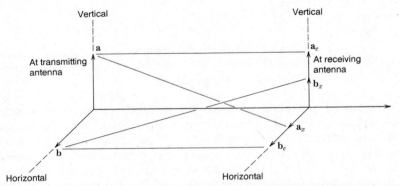

Figure 8.4 The transmitted fields **a** and **b** produce co-polarized components $\mathbf{a}_c$ and $\mathbf{b}_c$ and cross-polarized components $\mathbf{a}_x$ and $\mathbf{b}_x$ at the receiving antenna. With perfect antennas and in the absence of depolarization $\mathbf{a}_x$ and $\mathbf{b}_x$ would be zero.

crosstalk on an analog link and raise the BER on a digital link. This generation of unwanted cross-polarized components is called depolarization.

The measure of depolarization that is most useful for analyzing communications systems is the *cross-polarization isolation, xpi*. In terms of complex phasor field amplitudes, it is given by Eq. (8.3) for the *V* polarized channel and by Eq. (8.4) for the *H* polarized channel.

$$xpi_V = \mathbf{a}_c/\mathbf{b}_x \qquad (8.3)$$

$$xpi_H = \mathbf{b}_c/\mathbf{a}_x \qquad (8.4)$$

The *xpi* values are commonly expressed in decibels; for example

$$XPI_V = 20 \log_{10}|\mathbf{a}_c/\mathbf{b}_x| \text{ dB} \qquad (8.5)$$

Physically the *XPI* is the decibel ratio of wanted power to unwanted power in the same channel. The larger its value, the less interference there is and the better a communications system will perform.

XPI is difficult to measure; while at least two groups did measure it with the COMSTAR family of satellites [7, 8], most satellite propagation experiments transmit a single polarization (say *V*) and measure the co-polarized and cross-polarized components of the received signal (say $\mathbf{a}_c$ and $\mathbf{a}_x$). This leads to the *cross-polarization discrimination, xpd*. For the case where *V* polarization alone is transmitted (Figure 8.5),

$$xpd_V = \mathbf{a}_c/\mathbf{a}_x \qquad (8.6)$$

or in decibels

$$XPD_V = 20 \log_{10}|\mathbf{a}_c/\mathbf{a}_x| \text{ dB} \qquad (8.7)$$

In most practical situations *XPI* and *XPD* are equal [6, 9], and they are usually treated as the same thing and sometimes called "isolation."

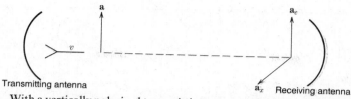

Figure 8.5 With a vertically polarized transmitting antenna, the ratio of the wanted received field component $\mathbf{a}_c$ to the unwanted component $\mathbf{a}_x$ is called the cross-polarization discrimination, *xpd*.

In practice real antennas do not transmit polarization pairs that are exactly orthogonal, and the isolation under clear-weather conditions is not infinite. Clear-weather isolations of 30 to 35 dB may be realized under operational conditions with precision antennas; off-the-shelf models will usually deliver at least 20 dB.

8.2 PROPAGATION EFFECTS THAT ARE NOT ASSOCIATED WITH HYDROMETEORS

In this section we will discuss propagation effects that are not caused by raindrops or ice crystals.

Atmospheric Absorption

At sufficiently high frequencies, electromagnetic waves interact with the molecules of atmospheric gases to cause attenuation. These interactions occur at resonance frequencies and are apparent in plots of zenith (90° elevation angle) attenuation versus frequency like that shown in Figure 8.6. The important resonances below 100 GHz are those of water vapor centered at 22.235 GHz and oxygen between 53.5 and 65.2 GHz [1]. The allocated bands for satellite communications were carefully chosen to avoid these frequencies, and thus gaseous absorption is unimportant in almost all commercial links. The resulting attenuation is commonly less than 1 dB, and this is usually within the overall uncertainty in a power budget. Readers for whom extremely small losses are important or who are concerned with specialized links operating in the water vapor or oxygen bands where the losses are not small should consult references 1 and 10.

Tropospheric Multipath and Scintillation Effects

At elevation angles below about 5°, the tropospheric portion of the propagation path is so long that turbulence may cause the rapid amplitude and phase variations called scintillations. See Figure 8.7 [11] for an example. In addition, waves may arrive simultaneously at a receiving antenna via several propagation paths and, by interfering with each other, give rise to multipath fading, also called frequency-selective fading. For these reasons, commercial links try to avoid low elevation angles. The reader who must design links to operate below 5° elevation should consult reference 1.

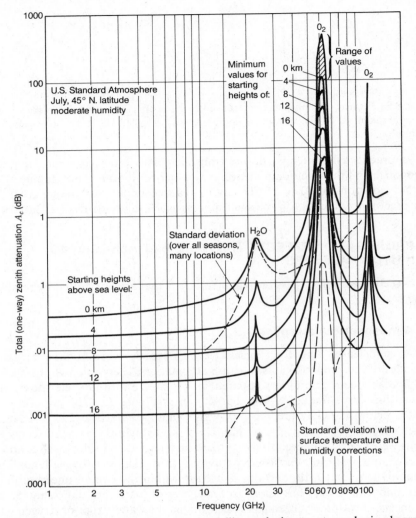

Figure 8.6 Total zenith attenuation on a satellite path due to atmospheric absorption. (Reprinted from Louis J. Ippolito, R. D. Kaul, and R. G. Wallace, *Propagation Effects Handbook for Satellite Systems Design* (NASA Reference Publication 1082(03)), National Aeronautics and Space Administration, Washington, DC, June 1983. Courtesy of NASA.)

Land and Sea Multipath

At higher elevation angles, multipath propagation could result through reflections from surrounding terrain; these are, however, eliminated by the highly directional antennas used at commercial earth stations. On the other hand, multipath could be a severe problem in the proposed land mobile satellite systems that will operate at frequencies below 4 GHz with low-gain antennas and in proposed low-cost maritime systems that will also use small antennas.

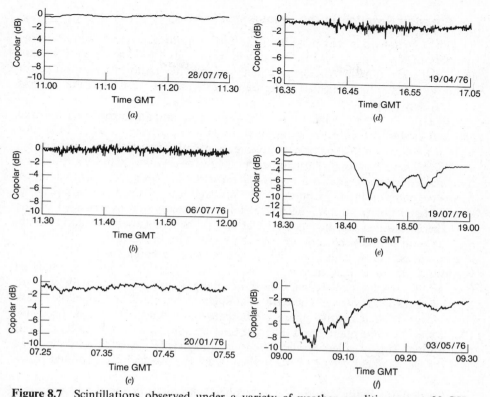

Figure 8.7 Scintillations observed under a variety of weather conditions on a 30-GHz downlink from ATS-6. (Reprinted with permission from T. Pratt and D. J. Browning, "Copolar Attenuation and Radiometer Measurements at 30 GHz for a Slant Path to Central England," *Proceedings ATS-6 Meeting* (ESA SP-131), ESTEC, Noordwijk, The Netherlands, pp. 15–20 (October 1977). Courtesy ESA.) (*a*) Clear-weather copolar signal with low scintillation. (*b*) Clear-weather copolar signal with high scintillation. (*c*) Copolar scintillation in cloud. (*d*) Copolar scintillation in cloud. (*e*) Copolar scintillation in rain. (*f*) Copolar scintillation in rain.

Multipath Effects in System Design

For analog FM systems, multipath fading is similar to rain fading in that its effects can be eliminated by providing a clear-weather margin sufficient to overcome the fades that are anticipated for a stated percentage of time. While the statistics of satellite link multipath fading are not well established, this procedure is routinely followed for terrestrial radio links and it should be possible to adapt it to satellite paths without difficulty. But the response of digital systems to multipath is different from analog, and terrestrial links designed according to the established procedures for analog systems have failed to provide the desired reliability. See reference 12 for an excellent review of the situation.

Faraday Rotation in the Ionosphere [13, 14]

The ionosphere is that portion of the earth's atmosphere that contains large numbers of electrons and ions. It extends from about 100 km to over 400 km above the earth. The ionosphere completely dominates radio propagation below about 40 MHz, but its effects on the frequencies used by most satellite systems are relatively minor.

Electrically the ionosphere is an inhomogeneous and anisotropic plasma, and an exact analysis of wave propagation through it is extremely difficult [13]. For a given frequency and direction of propagation with respect to the earth's magnetic field, there exist two characteristic polarizations. Waves with these polarizations, called *characteristic waves*, propagate with their polarization unchanged. Any wave entering the ionosphere can be resolved into two components with the characteristic polarizations. The phase shift and attenuation experienced by the characteristic waves can be calculated at any point along the propagation path, and the total field can be computed as the vector sum of the fields of the characteristic waves. This total field can be interpreted as an attenuated and depolarized version of the wave that entered the ionosphere. Thus when a linearly polarized (LP) satellite path signal reaches the ionosphere, it excites waves with the two characteristic polarizations. These travel at different velocities, and when they leave the ionosphere their relative phase is different from when they entered. The wave that leaves the ionosphere has a different polarization from the LP wave that was transmitted. This is called *Faraday rotation*, and its effect is essentially the same as if the electric field vector of the transmitted LP wave had been rotated by an angle ϕ.

For a path length Z m (in the ionosphere) the rotation angle ϕ is given by

$$\phi = \int \left(\frac{2.36 \times 10^4}{f^2} \right) ZNB_0 \cos \theta \, dz \text{ rad} \qquad (8.8)$$

Here θ is the angle between the geomagnetic field and the direction of propagation, N is the electron density in electrons/cubic meter, B_0 is the geomagnetic flux density in Teslas, and f is the operating frequency in Hz.

Typical values for satellite paths are as follows. Since ϕ varies inversely with f^2, these numbers are easy to scale to other frequencies.

	Average	Peak Seasonal	Solar Flare
Northern Europe at 11 GHz	0.3°	1°	1.6°
United States at 4 GHz	2°	6°	8°

The polarizations of an earth station antenna can be adjusted to compensate for the Faraday rotation observed under average conditions. The XPD that results when the polarization angle of an LP wave changes by an amount $\Delta\phi$ is given by

$$XPD = 20 \log_{10}(\cot \Delta\phi) \qquad (8.9)$$

Hence the 6° change from average conditions associated with a solar flare would reduce the XPD on a 4 GHz link to about 19.6 dB.

Ionospheric Scintillations [15]

Irregularities in the upper part of the ionosphere can cause scintillations with peak-to-peak fluctuations as large as 6 dB on 6 GHz downlinks to stations near the geomagnetic equator. These tend to occur about one hour after local sunset and are most frequent around the spring and fall equinoxes. They are generally not significant in the United States or in Europe.

8.3 RAIN AND ICE EFFECTS

At frequencies above 10 GHz rain is the dominant factor in satellite propagation. Rain propagation has been studied intensively since the late 1960s, and we can only summarize the literature in a few pages here. For an excellent tutorial and bibliography, consult reference 1. The effects of high-altitude ice crystals are less important, and we will discuss them after rain.

Characterizing Rain

Much of what meteorologists know about rain involves rain *accumulation*, since this is most important to farmers and hydrologists, among others. Unfortunately, rain accumulation is of little use to satellite-link designers, since it is the *rainfall rate* rather than the accumulation that determines propagation effects. If we imagine a large flat-bottomed container open to the rain (Figure 8.8), the rain rate R is the rate at which the water level in the container is rising. Rain rate is usually expressed in units of millimeters per hour (mm/h). Peak values of 100 to 150 mm/h may be expected for short periods during summer thunderstorms in the mid-Atlantic region of the United States.

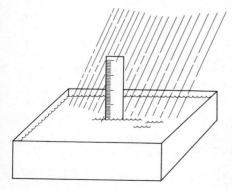

Figure 8.8 A simple device for measuring rain rate. The ruler measures the water level in a flat-bottomed container whose top is open to the rain. The rain rate is the rate at which the water level is rising.

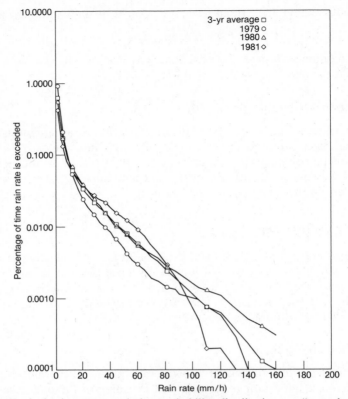

Figure 8.9 Typical rain rate cumulative probability distributions or "exceedance" curves. These were measured at the authors' Blacksburg, Virginia, earth station.

The long-term behavior of rain rate is described by what is properly called a *cumulative probability distribution* and popularly known as an *exceedance curve*. This gives the percentage of time (usually the percentage of one year) that the rain rate exceeds a given value. A typical plot for Blacksburg, Virginia, appears in Figure 8.9. Some measured values are as follows.

Percentage Time	Absolute Time Out of 1 Year	R(mm/hr) Exceeded for this Percentage Time
0.1	8.76 hr	8
0.01	52.56 min	43
0.001	5.26 min	106
0.0001	0.53 min	142

Mathematical models are available that give the exceedance curve for a given location based on the total rain accumulation and other meteorological data; see,

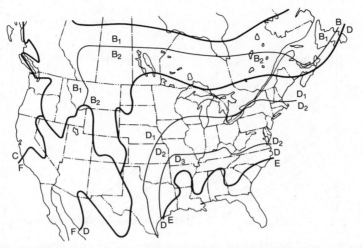

Figure 8.10   NASA rain rate climate regions for the continental United States and southern Canada. (Reprinted from Louis J. Ippolito, R. D. Kaul, and R. G. Wallace, *Propagation Effects Handbook for Satellite Systems Design* [NASA Reference Publication 1082(03)], National Aeronautics and Space Administration, Washington, DC, June 1983. Courtesy of NASA.)

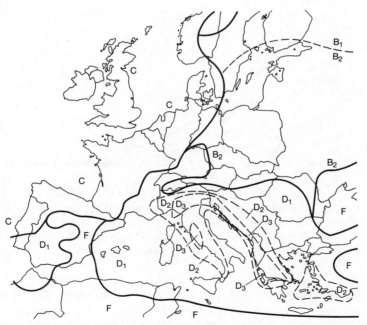

Figure 8.11   NASA rain rate climate regions for Europe. (Reprinted from Louis J. Ippolito, R. D. Kaul, and R. G. Wallace, *Propagation Effects Handbook for Satellite Systems Design* [NASA Reference Publication 1082(03)], National Aeronautics and Space Administration, Washington, DC, June 1983. Courtesy of NASA.)

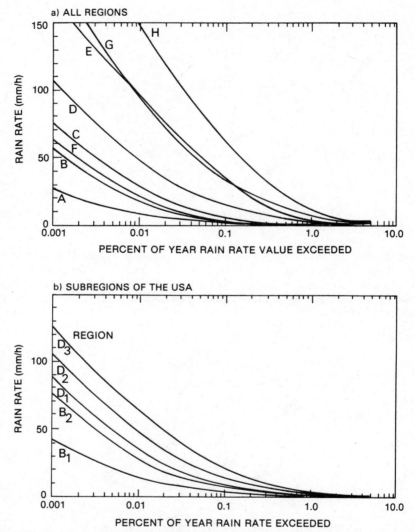

Figure 8.12 Rain rate cumulative probability distributions for the regions presented on the previous maps. (Reprinted from Louis J. Ippolito, R. D. Kaul, and R. G. Wallace, *Propagation Effects Handbook for Satellite Systems Design* [NASA Reference Publication 1082(03)], National Aeronautics and Space Administration, Washington, DC, June 1983. Courtesy of NASA.)

for example, reference 16. NASA [1] and CCIR [5, pp. 109ff] publish maps and tables for estimating the rain rate distribution at any location. These are not always consistent with each other. Figures 8.10 through 8.14 present some samples of the information that is available. Figures 8.10 and 8.11 show the rain climate regions for the United States, southern Canada, and Europe as presented in reference 1;

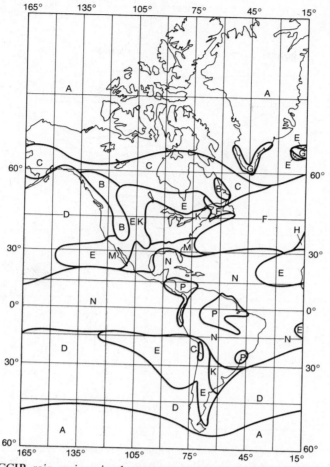

Figure 8.13 CCIR rain regions in the western hemisphere. (Reprinted with permission from International Radio Consultative Committee (CCIR), *Recommendations and Reports of the CCIR, 1982, Volume V, Propagation in Non-Ionized Media*, International Telecommunication Union, Geneva, Switzerland, 1982.)

Figure 8.12 and Table 8.1 provide the associated rain rate statistics. Figure 8.13 shows the CCIR rain regions in the western hemisphere, and Table 8.2 presents the corresponding percentage-of-time data. CCIR also publishes maps showing contours of rain rate exceeded for selected percentages of time; Figure 8.14 is an example.

Raindrop Distributions

Rain attenuation and depolarization occur because individual raindrops absorb energy from radio waves (i.e., the drops are heated) and because some energy

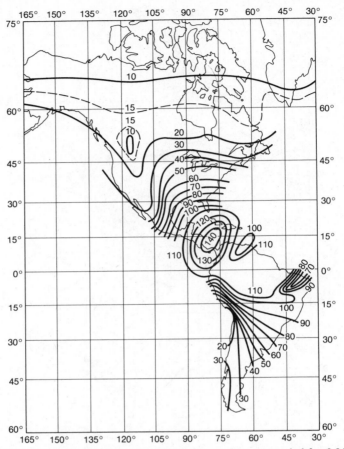

Figure 8.14 Contours of rainfall rate in millimeters per hour exceeded for 0.01 percent of the time as published by CCIR. (Reprinted with permission from International Radio Consultative Committee (CCIR), *Recommendations and Reports of the CCIR, 1982, Volume V, Propagation in Non-Ionized Media*, International Telecommunication Union, Geneva, Switzerland, 1982.)

in the waves is scattered out of the propagation path. These interactions depend on the number of raindrops encountered (and hence on the rain rate) and on their distribution of sizes and shapes.

Raindrop size and shape distributions have been measured in several ingenious experiments; see references 1 and 3 for references and more details. Generally the drop size distributions are exponential and take the form [3]

$$N(D) = N_0 \exp\left(\frac{-D}{D_m}\right) \text{mm}^{-1}\,\text{m}^{-3} \tag{8.10}$$

where D_m is the median drop diameter and $N(D)\,dD$ is the number of drops per cubic meter with diameters between D and $D + dD$ mm. The rain rate R is related

Table 8.1
Rain Rate Distribution Values Versus Percentage of Year That Rain Rate Is Exceeded

Percentage of Year	Rain climate region												Minutes per Year	Hours per Year
	A	B_1	B	B_2	C	D_1	$D=D_2$	D_3	E	F	G	H		
0.001	28.5	45	57.5	70	78	90	108	126	165	66	185	253	5.26	0.09
0.002	21	34	44	54	62	72	89	106	144	51	157	220.5	10.5	0.18
0.005	13.5	22	28.5	35	41	50	64.5	80.5	118	34	120.5	178	26.3	0.44
0.01	10.0	15.5	19.5	23.5	28	35.5	49	63	98	23	94	147	52.6	0.88
0.02	7.0	11.0	13.5	16	18	24	35	48	78	15	72	119	105	1.75
0.05	4.0	6.4	8.0	9.5	11	14.5	22	32	52	8.3	47	86.5	263	4.38
0.1	2.5	4.2	5.2	6.1	7.2	9.8	14.5	22	35	5.2	32	64	526	8.77
0.2	1.5	2.8	3.4	4.0	4.8	6.4	9.5	14.5	21	3.1	21.8	43.5	1052	17.5
0.5	0.7	1.5	1.9	2.3	2.7	3.6	5.2	7.8	10.6	1.4	12.2	22.5	2630	43.8
1.0	0.4	1.0	1.3	1.5	1.8	2.2	3.0	4.7	6.0	0.7	8.0	12.0	5260	87.7
2.0	0.1	0.5	0.7	0.8	1.1	1.2	1.5	1.9	2.9	0.2	5.0	5.2	10520	175
5.0	0.0	0.2	0.3	0.3	0.5	0.0	0.0	0.0	0.5	0.0	1.8	1.2	26298	438

Source: Louis J. Ippolito, R. D. Kaul, and R. G. Wallace, *Propagation Effects Handbook for Satellite Systems Design* (NASA Reference Publication 1082(03)), National Aeronautics and Space Administration, Washington, D.C., June 1983. Courtesy of NASA.

Table 8.2
Cumulative Rain Rate Statistics for the CCIR Rain Climate Regions

Percentage of time (%)	Rainfall Intensity Exceeded (mm/h)													
	A	B	C	D	E	F	G	H	J	K	L	M	N	P
1.0	—	1	—	3	1	2	—	—	—	2	—	4	5	12
0.3	1	2	3	5	3	4	7	4	13	6	7	11	15	34
0.1	2	3	5	8	6	8	12	10	20	12	15	22	35	65
0.03	5	6	9	13	12	15	20	18	28	23	33	40	65	105
0.01	8	12	15	19	22	28	30	32	35	42	60	63	95	145
0.003	14	21	26	29	41	54	45	55	45	70	105	95	140	200
0.001	22	32	42	42	70	78	65	83	55	100	150	120	180	250

Source: International Radio Consultative Committee (CCIR), *Recommendations and Reports of the CCIR, 1982, Volume V, Propagation in Non-Ionized Media*, International Telecommunication Union, Geneva Switzerland, 1982, p. 117. Reprinted with permission of the CCIR.

to $N(D)$ and to the terminal velocity $V(D)$ of falling drops in meters per second with diameter D by [3]

$$R = 0.6 \times 10^{-3} \pi \int D^3 V(D) N(D) \, dD \text{ mm/h} \tag{8.11}$$

Calculating Attenuation

The details of scattering and absorption by a single raindrop and the summation over the drop population that attenuation calculation requires are beyond the scope of this text. We will begin with the intermediate result that if the rain rate R is constant over a path of length L km, the attenuation A caused by the rain is given by

$$A = aR^b L \text{ dB} \tag{8.12}$$

The quantity aR^b is called the *specific attenuation* and has the units decibels per kilometer (dB/km). The coefficients a and b depend strongly on frequency and weakly on polarization, raindrop temperature, and other factors. Extensive tabulations are available, but for most practical purposes the following approximations work well [17].

$$a = \begin{cases} 4.21 \times 10^{-5} f^{2.42}, & 2.9 \leq f \leq 54 \text{ GHz} \\ 4.09 \times 10^{-2} f^{0.699}, & 54 \leq f \leq 180 \text{ GHz} \end{cases} \tag{8.13}$$

$$b = \begin{cases} 1.41 f^{-0.0779}, & 8.5 \leq f < 25 \text{ GHz} \\ 2.63 f^{-0.272}, & 25 \leq f < 164 \text{ GHz} \end{cases} \tag{8.14}$$

Values of f used in Eqs. (8.13) and (8.14) should be in GHz.

If the rain rate were constant along a satellite path of known length, then calculating the attenuation for a given ground rainfall rate by Eq. (8.12) through Eq. (8.14) would be straightforward. This is the usual situation for terrestrial radio links, but on satellite paths the rain rate changes with position and with time, and the length L of the portion of the path that contains rain is also variable. If this length is $L(t)$, and $R(y, t)$ is the rain rate in millimeters per hour (mm/h) at time t at a distance y km measured from the ground along the path, then the attenuation $A(t)$ is given by

$$A(t) = \int_0^{L(t)} a[R(y, t)]^b \, dy \text{ dB} \qquad (8.15)$$

Except in those experiments where a weather radar is available to measure $R(y, t)$, Eq. (8.15) is of little practical use. There is usually no relationship between simultaneous values of the attenuation $A(t)$ and the point rain rate $R(t)$ measured on the ground near a receiving antenna. This is because raindrops (particularly those caught in updrafts) require a surprisingly long time to fall from the upper part of a storm to a ground-based rain gauge. Typically the behavior of the attenuation at time t resembles the behavior of the ground rain rate of time $t + T$, where the delay T is variable and may be as long as 4 to 8 min. For this reason, predictive models for rain attenuation are statistical rather than instantaneous.

Since attenuation is caused by rain, both A and R exhibit the same kind of statistical behavior and both can be described by exceedance plots. Figure 8.15 shows such a plot for attenuation measured at the authors' earth station on three frequencies during the same one-year period.

The idea behind statistical models of rain attenuation is this. We can define a path average rain rate $R_p(t)$ so that

$$A(t) = \int_0^{L(t)} a[R(y, t)]^b \, dy = a[R_p(t)]^b L(t) \text{ dB} \qquad (8.16)$$

All the rain that passes through the propagation path should ultimately reach the ground. Hence the exceedance curve of the path average rain rate should be the same as the exceedance curve of the ground (point) rain rate. Therefore the attenuation value $A(P)$ equaled or exceeded P percent of the time should be proportional to the value of ground rain rate $R(P)$ equaled or exceeded for the same P percent of time. The proportionality factor between $A(P)$ and $a[R(P)]^b$ has units of kilometers (km) and is called the *effective path length*, L_{eff}. Thus

$$A(P) = a[R(P)]^b L_{eff} \text{ dB} \qquad (8.17)$$

The effective path length may be found from measured data by identifying and tabulating corresponding values of $A(P)$ and $R(P)$ with P as a parameter. Attenuation and rain rate values so paired are called *equal-probability values*. Figure 8.16 illustrates the process for finding them.

Most predictive models provide a relationship between $A(P)$ and $R(P)$ in the form of Eq. (8.17), but many avoid calling the constant of proportionality "effective path length." The full reasons for this usage are beyond the scope of this text, but basically they involve the dependence of effective path length on some factors that

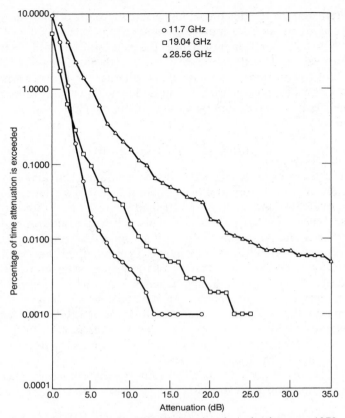

Figure 8.15 Cumulative attenuation distribution for the calendar year 1978 as measured at the authors' earth station on satellite downlinks at three frequencies. The 11.7-GHz path was at 33° elevation; the other two were at 45°.

are unrelated to path geometry. We will present two models here; for a compendium of all of the current ones the reader should consult reference 1.

The *SAM* model [18] was developed with NASA support to provide a simplified technique for hand calculation. It assumes that rain extends from the earth station's elevation H_0 km above sea level to an effective storm height H_e km. See Figure 8.17 for an illustration. For an elevation angle El, the path length L in rain is given by

$$L = \frac{H_e - H_0}{\sin (El)} \text{ km} \qquad (8.18)$$

Temperature decreases with height in the lower atmosphere, and the locus of points where the temperature reaches 0°C is called the *zero degree isotherm*. Its height is H_i; generally, above H_i no liquid water exists. (Supercooled water could be present as could warm raindrops carried above H_i by an updraft and not yet

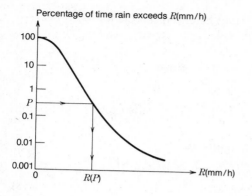

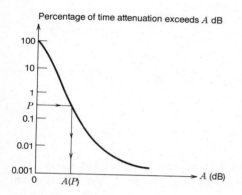

Figure 8.16 The procedure for finding equal-probability values of attenuation A and rain rate R. For each precent of time P we enter the graphs and find the corresponding values of rain rate $R(P)$ and attenuation $A(P)$.

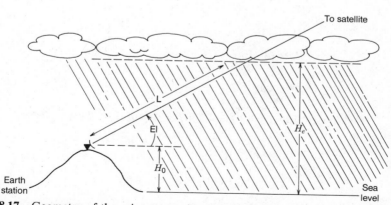

Figure 8.17 Geometry of the rain attenuation problem as used in the SAM model. H_e is called the storm height and H_0 is the earth station's elevation above sea level. The total propagation path length in rain is L.

frozen, but these situations are not important here.) For prediction purposes, typical summer values of H_i may be estimated from the earth station latitude Λ_e degrees by

$$H_i = \begin{matrix} 4.8, & |\Lambda_e| \le 30° \\ 7.8 - 0.1|\Lambda_e|, & |\Lambda_e| > 30° \end{matrix} \tag{8.19}$$

In the SAM model the effective rain height H_e is related to H_i and the rain rate R by

$$H_e = \begin{matrix} H_i, & R \le 10 \text{ mm/h} \\ H_i + \log_{10}(R/10), & R > 10 \text{ mm/h} \end{matrix} \tag{8.20}$$

Once $R(P)$ is determined and L is calculated from Eqs. (8.20), (8.19), and (8.18), the equal probability attenuation $A(P)$ is given by

$$A(P) = a[R]^b L, \qquad R \le 10 \text{ mm/h}$$

$$= a[R]^b \left\{ \frac{1 - \exp\left[-\gamma b \, \log_e\left(\dfrac{R}{10}\right) L \cos(El)\right]}{\gamma b \, \log_e\left(\dfrac{R}{10}\right)\cos(El)} \right\}, \qquad R > 10 \text{ mm/h} \tag{8.21}$$

where the empirical quantity γ is $1/22$.

Example 8.3.1

As an example of using the SAM model, consider a 28 GHz path at a 45° elevation angle from the authors' Blacksburg, Virginia, earth station at $\Lambda_e = 37.229°$ N latitude with $H_0 = 0.640$ km. The rain rate R that occurs for $P = 0.01$ percent of the time in an average year is 43 mm/h. Equations (8.13) and (8.14) yield $a = 0.1338$ and $b = 1.062$. Substituting into the equations we have

$$H_i = 7.8 - 0.1 \times 37.229 = 4.0771 \text{ km}$$

$$H_e = 4.0771 + \log_{10}(43/10) = 4.7106 \text{ km}$$

$$L = (4.7106 - 0.640)/\sin(45°) = 5.7567 \text{ km}$$

$$\gamma b \, \log_e(R/10)\cos(45°) = \frac{1}{22} \times 1.062 \times 1.4586 \times 0.707 = 0.0498$$

$$L\gamma b \, \log_e(R/10)\cos(45°) = 5.7567 \times 0.0498 = 0.2866$$

$$A = 0.1338(43)^{1.062} \frac{1 - e^{-.2866}}{(0.0498)}$$

$$= 0.1338 \times 54.29 \times 5.0038$$

$$= 36.3 \text{ dB}$$

The long-term value measured by the authors for 0.01 percent of the time is 32 dB. The year-to-year variation is several decibels each way.

The *CCIR model* [5, pp. 334ff] is based on predicting the attenuation expected for 0.01 percent of the time and then scaling it to other percentages. We will illustrate it here for time percentages between 0.001 and 0.1 percent.

For earth station latitude Λ_e, the rain height h_R is calculated by

$$h_R = 5.1 - 2.15 \log_{10}\{1 + 10^{[(|\Lambda_e| - 27)/25]}\} \text{ km} \tag{8.22}$$

The slant-path length L_S is calculated from h_R, the station height above sea level h_0, and the path elevation angle El by

$$L_S = \frac{2(h_R - h_0)}{[\sin^2(El) + 2(h_R - h_0)/8500]^{\frac{1}{2}} + \sin(El)} \quad \text{km for } El < 10°$$

$$L_S = \frac{h_R - h_0}{\sin(El)} \quad \text{km for } El \geq 10° \tag{8.23}$$

The path length reduction factor introduced to account for the spatial nonuniformity of the rain rate is called r_p. It is calculated from the horizontal projection of the slant path L_G in kilometers (km) by

$$r_p = \frac{90}{90 + 4L_G} \tag{8.24}$$

where

$$L_G = L_S \cos(El) \tag{8.25}$$

If R_p is the rainrate exceeded for 0.01 percent of an average year, and a and b are the coefficients of Eq. (8.12), then the attenuation that will be exceeded for 0.01 percent of the time as predicted by the CCIR model is

$$A_{0.01} = a(R_p)^b L_S r_p \text{ dB} \tag{8.26}$$

The attenuation value A_p predicted for any other percentage of the time P may be calculated from $A_{0.01}$ by

$$A_p = A_{0.01}\left(\frac{P}{0.01}\right)^{-\alpha} \text{ dB} \tag{8.27}$$

where

$$\alpha = \begin{array}{ll} 0.33 & \text{for } 0.001 \leq P \leq 0.01 \\ 0.41 & \text{for } 0.01 < P \leq 0.1 \end{array} \tag{8.28}$$

For percentages of time outside the 0.001 to 0.1 percent range, a relatively complicated extrapolation procedure is available [5].

Example 8.3.2

As an example of using the CCIR model, we will calculate the 0.01 percent time attenuation for the path of Example 8.3.1. The rain height h_R is given by

$$h_R = 5.1 - 2.15 \log_{10}\left\{1 + 10^{[(37.229-27/25)]}\right\} = 3.91 \text{ km}$$

$$L_S = \frac{3.91 - 0.64}{\sin(45°)} = 4.63 \text{ km}$$

$$L_G = 4.63 \cos(45°) = 3.27 \text{ km}$$

$$r_p = \frac{90}{90 + 4 \times 3.27} = 0.87$$

$$A_{0.01} = 0.1338(43)^{1.062} \times 4.63 \times 0.87 = 29.3 \text{ dB}$$

Thus for this case the CCIR model underestimates the measured value and the SAM model overestimates it.

Scaling Attenuation with Elevation Angle and Frequency

While the two models presented both permit the calculation of attenuation for any elevation angle and frequency, it is often useful, given an attenuation value for one path and frequency, to estimate quickly the effects of changing one or both of these factors. There are two rules of thumb for doing this. For a uniform rain environment and a flat earth, attenuation in decibels is proportional to path length. If the rain height is constant, then path length is proportional to the cosecant of the elevation angle El. Thus the attenuations in decibels at the same frequency at elevation angles El_1 and El_2 are approximately related by

$$\frac{A(El_1)}{A(El_2)} = \frac{\csc(El_1)}{\csc(El_2)} \tag{8.29}$$

This formula breaks down at low elevation angles ($El < 10°$) where its implicit flat earth and uniform rain assumptions fail to hold.

Many sophisticated formulas have been published for scaling attenuation with frequency, but a good approximation attenuation in decibels simply varies in proportion to the square of the frequency in GHz. Thus if $A(f_1)$ and $A(f_2)$ are the attenuations that would be measured on the same path at frequencies f_1 and f_2 GHz, they are approximately related by

$$\frac{A(f_1)}{A(f_2)} = \frac{f_1^2}{f_2^2} \tag{8.30}$$

CCIR [5, p. 337] recommends using

$$\frac{A(f_1)}{A(f_2)} = \frac{g(f_1)}{g(f_2)} \tag{8.31}$$

where

$$g(f) = \frac{f^{1.72}}{[1 + 3 \times 10^{-7}(f^{1.72})^2]} \tag{8.32}$$

where the attenuations are in dB and the frequencies are in GHz.

Calculating *XPD*

Rain depolarization occurs because the larger raindrops in a storm are not spherical but instead flatten out and are shaped more like oblate spheroids ("M & M's" to U.S. readers) and fall with their major axes nearly horizontal. If all the drops in a rainstorm were aligned, then waves propagating with their electric field vectors parallel to the drops' minor axes (for all practical purposes vertically polarized waves) would experience the minimum attenuation for that rain rate, and waves propagating with their electric field vectors parallel to the major axes (i.e., horizontally polarized waves) would experience the maximum attenuation. In these two cases, no depolarization would occur. The difference between the attenuations experienced by waves with horizontal and vertical polarization is small—only a few decibels. It is called the *differential attenuation.*

Imagine the case of a wave whose linear polarization is intermediate between horizontal and vertical. We can resolve it into its vertically polarized and horizontally polarized components as in Figure 8.18. These components propagate through the rain with their polarizations unchanged, but the horizontal component is attenuated more than the vertical component. If at any point we recombine the vertical and horizontal components to reconstruct the wave, we find that its polarization has rotated toward the vertical and a cross-polarized component is now present.

This explanation of depolarization was apparently developed independently by at least two research groups [19, 20] and is of course a simplification of a complicated problem in electromagnetic wave scattering. For details of the process the reader should consult the extensive publications of T. Oguchi, the pioneer researcher in this field [21].

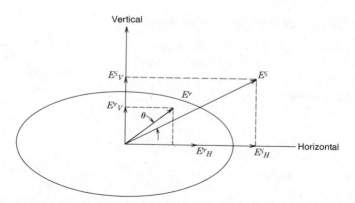

Figure 8.18 A simplified explanation of rain depolarization. An incident electromagnetic wave with electric field vector **E**^{i} strikes a raindrop. We resolve it into a horizontal component E_H^i and a vertical component E_V^i. The horizontal component is attenuated more than the vertical component because it encounters more water. Thus when we recombine the horizontal and vertical field components E_H^r and E_V^r that arrive at the receiver, we find that the received wave **E**^{r} has had its polarization rotated toward the vertical by the angle θ.

Attenuation depends mainly on the volume of water that is present in the propagation path, and hence it may be predicted with acceptable accuracy from the rainrate. But depolarization is sensitive to the size, shape, and orientation distributions of the raindrops, and these dependences make depolarization more difficult to model than attenuation. See reference 22 for a further discussion of this point. Statistically, XPD may be related to attenuation A (dB) by

$$XPD = U - V \log_{10}(A) \text{ dB} \tag{8.33}$$

for A values in the nominal range $3 \leq A \leq 30$ dB. (Taking the logarithm of a decibel quantity may startle the reader, but that is the correct procedure. See reference 23 for an analytical justification.) This equation is not useful for predicting instantaneous values of XPD. It provides an estimate of the average value of XPD that will be associated with a given attenuation.

There is some uncertainty about the best values to use for U and V in Eq. (8.33), and the problem is an active area of research at the present time (1984). There is, for example, disagreement about whether U should vary with frequency in GHz as $30 \log_{10}(f)$ or as $20 \log_{10}(f)$. For quick estimates with linear polarization, the so-called "CCIR approximation" [5] is

$$U = 30 \log_{10}(f) - 40 \log_{10} \cos(El) - 20 \log_{10} [\sin(2\tau)] \tag{8.34}$$

$$V = 20 \tag{8.35}$$

Here El is the path elevation angle and τ is the angle between the received electric field and local horizontal. For circular polarization use $\tau = 45°$. The frequency f must be expressed in GHz. For a detailed discussion of this and other models for calculating XPD from attenuation, see reference 24.

Rain Effects on Antenna Noise Temperature

In addition to the attenuation and depolarization that it causes, rain also degrades the performance of a satellite communications system by increasing the earth station antenna noise temperature. In clear weather the antenna sees the cold background of space, but in rain it receives thermal radiation from the raindrops.

The *increase* in antenna noise temperature due to rain, T_b, may be estimated by

$$T_b = 280(1 - e^{-A/4.34}) \text{ K} \tag{8.36}$$

where A is the rain attenuation in decibels [2]. This should be added to the clear weather antenna noise temperature when calculating (C/N) in rain. The number 280 in the equation is called the *effective temperature* of the rain; it differs from the true temperature because of scattering and antenna affects. Values ranging from 270 to 290 K are used. Chapter 9 presents a more exact procedure for calculating the increase in antenna noise temperature caused by rain.

Under some conditions the increase in antenna noise temperature can have more effect on a satellite link than does the associated attenuation. As an example, note that a 1-dB rain fade will cause about a 58 K increase in antenna temperature. If the clear weather system (antenna plus receiver) noise temperature is 232 K, this increase will raise the noise power by 1 dB. The rain fade will be accompanied by a 2-dB drop in (C/N): 1 dB for the fade and 1 dB for the increase in noise level.

Ice Crystal Effects

Intense depolarization in the absence of attenuation was first noted in 1976 during the European phase of NASA's ATS-6 experiment. It is caused by high-altitude ice crystals that frequently are associated with rainstorms. Thus, ice depolarization often precedes or follows rain, and the presence of ice above rain is one factor that causes wide variations in the instantaneous value of XPD that may accompany a given attenuation. See references 25 and 26 for detailed discussions of ice depolarization.

Because high-altitude ice is undetectable by terrestrial weather instruments and because ice depolarization is not accompanied by attenuation, there are no predictive models for estimating ice depolarization on a satellite link. This has little practical consequence, for while individual ice-depolarization events are dramatic, in North America and Europe satellite-link performance is dominated by rain depolarization rather than ice. At most, ice causes about a 2 dB reduction in the limiting XPD value that rain would cause for a given percentage of time [27].

Examples of Rain and Ice Effects

Figure 8.19 displays typical rain effects on satellite links. It was measured on the CTS 11.7-GHz circularly polarized downlink and the COMSTAR linearly polarized 19.04- and 28.56-GHz downlinks.

8.4 ELIMINATING OR ALLEVIATING PROPAGATION EFFECTS

Attenuation

Conceptually, the simplest way of overcoming attenuation is to provide so much signal at the receiver that the fade does not matter. As discussed in Chapter 4, the fade margin is the decibel difference between the link (C/N) that is achieved in clear weather and the minimum value of link (C/N) that is necessary for satisfactory operation. For a link reliability r (expressed as a percentage of the time), the required rain margin M_r in decibels is equal to the attenuation $A(P)$ that will be equaled or exceeded for P percent of the time, where

$$P = 100\% - r \tag{8.37}$$

plus ΔN, the decibel increase in the link thermal noise due to rain in the propagation path. The last is computed from the T_b of the previous section.

$$M_r = A(P) + \Delta N \text{ dB} \tag{8.38}$$

In theory M_r can be achieved through a combination of increased receiving antenna gain and increased transmitter EIRP. But for large rain attenuations, the necessary power and gain may be unavailable, and it may be impossible to achieve the desired reliability with a single downlink. The choice then is between abandoning the reliability requirement or going to a *site-diversity* system.

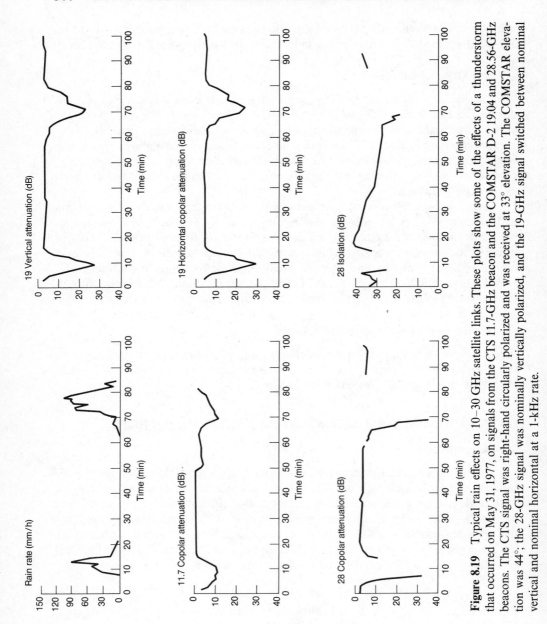

Figure 8.19 Typical rain effects on 10–30 GHz satellite links. These plots show some of the effects of a thunderstorm that occurred on May 31, 1977, on signals from the CTS 11.7-GHz beacon and the COMSTAR D-2 19.04 and 28.56-GHz beacons. The CTS signal was right-hand circularly polarized and was received at 33° elevation. The COMSTAR elevation was 44°; the 28-GHz signal was nominally vertically polarized, and the 19-GHz signal switched between nominal vertical and nominal horizontal at a 1-kHz rate.

Site Diversity

In site-diversity operation (hereafter abbreviated as "diversity"), two earth stations spaced a few kilometers apart are available to communicate with the same transponder. Since heavy rain occurs in geographically small "cells," there is a probability (which increases with station separation) that equally heavy rain will not affect both stations at the same time. Thus both sites will probably not experience the same rain attenuation at the same time. If we refer to the two sites by the numbers 1 and 2 and if the attenuations they experience at time t are $A_1(t)$ and $A_2(t)$ dB, then the *joint attenuation* $A_j(t)$ is defined by

$$A_j(t) = \text{minimum}[A_1(t), A_2(t)] \text{ dB} \tag{8.39}$$

The average single-site attenuation $A_S(t)$ is the mean of $A_1(t)$ and $A_2(t)$.

$$A_S(t) = [A_1(t) + A_2(t)]/2 \text{ dB} \tag{8.40}$$

An ideal diversity system that monitored the received downlink signals at both sites and always selected the stronger one would experience an attenuation of $A_j(t)$ and the diversity system would perform better than either site alone. How much better is measured by two statistical quantities, *diversity gain* and *diversity improvement*

Diversity gain, $G_D(P)$, is the decibel difference between the average single-site attenuation $A_S(P)$ equaled or exceeded P percent of the time and the joint attenuation $A_j(P)$ equaled or exceeded P percent of the time.

$$G_D(P) = A_S(P) - A_J(P) \text{ dB} \tag{8.41}$$

Diversity improvement $I_D(A)$ is the ratio between the percentage of time P_S that the average single-site attenuation A_S exceeds A dB to the percentage of time P_J that the joint attenuation A_J exceeds A dB.

$$I_D(A) = \frac{P_S(A)}{P_J(A)} \tag{8.42}$$

See Figure 8.20 for an illustration of these two quantities. Diversity gain determines system margin, and it is the measure of diversity system performance that we will use here. Diversity improvement is usually preferred by engineers designing terrestrial microwave links that may employ diversity to overcome multipath fading.

The margin M_r required for rain fades in a site-diversity system is calculated from

$$M_r = A(P) + \Delta N - G_D(P) \text{ dB} \tag{8.43}$$

where r and P are related by Eq. (8.37) and $A(P)$ is the expected attenuation for a single site. The increased noise in decibels contributed by the increase in antenna temperature is ΔN.

Diversity gain may be estimated by two empirical models developed by Hodge [28, 29], which are described in detail in reference 1. In the earlier and simpler of the two, G_D is related to A by

$$G_D = a'(1 - e^{-b'd}) \text{ dB} \tag{8.44}$$

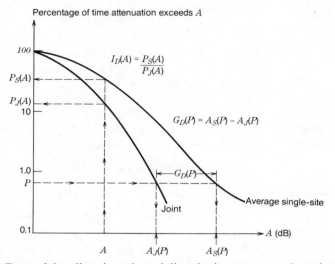

Figure 8.20 Determining diversity gain and diversity improvement. At a given percentage of time, P, the diversity gain $G_D(P)$ is the decibel difference between the average single-site attenuation exceeded $A_S(P)$ and the joint attenuation $A_J(P)$ exceeded. At a given attenuation A, the diversity improvement $I_D(A)$ is the ratio of the percentage of time $P_S(A)$ that the single-site attenuation exceeds A to the percentage of time $P_J(A)$ that the joint attenuation exceeds A.

where the coefficients a' and b' are given by

$$a' = A - 3.6(1 - e^{-0.24A}) \qquad (8.45)$$

$$b' = 0.46(1 - e^{-0.26A}) \qquad (8.46)$$

and d is the site separation distance in kilometers.

The later and more detailed model added the effects of frequency f in GHz, elevation angle El, and baseline-to-path angle Δ. The baseline is the straight line joining the two sites. The diversity gain is the product of terms containing the dependence on each factor

$$G_D = G_d G_f G_E G_\Delta \, \text{dB} \qquad (8.47)$$

where

$$G_d = a(1 - e^{-bd}) \, \text{dB} \qquad (8.48)$$

$$a = 0.64A - 1.6(1 - e^{-0.11A}) \qquad (8.49)$$

$$b = 0.585(1 - e^{-0.98A}) \qquad (8.50)$$

$$G_f = 1.64e^{-0.025f} \qquad (8.51)$$

$$G_E = 0.00492 \, El + 0.834 \qquad (8.52)$$

$$G_\Delta = 0.00177\Delta + 0.887 \qquad (8.53)$$

In these equations Δ and El are measured in degrees and f is in GHz.

Diversity gain G_D depends most strongly on the site separation distance d. Logically it should range from 0 dB at 0 km separation—where both sites experience exactly the same rain—to a value equal to the single-site attenuation when d became large enough that the rain at the two sites was completely uncorrelated. In practice, measurements show a saturation effect in which G_D reaches limiting values that are several decibels below the associated single-site attenuations. This is thought to be due to the effects of widespread stratiform rain in which more intense rain cells are embedded. There is also an average spacing of 10 to 30 km between thunderstorms.

While diversity has been studied extensively, it has not been widely used in commercial satellite systems. This is because only a few systems are operating at 11/14 GHz, and all the rest are at 4/6 GHz, where rain attenuation is not a problem. Most of the systems now at 11/14 GHz have chosen to accept a lower reliability or a higher single-site margin in preference to the added expense of a diversity terminal.

Depolarization

Dual-polarized links must contend with both attenuation and depolarization. The amount of attenuation that can be tolerated is easy to calculate from the minimum (C/N) and the margin, but determining minimum allowable polarization isolation values is a difficult problem. As far as one channel is concerned, the cross-polarized interference from the other channel is simply co-channel interference. The extent to which this interference degrades link performance depends on the details of the modulation process employed and on the information carried by the two channels. As a first approximation for CSSB systems, the XPD value can be used as an estimate for the (S/X) term in Eq. (6.17). See references 6 and 30 for more information on including depolarization in FM and QPSK link analysis. Users perceive depolarization as crosstalk in analog systems and by an increased bit error rate in digital systems. Unlike attenuation, depolarization cannot be alleviated by adding margin to the link power budget. Two potential remedies are site-diversity reception and adaptive depolarization compensation (cancellation). Both are still in the experimental stage.

The idea of diversity operation to overcome depolarization is the same as that for attenuation—namely, that two propagation paths spaced a few kilometers apart will not experience severe rain or ice crystal effects simultaneously. At the cost of maintaining two earth stations for the same link, performance can be improved by monitoring depolarization at both sites and selecting the best one. This leads to the concept of *polarization diversity gain*, the decibel difference between the single-site and joint XPD values observed for the same percentage of time. The possible use of site diversity to improve the performance of dual-polarized links is at too early a stage in its development to justify a detailed discussion here. Further information is available in reference 31.

Another remedy for depolarization is *cancellation* [32], where a suitably attenuated and phase-shifted replica of the unwanted signal component is introduced at the receiver to cancel out the cross-polarized signal that was generated in the propagation path. Since the amplitude and phase of the depolarized signal vary with time, some sort of adaptive control is required to minimize the crosstalk. Chu [33, 34] has discussed the cancellation problem in detail and experiments conducted by Comsat Laboratories with a 4-GHz downlink have yielded good results [35].

8.5 SUMMARY

The atmosphere introduces a time-varying loss called attenuation into link power budgets. This affects all satellite links; those employing orthogonal polarizations for frequency reuse are also degraded by depolarization, the conversion of some of the energy in an electromagnetic wave from one polarization to another.

Propagation effects of the clear atmosphere are absorption, tropospheric scintillation, multipath, ionospheric scintillation, and Faraday rotation. The first three are relatively unimportant in commercial satellite applications for elevation angles above about 15° and frequencies below 10 GHz. Faraday rotation varies inversely with frequency squared, and its depolarizing effects at 4 and 6 GHz are usually insignificant. Ionospheric scintillation can cause signal fluctuations as large as 6 dB for stations near the geomagnetic equator.

At frequencies above 10 GHz rain strongly attenuates and depolarizes radio waves and constitutes a major impediment to satellite link performance. For a specified earth station location, the rain attenuation that will be exceeded for a given percentage of time can be predicted on a statistical basis and used to determine the fade margin that a given satellite link will need in order to provide a specified reliability. Depolarization may also be predicted, but with less accuracy. Depolarization may be overcome with a compensation scheme that cancels the unwanted cross-polarization signal. Diversity operation (using two nearby earth stations to communicate with the same satellite and dynamically selecting the station that has the best propagation conditions at a given time) offers an alternative both to depolarization compensation and to large fade margins.

REFERENCES

1. Louis J. Ippolito, R. D. Kaul, and R. G. Wallace, *Propagation Effects Handbook for Satellite Systems Design* (NASA Reference Publication 1082(03)), National Aeronautics and Space Administration, Washington, DC, June 1983.
2. Warren L. Flock, *Propagation Effects on Satellite Systems at Frequencies Below 10 GHz* (NASA Reference Publication 1108), National Aeronautics and Space Administration, Washington, DC, December 1983.
3. Martin P. M. Hall *Effects of the Troposphere on Radio Communication*, Peter Peregrinus, Ltd., Stevenage, UK, 1979.
4. Petr Beckmann, *The Depolarization of Electromagnetic Waves*, The Golem Press, Boulder, CO, 1968.

5. International Radio Consultative Committee (CCIR), *Recommendations and Reports of the CCIR, 1982, Volume V, Propagation in Non-Ionized Media,* International Telecommunication Union, Geneva, Switzerland, 1982.

6. C. W. Bostian, W. L. Stutzman, and J. M. Gaines, "Depolarization Modeling for Earth–Space Radio Paths at Frequencies Above 10 GHz," *Radio Science,* **17,** 1231–1241 (September–October, 1982).

7. D. C. Cox and H. W. Arnold, "Comparison of Measured Cross-Polarization Isolation and Discrimination for Rain and Ice on a 19-GHz Space-Earth Path," *Radio Science,* **19,** 617–628 (March–April, 1984).

8. Prabha N. Kumar, "Depolarization of 19 GHz Signals," *Comsat Technical Review,* **12,** 271–293 (Fall 1982).

9. P. A. Watson and M. Arbabi, "Cross-Polarization Isolation and Discrimination," *Electronics Letters,* **9,** 516–519 (November 1, 1973).

10. H. J. Liebe, "Modeling Attenuation and Phase of Radio Waves in Air at Frequencies Below 1000 GHz," *Radio Science,* **16,** 1183–1199 (November–December, 1981).

11. T. Pratt and D. J. Browning, "Copolar Attenuation and Radiometer Measurements at 30 GHz for a Slant Path to Central England," *Proceedings ATS-6 Meeting* (ESA SP-131), ESTEC, Noordwijk, The Netherlands, 15–20, October 1977.

12. Curtis A. Siller, Jr., "Multipath Propagation," *IEEE Communications Magazine,* **22,** 6–15 (February 1984).

13. Kenneth Davies, *Ionospheric Radio Propagation* (National Bureau of Standards Monograph 80), U.S. Government Printing Office, Washington, DC, April 1, 1965.

14. T. Murakami and G. S. Wickizer, "Ionospheric Phase Distortion and Faraday Rotation of Radio Waves," *RCA Review,* **30,** 475–503 (September 1979).

15. Roger R. Taur, "Ionospheric Scintillation at 4 and 6 GHz," *Comsat Technical Review,* **3,** 145–163 (Spring 1973).

16. P. L. Rice and N. R. Holmberg, "Cumulative Time Statistics of Surface Point Rainfall Rates," *IEEE Transactions on Communications,* **COM-21,** 1131–1136 (October, 1973).

17. R. L. Olsen, D. V. Rogers, and D. B. Hodge, "The aR^b Relation in the Calculation of Rain Attenuation," *IEEE Transactions on Antennas and Propagation,* AP-26, 318–329 (March 1978).

18. W. L. Stutzman and W. K. Dishman, "A Simple Model for the Estimation of Rain-Induced Attenuation Along Earth–Space Paths at Millimeter Wavelengths," *Radio Science,* **17,** 1465–1476 (November–December 1982).

19. David T. Thomas, "Cross-Polarization Distortion in Microwave Radio Transmission Due to Rain," *Radio Science,* **6,** 833–839 (October 1971).

20. P. A. Watson and M. Arbabi, "Rainfall Crosspolarization at Microwave Frequencies," *Proceedings of the IEE* (London), **120,** 413–418 (April 1973).

21. T. Oguchi, "Electromagnetic Wave Propagation and Scattering in Rain and Other Hydrometeors," *Proceedings of the IEEE,* **71,** 1029–1078 (September 1983).

22. C. W. Bostian, W. L. Stutzman, and J. M. Gaines, "A Review of Depolarization Modeling for Earth–Space Radio Paths at Frequencies Above 10 GHz," *Radio Science*, **17**, 1231–1241 (September–October 1982).

23. W. L. Nowland, R. L. Olsen, and I. P. Shkarofsky, "Theoretical Relationship Between Rain Depolarization and Attenuation," *Electronics Letters*, **13** (22), 676–677 (October 1977).

24. W. L. Stutzman and D. L. Runyon, "The Relationship of Rain-Induced Crosspolarization Discrimination to Attenuation for 10 to 30 GHz Earth–Space Radio Links," *IEEE Transactions on Antennas and Propagation*, **AP-32** (July 1984).

25. C. W. Bostian and J. E. Allnutt, "Ice Crystal Depolarization on Satellite–Earth Microwave Radio Paths," *Proceedings of the IEE* (London), **126**, 148–153, 951–960 (October 1979).

26. D. C. Cox, "Depolarization of Radio Waves by Atmospheric Hydrometeors in Earth–Space Paths: A Review," *Radio Science*, **16**, 781–812 (September–October 1981).

27. W. L. Stutzman, C. W. Bostian, A. Tsolakis, and T. Pratt, "The Impact of Ice Along Satellite-to-Earth Paths on 11 GHz Depolarization Statistics," *Radio Science*, **18**, 720–724 (September–October 1983).

28. D. B. Hodge, "An Empirical Relationship for Path Diversity Gain," *IEEE Transactions on Antennas and Propagation*, *AP-24*, 250–251 (March 1976).

29. D. B. Hodge, "An Improved Model for Diversity Gain on Earth–Space Propagation Paths," *Radio Science*, **17**, 1393–1399 (November–December 1982).

30. J. Neessen and F. Zelders, "The Impact of Cross Polarization Phenomena on the Fading Margin in Satellite Communications Systems," *Alta Frequenza*, **49**, 338–343 (September–October, 1980).

31. R. E. Marshall and C. W. Bostian, "Using Site-Diversity Reception to Overcome Rain Depolarization in Millimeter Wave Satellite Communications Systems," *IEEE Transactions on Antennas and Propagation*, **AP-30**, 990–991 (August 1982).

32. M. Yamada, H. Yuki, and K. Inagaki, "Compensation Techniques for Rain Depolarization in Satellite Communications," *Radio Science*, **17**, 1221–1226, (September–October 1982).

33. T. S. Chu, "Restoring the Orthogonality of Two Polarizations in Radio Communications Systems," Part I, *Bell System Technical Journal*, **50**, 3063–3069, (November 1971).

34. T. S. Chu, "Restoring the Orthogonality of Two Polarizations in Radio Communications Systems," Part II, *Bell System Technical Journal*, **52**, 319–327 (March 1983).

35. R. R. Persinger, R. W. Gruner, J. E Effland, and D. F. DiFonzo, "Operational Measurement of a 4/6 GHz Adaptive Polarization Compensation Network Employing Up/Down Link Correlation Algorithms," *Proceedings of the International Conference on Antennas and Propagation* (Part 2), IEE Conference Publication 195, April 1981.

PROBLEMS

The first three problems assume a satellite system with an uplink center frequency of 27.5 GHz and a downlink center frequency of 17.7 GHz. Both earth stations see the satellite at an elevation angle of 35°.

1. Calculate the uplink and downlink specific attenuations in dB/km.

2. For an earth station in Florida at 27° N latitude and sea-level elevation (NASA handbook rain climate region E) determine
    a. The rain rates that will be exceeded for 0.001, 0.01, 0.1, and 1 percent of a year.
    b. The uplink and downlink rain attenuations predicted by the SAM model for the rain rates in Part a.
    c. The uplink and downlink rain attenuations predicted by the CCIR model for 0.001, 0.01, and 0.1 percent of a year.
    d. The increase in earth station antenna noise temperature that will accompany the attenuations calculated in Part b.
    e. The downlink XPD that would be associated with the attenuations calculated in Part b for circular polarization. Use the CCIR approximation of Eq. (8.34).
    f. The diversity gain G_D that could be achieved on the downlink for a site separation of 10 km based on the simpler Hodge model of Eqs. (8.44) through (8.46) and single-site attenuation values given by the answers to Part b.

3. Work Problem 2 for a sea-level earth station near Portland, Oregon, at 45° N latitude in NASA handbook rain climate region C.

4. Assuming that the three-year average rain rate distribution shown in Figure 8.9 represents the rain that fell in 1978, calculate the effective path lengths associated with attenuation values of 5, 10, and 15 dB in Figure 8.15.

5. Calculate the ratio of 28.56 GHz attenuation exceeded for 0.01 percent of the time to the value of 19.04 GHz attenuation exceeded for the same percentage of the time as shown in Figure 8.15. Compare this ratio to those predicted by Eq. (8.30) and (8.31).

6. Using the data of Figure (8.15), scale the attenuation exceeded for 0.01 percent of the time on the 28.56-GHz path to 11.7 GHz using the equations in the text. Remember to scale for both the frequency difference and the elevation angle difference. Compare your result to the observed 11.7-GHz attenuation.

7. Using the "CCIR approximation" of Eqs. (8.34) and (8.35), calculate the XPD associated with attenuations of 10 and 20 dB for an 11.7-GHz circularly polarized downlink with a 33° elevation angle.

8. The attenuation that would be observed on a satellite path may be estimated by measuring the increase in antenna noise temperature and calculating the associated attenuation A from Eq. (8.36). Instruments that do this are called

radiometers. They are useful for attenuation values between 0 and about 10 dB but not very good for larger ones. Using Eq. (8.36), show why this occurs. The effect is called radiometer saturation.

9. If the rain rate distributions along the two paths of a site-diversity system were statistically independent, the probability that the joint attenuation A_j would exceed some value A_0 would be the square of the probability that the single-site attenuation would exceed A_0. Assume a site-diversity system operating at 28.56 GHz and having identical single-site cumulative attenuation distributions like the 28.56 GHz curve in Figure 8.15. Calculate the diversity gain that would be achieved at $P = 0.01$ percent if the rain rate distributions on the two paths were statistically independent and compare it to the predictions of the simpler Hodge model [Eq. (8.44)] for site separations d of 3, 10, and infinity kilometers.

9

EARTH STATION TECHNOLOGY

9.1 EARTH STATION DESIGN

Earth Station Design for Low System Noise Temperature

An *earth station* is any transmitting or receiving system that sends signals to or receives signals from a satellite. The earth station may be located on a ship at sea or on an aircraft, but it is still called an earth station since it forms the earth-based end of the earth–space link. The most visible part of an earth station is usually the antenna, which may be as large as 30-m diameter in the Intelsat network, or as small as 0.7 m for reception of direct broadcast satellite television (DBS-TV). The one feature that all earth stations have in common is the need to achieve a low system noise temperature in the receiving channel.

In Chapters 4 and 5, we showed that the downlink carrier-to-noise ratio in the receiving channel is nearly always the limiting factor that sets the communications capacity and performance of the link. Typically, the (C/N) is in the range of 5 to 25 dB, measured in the IF predetection bandwidth of the receiver. Recalling Eq. (4.20), (C/N) is found from

$$(C/N) = \frac{P_t G_t}{kB} \left[\frac{\lambda}{4\pi R} \right]^2 \frac{G_r}{T_s} \tag{9.1}$$

where G_r/T_s is the *G/T ratio* of the earth station. For a given satellite and signal transmission, the only parameters of the earth station that affect the (C/N) are G_r, the antenna gain when receiving the wanted transmission, and T_s, the system noise temperature at the frequency of the transmission. (G_r and T_s are not necessarily constant with frequency.) Since (C/N) is directly proportional to G/T, this is a useful parameter with which to characterize earth stations.

Earth station G/T ratios are usually expressed in dBK^{-1}, obtained by converting system noise temperature into dBK (dBKelvins, or dB greater than a temperature of 1 K). Thus

$$G/T = G_r \text{ in dB} - 10 \log_{10}(T_s \text{ in Kelvins}) \text{ dBK}^{-1} \qquad (9.2)$$

G/T may have a value from as low as -10 dBK^{-1} up to 46 dBK^{-1}. Most satellite communication systems are designed around a specified earth station G/T, and there may be a minimum earth station G/T that can be used in the system. For example, Intelsat Standard A earth stations are required by Intelsat to have a G/T of 40.7 dB or greater at 4000 MHz and 5° elevation angle [1]. An earth station owner must prove to Intelsat that this G/T is achieved before the earth station is permitted to join the network.

The specified G/T can be obtained from an infinite number of combinations of G and T. In practice, however, the range of values of T_s, the system noise temperature, is somewhat limited. While it is desirable to reduce T_s toward zero so that G/T becomes large, this cannot be done in practice. External noise sources such as the atmosphere, the receiving system waveguides, and the low noise amplifier of the receiver all contribute to T_s and set a lower limit to its value. Typically, T_s is rarely below 70 K, and may be as high as 2000 K. The gain of the receiving antenna can be increased by using a larger aperture area, with an upper limit on G_r of about 65 dB. This limits G/T to about 46 dB with the largest practical antenna and the lowest practical noise temperature.

The optimum G/T for a given application is invariably a compromise between the cost of a large antenna, to increase G, and the cost of lower system noise to decrease T. Once the antenna becomes large, increasing G becomes extremely expensive, making it worthwhile to minimize T. Thus, large antenna earth stations may use cryogenically cooled low noise amplifiers in an attempt to obtain the lowest possible system noise temperature. With a small earth station antenna, it is usually more cost effective to increase the antenna aperture area than to lower the system noise temperature.

The high cost of large-aperture antennas (several million dollars for a 25-m diameter steerable dish) led to a great deal of research into design techniques aimed at improving the efficiency of large antennas during the period from 1960 to 1980. Some notable advances in antenna technology stemmed from the requirements of large earth stations—the shaped Cassegrain antenna, beam waveguide feed, corrugated horn, and low cross-polarization design methods can all be attributed to this research effort. Similar advances have occurred in low-noise amplifiers. In 1962 when the first TELSTAR experiment was tried, the low-noise amplifiers used at the earth stations were masers immersed in liquid helium. The need for more practical LNA's led to the development of GaAsFET amplifiers at 4 and 11 GHz, with noise temperatures down to 70 K without cooling.

In this chapter we discuss the design of large and small earth stations, and how the antenna gain can be traded for system noise temperature to obtain a specified G/T. There are significant differences in approach depending on the number of voice or data channels an earth station is to carry, and the flux density the satellite achieves at the earth's surface. The system noise temperature is made up of contributions from many sources; minimizing T_s requires considerable care in the design of the earth station antenna, as well as the selection of a suitable LNA or receiver. These aspects are considered in Sections 9.2, 9.3, and 9.5.

Large antennas invariably produce narrow beams with the result that satellites that move their position by more than a fraction of a degree must be tracked. Most large antennas are equipped with automatic tracking (called *autotrack*) facilities so that the motion of the satellite can be followed. Smaller antennas with broader beams may require only a reposition capability. Tracking and beam steering are discussed in Section 9.4.

The antenna of an earth station is its most visible part, but receivers are needed at all earth stations and many also need transmitters. Earth stations that multiplex many telephone or data signals together and provide links to many countries simultaneously require extensive baseband and IF equipment to implement the required multiplexing and may also have a great many transmitters and receivers. These are discussed in Sections 9.6 and 9.7. Finally, *frequency coordination* is required to ensure that interference at or from an earth station is held to an acceptable level. This is discussed in Section 9.8.

Figure 9.1 and 9.2 show some examples of earth station antennas.

Figure 9.1 Examples of large earth stations. Foreground: 19-m 14/11 GHz Standard C Cassegrain antenna with beam waveguide and wheel and track mount. Background: 32-m 6/4 GHz Standard A Cassegrain antenna with kingpost mount. (Photograph courtesy of Marconi Research Centre and Marconi Communications Systems Ltd., U.K.)

Figure 9.2a A 13-m 14/11 GHz Cassegrain antenna installed in London's Dockland re-development area. (Photograph courtesy of Marconi Research Centre, U.K.)

Figure 9.2b 5.6 × 2.8 m 14/11 GHz offset Gregorian antenna. Fully transportable for broadcast uplinks, this antenna meets the FCC 29 - 25 log θ sidelobe envelope requirement. (Photograph courtesy of GEC-McMichael Ltd., U.K.)

Large Earth Station Antennas

Large antennas are used at earth stations that require a high G/T ratio, and are capable of carrying large numbers of telephone, television, or data channels simultaneously. A high volume of traffic is needed to recover the cost of a large earth station antenna, which implies operation with wideband carriers.

As an example, consider the case of an earth station designed to operate within a zone coverage region of an INTELSAT V satellite. The satellite is designed for operation with Standard A earth stations having a minimum G/T of 40.7 dBK^{-1}, using FDM/FM/FDMA. The satellite cannot be used with earth stations having a G/T much below 40 dBK^{-1}, as illustrated by the following calculations. Suppose we send just one FM carrier from a 4-GHz transponder of an INTELSAT V satellite so that it occupies the full 72 MHz of a wideband transponder and has an EIRP of 26 dBW in the direction of the receiving earth station. The received (C/N) is given by Eq. (9.1) (ignoring any interference or uplink noise contribution)

$$(C/N) = (26 + 228.6 - 78.6) - 196 + G/T \text{ dB}$$
$$= G/T - 20 \text{ dB} \tag{9.3}$$

INTELSAT V links operate with a 7-dB margin over a minimum (C/N) of 11 dB, so for an 18 dB (C/N) we must have a G/T of 38 dB. This is the minimum G/T that can be used for any earth station to communicate via an INTELSAT V zone beam transponder, when the transponder is fully loaded. Even if we use several narrower bandwidth FM carriers, the transponder transmitter power must be shared among the carriers in proportion to their bandwidth, and the incident flux density at the earth's surface remains the same. Thus only Standard A earth stations can join the Intelsat global network using the standard FM/FDMA carriers, even when the earth station wants to send only one or two voice signals. This makes small-capacity systems very costly and has led to the establishment of different standards for such applications.

Intelsat has made provision for smaller earth stations (*Standard B*) in its network by using some transponders on its satellites in a TDMA mode [1]. These transponders can be operated in or close to saturation, increasing the transmitted power and allowing a reduction in antenna diameter. Typical Standard B earth stations have 19-m diameter antennas, but achieve the same low noise temperatures as Standard A antennas. Table 9.1 compares the specification of Intelsat Standard A, B, and C earth stations. Standard C earth stations use smaller antennas and operate in the 14/11 GHz band using the spot beams of INTELSAT V and VI spacecraft.

9.2 BASIC ANTENNA THEORY

Earth station antennas are required to supply gains in excess of 25 dB in nearly all cases. This requires the use of an *aperture antenna* that has a significant gain and area. Examples of aperture antennas are waveguide horns and reflector antennas. In this section we review the basic theory of aperture antennas before proceeding

Table 9.1
Summary of Intelsat Standard A, B, and C Earth Station
Characteristics

Standard	A	B	C
Frequencies (GHz)	6/4	6/4	14/11
Polarization	Circular	Circular	Linear
G/T dBK^{-1}	40.7	31.7	$39 + 20 \log\left(\dfrac{f}{11.2}\right)$
Typical dish diameter (m)	30	11–13	19
Antenna midband receive gain (dB)	61	51.5	65
Antenna midband transmit gain (dB)	64	54.1	66.4
Main reflector rms surface tolerance (mm)	1.0	0.8	0.6
Typical LNA noise temperature (K)	40	40	120

Source: Reproduced from Intelsat Publications BG-28-72E Rev. 1 and BG-28-73E Rev. 1, International Telecommunications Satellite Organization, Washington, D.C., December 15, 1982, by permission of the International Telecommunications Satellite Organization.

to discuss the special requirements imposed by the earth station application. We concentrate on reflector antennas because that is the most widely used configuration in earth stations. The theory is equally applicable to antennas used on spacecraft, although there are different requirements related to the need for area coverage in that case that lead to differences in antenna design. Many spacecraft antennas have large array feeds to create shaped beam patterns, and some use phased arrays. We concentrate on the earth station antenna that is optimized for G/T ratio and is pointed at a single source of radiation, the satellite.

It is common practice to analyze aperture antennas in the transmitting mode, and then use reciprocity to equate the receiving characteristics to those calculated for the antenna when transmitting. We will follow this approach in reviewing the basic theory of aperture antennas and in calculating the gain and radiation pattern. When considering tracking antennas, it is useful to examine their operation in a receiving mode, since tracking is usually accomplished on reception rather than transmission.

An aperture antenna achieves gain and a narrow beam by creating an electromagnetic field over the aperture that has uniform phase—a plane wavefront. It is necessary to control the amplitude distribution of the aperture field, both to maximize gain and to minimize losses; the aperture field distribution determines the radiation pattern in the far field and also affects the gain, so we first examine the relationship between the aperture field and the far-field radiation pattern for several cases.

Linear Apertures

Consider first the linear aperture distribution shown in Figure 9.3. The aperture is assumed to have a small width dy and a length a. The field in the aperture is constant in the y direction, and its amplitude varies as $E(x)$ in the x direction. The field at a distant point P in the (x, z) plane can be calculated by adding together at P all contributions from small elements dx, dy of the field in the aperture. Hence the field at P is given by

$$E(R, \theta) = A \int_{-a/2}^{+a/2} \frac{E(x)}{R_1} \exp\left(-jkR_1\right) dx \tag{9.4}$$

where R_1 is the distance from the observation point P to the element dx in the aperture, and $k = 2\pi/\lambda$. Uniform phase across the aperture is assumed. A is a constant that we can ignore in this analysis. If we make R_1 very large compared

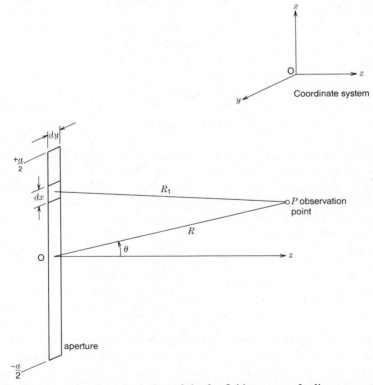

Figure 9.3 Geometry for the calculation of the far-field pattern of a linear aperture.

to a, the variation in the $1/R_1$ term is small and we can replace $1/R_1$ by $1/R$, where R is the distance from P to the center of the aperture. When $R \gg a$, we are in the *far field* and we can replace R_1 within the exponential term of Eq. (9.4) by the approximation

$$R_1 \simeq R + x \sin \theta \tag{9.5}$$

Then Eq. (9.4) becomes

$$E(R, \theta) = A \frac{\exp(-jkR)}{R} \int_{-a/2}^{a/2} E(x) \exp(-jkx \sin \theta) \, dx \tag{9.6}$$

The term $\exp(-jkR)/R$ outside the integral represents the path loss and phase shift over a distance R and can be ignored when evaluating the antennna pattern. Thus the simple linear aperture case gives us a far field distribution in the plane of the aperture of

$$E(\theta) = \int_{-a/2}^{+a/2} E(x) \exp(-jkx \sin \theta) \, dx \tag{9.7}$$

Equation (9.7) can be evaluated analytically when the field distribution $E(x)$ is constrained to certain simple forms. For example, if we have *uniform illumination* of the aperture, $E(x) = 1$, evaluation of the integral in Eq. (9.7) gives

$$E(\theta) = \frac{a \sin\left(\dfrac{\pi a}{\lambda} \sin \theta\right)}{\dfrac{\pi a}{\lambda} \sin \theta} = \frac{a \sin X}{X} \tag{9.8}$$

where

$$X = \frac{\pi a}{\lambda} \sin \theta$$

The maximum value of $E(\theta)$ is a when $X = 0$, that is, when $\theta = 0$, and the far-field pattern has a $\sin X/X$ shape, shown in Figure 9.4. The pattern has a main lobe with a half power (3 dB) width of $51\lambda/a$ degrees, and first sidelobe amplitude of 13.2 dB, when $a \gg \lambda$, which is usually the case for aperture antennas [2].

Equation (9.6) has the general form of a Fourier transform. The Fourier transform is defined as

$$F(f) = \int_{-\infty}^{+\infty} f(t) \exp(-j2\pi ft) \, dt \tag{9.9}$$

and the inverse Fourier transform is

$$f(t) = \int_{-\infty}^{\infty} F(f) \exp(j2\pi ft) \, df \tag{9.10}$$

The limits in Eqs. (9.9) and (9.10) extend to infinity, so to make Eq. (9.7) for the far-field pattern match the Fourier transform, we need to extend the limits from $\pm a/2$ to infinity. We can do this because we assume that the field is zero beyond the ends of the aperture.

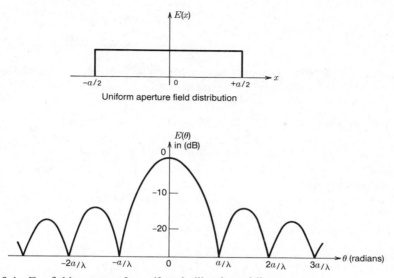

Figure 9.4   Far-field pattern of a uniformly illuminated linear aperture.

The value of the Fourier transform approach is that it can provide a general guide to the shape of far-field patterns that result from certain aperture field distributions. For example, a rectangular field distribution $E(x) = 1$ was used to obtain the $\sin X/X$ shape far-field pattern in Eq. (9.8). This is a familiar result from Fourier transform theory. If we wish to create a far-field pattern $E(\theta) = 1$ over a range of angles θ_1 to θ_2, we can use the inverse Fourier transform to show that the required aperture field distribution must have a shape $\sin X/X$, which will extend from $-\infty$ to $+\infty$. That means that we cannot produce a sector-shaped beam from a finite-sized aperture. Truncation of the aperture always results in a far-field pattern with sidelobes.

It is useful to examine the far-field pattern that results from some particular aperture distributions, in order to understand how beamwidth and sidelobe level are related to the aperture field. Table 9.2 gives some sample results for easily integrated forms of $E(x)$. The *illumination efficiency* is the ratio of the on-axis directivity for the given aperture illumination to that produced by a uniform aperture distribution with the same total radiated power. Uniform illumination produces the highest illumination efficiency of 1, and all tapered distributions have lower gain.

In designing an antenna for a satellite communications earth station we invariably have a set of conflicting requirements. We want maximum gain, a narrow main beam, and low sidelobes to minimize interference. The gain, beamwidth, and sidelobe levels in the far-field pattern are all controlled by the aperture distribution of the antenna. In general, we must have uniform phase across the aperture to achieve maximum on-axis gain, which leaves us only one parameter, the amplitude distribution, with which to optimize three requirements.

Table 9.2
Characteristics of Linear Apertures

Illumination function	Illumination efficiency	3-dB Beamwidth (rad)	Position of first null (rad)	First sidelobe level (dB)
Parabolic: $f(x) = 1 - (1 - \Delta)\left(\dfrac{2x}{a}\right)^2$, $\quad |x| \leq a/2$, $\quad$ aperture width $= a$				
$\Delta = 1.0$	1.00	$0.88\lambda/a$	λ/a	-13.2
(uniform illumination)				
0.8	0.994	$0.92\lambda/a$	$1.06\lambda/a$	-15.8
0.5	0.970	$0.97\lambda/a$	$1.14\lambda/a$	-17.1
0	0.833	$1.15\lambda/a$	$1.43\lambda/a$	-20.6
Cosine raised to a power: $f(x) = \cos^n\left(\dfrac{\pi x}{a}\right)$, $\quad |x| \leq a/2$, $\quad$ aperture width $= a$				
$n = 0$	1.00	$0.88\lambda/a$	λ/a	-13.2
(uniform illumination)				
1	0.81	$1.2\ \lambda/a$	$1.5\lambda/a$	-23
2	0.667	$1.45\lambda/a$	$2.0\lambda/a$	-32
3	0.575	$1.66\lambda/a$	$2.5\lambda/a$	-40
Triangular: $f(x) = 1 - \left|\dfrac{2x}{a}\right|$, $\quad |x| \leq a/2$, $\quad$ aperture width $= a$				
	0.75	$1.28\lambda/a$	$2.0\lambda/a$	-26.4
Consine squared on a pedestal: $f(x) = A + B\cos^2\left(\dfrac{\pi x}{a}\right)$, $\quad |x| \leq a/2$, $\quad$ aperture width $= a$				
$A = 0.33$ $B = 0.67$	0.88	$1.10\lambda/a$	—	-25.7
$A = 0.08$ $B = 0.92$	0.74	$1.34\lambda/a$	—	-42.8

Source: Reprinted with permission from S. Silver, Ed., *Microwave Antenna Theory and Design*, Vol. 12. M.I.T. Radiation Lab. Series, McGraw-Hill, New York, 1949, Table 6.1, p. 187, and M. I. Skolnik, *Introduction to Radar Systems*, (2nd ed.) McGraw-Hill, New York, 1980, Table 7.1 p. 232.

Synthesis of the aperture distribution from the desired far-field pattern is rarely attempted; instead we search for "good" aperture distributions that lead to desirable features in the far-field patterns. Several basic rules can be stated that govern the relationship between the aperture distribution and the far-field pattern when the aperture dimension D is greater than one wavelength, λ, that is, $D > \lambda$.

The relationships can be proved via the Fourier transform and are evident in the results given in Table 9.2.

1. The gain of the antenna increases linearly with the aperture dimension, measured in wavelengths, $G \propto D/\lambda$.
2. The half-power beamwidth decreases linearly with the aperture dimension, measured in wavelengths, $\theta_{3dB} \propto \lambda/D$.
3. The sidelobe levels decrease as the amplitude distribution is tapered more severely.
4. Gain reduces and the half-power beamwidth increases as the amplitude distribution is tapered more severely.

Uniform illumination of an aperture results in the highest gain for a given aperture dimension but also gives high sidelobes in the far-field pattern. We can lower the sidelobe levels by tapering the amplitude distribution, but this will lower the gain and increase the half-power beamwidth at the same time. Selecting a "best" aperture distribution depends heavily on what is required from the antenna. In a large earth station antenna the emphasis is on maximum gain, and a near-uniform amplitude distribution is used. The resulting high sidelobe levels can be tolerated because of the narrow angular scale of the beam. In a small earth station antenna the sidelobes are spread much more widely and the aperture amplitude distribution must be carefully tapered to keep the sidelobes within the applicable specification. The design of large and small earth station antennas is discussed in detail in Sections 9.4 and 9.5.

In practical antennas we do not know $E(x)$ as an analytical function, and we may also have a phase variation as well as an amplitude taper across the aperture. In general, the far-field pattern will be given by

$$E(\theta) = \int_{-a/2}^{+a/2} |E(x)| \exp\left[-jkr \sin \theta + \Phi(x)\right] dx \qquad (9.11)$$

Where $|E(x)|$, $\Phi(x)$ describes the complex field distribution in the aperture. We may not be able to carry out the integration analytically, but we can obtain the far-field pattern by numerical integration using a computer. Sufficiently small increments dx across the aperture must be used in the numerical integration to avoid ambiguities in the far field pattern (these appear as *grating lobes*), but otherwise the procedure is straightforward.

Rectangular Apertures

Practical antennas have apertures with finite area extending in two dimensions. For the case of a rectangular aperture with dimensions a and b in the x and y directions, the general form of the far field pattern is [3]

$$G(\theta, \phi) = \int_{-a/2}^{+a/2} \int_{-b/2}^{+b/2} E(x, y) \exp\left[jk \sin \theta (x \cos \phi + y \sin \phi)\right] dx \, dy \quad (9.12)$$

where $E(x, y)$ is the aperture illumination function and θ and ϕ are the angles of a spherical coordinate set with its origin at the center of the aperture, as shown in Figure 9.5.

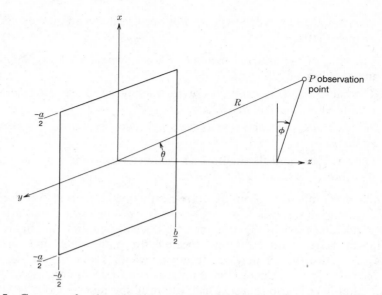

Figure 9.5 Geometry for the calculation of the far-field pattern of a rectangular aperture.

In the *principal planes*, the two integrations in Eqs. (9.12) can be separated so that the pattern is just that produced by the aperture field distribution in that plane. The illumination efficiency, η_I, of the aperture is given by the product of the illumination efficiencies for the two aperture field distributions

$$\eta_I = \eta_x \times \eta_y \tag{9.13}$$

Example 9.2.1

Consider the case of a rectangular waveguide horn excited in the TE_{10} mode, as illustrated in Figure 9.6. The E field is uniform in the x direction, across the horn dimension a, and varies as $\cos(\pi y/b)$ in the y direction. The resulting patterns in the xz and yz planes are [3, p343]

$$E(\theta)_{xz} = \frac{\sin\left(\dfrac{\pi a}{\lambda} \sin\theta\right)}{\dfrac{\pi a}{\lambda} \sin\theta}$$

and

$$E(\theta)_{yz} = \frac{\pi^2}{4}\left[\frac{\cos\left(\dfrac{\pi b}{\lambda} \sin\theta\right)}{\dfrac{\pi^2}{4} - \left(\dfrac{\pi b}{\lambda} \sin\theta\right)^2}\right]$$

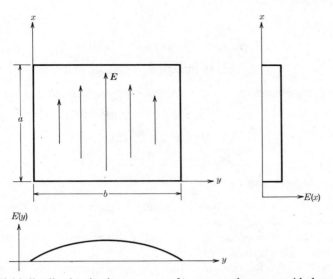

Figure 9.6 Field distribution in the aperture of a rectangular waveguide horn supporting the TE_{10} mode.

The aperture field distribution $E(x)$ is uniform, giving an illumination efficiency $\eta_x = 1$. The distribution $E(y)$ is of the form $\cos(\pi y/b)$, which from Table 9.2 gives an illumination efficiency η_y of 0.81. Thus the horn illumination efficiency η_I is

$$\eta_I = \eta_x \times \eta_y = 1 \times 0.81 = 0.81$$

Thus the gain of the horn is given by

$$G = \eta_I \cdot \frac{4\pi A}{\lambda^2} = 0.81 \times \frac{4\pi ab}{\lambda^2}$$

The above analysis assumes uniform phase across the horn aperture. This is approximately the case for small horns with long flare length, but large horns have significant phase curvature due to the difference between the flare length and axis length from the horn throat to the aperture plane. Phase error in the aperture results in lower gain and higher sidelobes in the far-field pattern.

If we assume a small horn with $a = 2\lambda$ and $b = 3\lambda$, we can calculate the gain and beamwidths in the principal planes and the first sidelobe levels. Using the data in Table 9.2

$$\text{gain} = 0.81 \times \frac{4\pi ab}{\lambda^2}$$

$$= 0.81 \times 4\pi \times 6 = 61 \quad \text{or} \quad 17.9 \text{ dB}$$

3-dB Beamwidths

$$E \text{ plane } (x \text{ direction}) = (51\lambda/a) = 25.5°$$
$$H \text{ plane } (y \text{ direction}) = (69\lambda/b) = 23°$$

The first sidelobe-levels in the principal planes are, from Table 9.2

$$E \text{ plane:} \quad 13.2 \text{ dB}$$
$$H \text{ plane:} \quad 23 \text{ dB}$$

Circular Apertures

We can use the same analysis procedure to calculate the far-field pattern of a circular aperture. We define an aperture distribution $E(r, \psi)$ where r is the radial distance from the center of the aperture and ψ is the angle in the aperture plane from a reference direction (usually the x direction) as shown in Figure 9.7. At a distant point P, in the far field, a distance R from the center of the aperture, the field is given by

$$E(R, \theta, \phi) = \frac{A}{R_1} \int_0^{2\pi} \int_0^{r_0} rE(r, \psi) \exp\left(-j\frac{2\pi R_1}{\lambda}\right) dr\, d\psi \qquad (9.14)$$

where r_0 is the radius of the aperture. Making the assumption that $1/R_1 \simeq 1/R$ and normalizing the field by ignoring the $A \exp(jkR)/R$ term, gives the general result

$$E(\theta, \phi) = \int_0^{2\pi} \int_0^{r_0} rE(r, \psi) \exp\left[-jkr \sin\theta \cos(\phi - \psi)\right] dr\, d\psi \qquad (9.15)$$

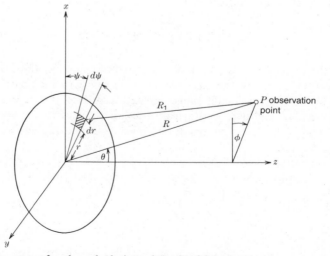

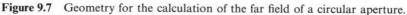

Figure 9.7 Geometry for the calculation of the far field of a circular aperture.

where (R, θ, ϕ) are the polar coordinates of the far-field point, with origin at the center of the aperture.

Evaluating this integral analytically is difficult for all but the simplest aperture field distributions. In the simplest case of uniform illumination of the aperture, $E(r, \psi) = 1$, Eq. (9.15) gives

$$E(\theta) = \pi r_0^2 \frac{2J_1(u)}{u} \tag{9.16}$$

where

$$u = \frac{2\pi r_0}{\lambda} \sin \theta$$

and the pattern is circularly symmetric about the beam axis. In deriving the far-field pattern for the uniformly illuminated circular aperture, we have ignored several amplitude factors. To normalize $E(\theta)$ in Eq. (9.16) to the gain of the antenna, $G(\theta)$, we must replace the πr_0^2 factor by $\sqrt{4\pi A/\lambda^2} = 2\pi r_0/\lambda = \pi D/\lambda$, where D is the diameter of the aperture. Table 9.3 gives results for illumination efficiency, 3-dB beamwidth, and first sidelobe level for some analytical aperture distributions.

The cosine illumination function closely approximates many tapered aperture distributions found in practice. Thus antennas with circular apertures often have illumination efficiency around 0.75, half-power beamwidth of $72\lambda/D°$, where D is the aperture diameter, and first sidelobe slightly greater than 24 dB below the peak of the main lobe [4]. The gain of a circular aperture with diameter D is given by

$$G = \eta_I \frac{4\pi A}{\lambda^2} = \eta_I \left(\frac{\pi D}{\lambda}\right)^2 \tag{9.17}$$

When other losses are taken into account, the aperture efficiency η_a will be less than the illumination efficiency η_I, and is typically 55 to 60 percent for a front-fed

Table 9.3
Characteristics of Circular Apertures

Illumination function	Illumination efficiency	3-dB Beamwidth (rad)	Position of first null (rad)	First sidelobe level (dB)
$f(x) = (1 - r^2)^n$,		$r \leq 1$,	aperture diameter $= D$	
$n = 0$	1.00	$1.02\lambda/D$	$\sin^{-1} 1.22\lambda/D$	-17.6
$n = 1$	0.75	$1.27\lambda/D$	$\sin^{-1} 1.63\lambda/D$	-24.6
$n = 2$	0.56	$1.47\lambda/D$	$\sin^{-1} 2.03\lambda/D$	-30.6

Source: Reprinted with permission from S. Silver, Ed., *Microwave Antenna Theory and Design*, Vol. 12, M.I.T. Radiation Lab. Series, McGraw Hill, 1949, Table 6.2, p. 195.

paraboloid. Then

$$G \simeq 0.6 \left(\frac{\pi D}{\lambda}\right)^2 \tag{9.18a}$$

and

$$\theta_{3dB} \simeq \frac{72\lambda}{D} \text{ degrees} \tag{9.18b}$$

Substituting for λ/D in the gain expression gives

$$G = 0.6 \left(\frac{72\pi}{\theta_{3dB}}\right)^2$$

$$= \frac{30,698}{(\theta_{3dB})^2} \tag{9.19}$$

This is the basis for the rule of thumb quoted in Chapter 3 for computing antenna gain from the 3-dB beamwidth.

9.3 DESIGN OF LARGE ANTENNAS

Large antennas are expensive to construct and install, with costs exceeding $1M for 30-m diameter fully steerable antennas. This makes it worthwhile to achieve the highest possible aperture efficiency and the lowest possible noise temperature so that G/T is maximized. The cost of large fully steerable antennas has been quoted as [5]

$$\text{cost} = y(D)^{2.7} \tag{9.20}$$

Another source [6] gives the cost as

$$\text{cost} = \$(42 + 3.13D + 0.191D^2) \times 1000 \tag{9.21}$$

where D is the diameter of the antenna aperture in feet.

The constant y in Eq. (9.20) depends on the currency used and inflation, but might typically be around five U.S. dollars in the early 1980s. This gives a figure of $1.25M for a 100-ft diameter steerable antenna using Eq. (9.20) and a figure of $2.26M using Eq. (9.21), which lie within the range of expected values for 100-ft earth station antennas. The power 2.7 reflects the fact that an antenna grows in three dimensions as the dish diameter increases. The need for a stronger support structure as the reflector size is increased forces up the cost more nearly as D^3 than D^2.

Taking our 100-ft diameter reflector as a baseline, consider the cost of increasing the antenna gain by 1 dB, or a factor of 1.258. Since $G \propto 4\pi A/\lambda^2$, we need to increase the reflector diameter by $\sqrt{1.258}$ to 112.2 ft. Using Eq. (9.20), the cost of a 112-ft antenna is $1.7M with $y = 5$, an increase of $450,000 to obtain just 1 dB extra gain. Antennas of this size must therefore achieve very high efficiency and low system noise temperature to keep the diameter of the reflector as low as possible for a given G/T.

The earliest earth station antennas used with the TELSTAR satellite experiment had apertures of about 25-m diameter. Bell Labs built a large *horn-reflector* antenna located at Andover, Maine, which was a scaled-up version of the 6/4 GHz hoghorns used in microwave line of sight links [7]. The horn-reflector or *hoghorn* has very low far-out sidelobes and provides better rejection of interference than open reflector antennas by virtue of the screening around the aperture. Andover was chosen as the location to minimize interference from microwave links operating in the 4-GHz band, because of concerns about potential interference levels. Experience showed that provided the earth station site was chosen with care, interference was not a severe problem and the very high cost of the hoghorn was not justified. The French PT&T authority built a hoghorn similar to the Bell Labs design at Plemeau-Bodeau in northwest France for the TELSTAR experiment.

The British Post Office constructed a 26-m diameter *front-fed* paraboloidal antenna at Goonhilly Downs, in the southwest corner of England, using a design developed from the large aperture antennas used for radio astronomy [8]. This is the only front-fed antenna that has ever been used for a Standard A earth station, but has the notable distinction of still being in service in 1984, 23 years after it was built for one experiment with TELSTAR. The hoghorns at Maine and Plemeau-Bodeau were replaced by Standard A Cassegrain antennas.

The front-fed paraboloid is widely used as a configuration for smaller earth stations, but has an excessively long waveguide run from the rear of the reflector out to the feed at the focus of the dish in a large antenna. The loss in this waveguide gives a major contribution to the system noise temperature, reducing the G/T ratio. The preferred design for all large earth station antennas built since the original TELSTAR experiment has been the *Cassegrain* configuration, which requires much shorter waveguide runs than the corresponding front-fed configuration. The *Gregorian* configuration is also used by some manufacturers. The geometry of front-fed, Cassegrain and Gregorian configurations is illustrated in Figure 9.8. The names Cassegrain and Gregorian are derived from the seventeenth-century astronomers William Cassegrain and James Gregory, who designed optical reflecting telescopes using these geometries. Isaac Newton also constructed reflecting telescopes using a flat plate for the secondary reflector, but this configuration has not been widely adopted for microwave antennas.

The Cassegrain antenna is popular for large earth stations for several reasons:

1. The gain can be increased by approximately 1 dB, relative to a front-fed reflector, by *shaping* of the dual reflector system.
2. Low antenna noise temperatures can be achieved by controlling *spillover* and using short waveguide runs or *beam waveguide* feeds.
3. Beam waveguide feeds place the low noise amplifier of the receiver, and other complex feed components, in a convenient, stationary position.

The design procedure for a Cassegrain antenna is discussed in the next section.

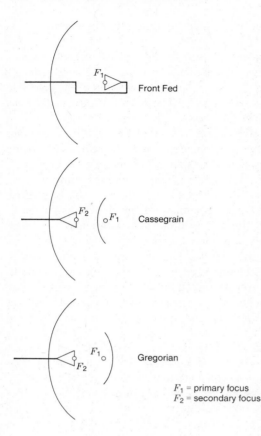

F_1 = primary focus
F_2 = secondary focus

Figure 9.8 Front-fed, Cassegrain, and Gregorian configurations of reflector antennas.

Large Cassegrain Antennas

The basic geometry of the Cassegrain antenna is shown in Figure 9.9. The main reflector is a paraboloid, which reflects incoming radiation toward the prime focus. The hyperboloid secondary reflector, often called a *subreflector* or *subdish*, has one focus coincident with the paraboloid reflector's focus, and a feed system is placed at the second focus, such that the phase center of the feed coincides with the hyperboloid's second focus. The paraboloid converts an incoming plane wave to a spherical wave converging toward the prime focus, which is then reflected by the subreflector to form a spherical wave converging on the feed, as illustrated in Figure 9.9.

The design of a Cassegrain antenna is conventionally carried out by assuming that the feed is transmitting. By reciprocity, the same antenna pattern and aperture efficiency will be achieved when the antenna is receiving. Thus we need a feed system with a radiation pattern such that the subreflector intercepts most of the energy transmitted by the feed. Typically, the subreflector edge is *illuminated* with a power level 10 to 15 dB below that at the center. This keeps the spillover from

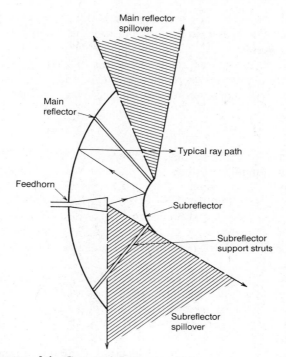

Figure 9.9   Geometry of the Cassegrain antenna.

the feed, past the edge of the subreflector, to a low level. Spillover reduces the efficiency of the antenna and increases its noise temperature, so it must be kept as low as possible. With a *corrugated horn* as the feed, spillover can be reduced to around 3 percent in a well-designed Cassegrain antenna [9]. Figure 9.9 illustrates spillover losses in a Cassegrain antenna. Spillover also occurs when the subreflector radiates toward the main reflector. It can be kept to a minimum by using a large-diameter subreflector that gives accurate control of the subdish radiation pattern, and by keeping the subreflector edge illumination low.

The gain of the antenna is the product of many factors that contribute to its aperture efficiency. In general, the gain is given by

$$G = \eta \cdot \frac{4\pi A}{\lambda^2} \tag{9.22}$$

where A is the aperture area of the antenna and η is the aperture efficiency. Aperture efficiency can be computed from efficiencies of various parts of the antenna. Major contributors to this factor are the *illumination efficiency* of the main aperture, η_I, the *spillover efficiency* (1 − spillover loss) for the subreflector and the main reflector, η_{ss} and η_{ms}, and *blockage efficiency*, η_b, which accounts for radiation lost by scattering from the subreflector and its support legs after reflection by the main dish.

Other losses contribute to aperture efficiency to a lesser degree. The most important ones are *ohmic losses* in the feed and waveguide components and *surface errors* on the main reflector.

Many other losses such as polarization loss, reflections at the feed, phase errors across the feed, gaps between reflector panels, and so on can be introduced to refine the calculation of aperture efficiency. For the efficiency factors given above we have

$$\eta = \eta_I \times \eta_{ss} \times \eta_{ms} \times \eta_b \times \ldots \tag{9.23}$$

Typical values for each of these efficiencies are given in Table 9.4 for the 27.4-m Cassegrain antenna operated by Cable and Wireless at Bahrein [10]. The overall efficiency is 70.8 percent, including 8.81 percent losses due to phase errors in the aperture resulting from horn phase error, reflector surface error, and misalignments of the feed and subreflector [10]. Any radiation in the polarization orthogonal to that desired will not be accepted by the distant receiving antenna and represents a polarization loss that should be included in the efficiency calculation.

A calculation such as that illustrated in Table 9.4 is an essential part of the design process of a large antenna. The gain is calculated from Eq. (9.22) using the aperture efficiency factor derived in Table 9.4. For this example, the gain of a 27.4-m diameter circular aperture at 4.0 GHz with an aperture efficiency of 70.8 percent is 59.7 dB.

The efficiency factors used in this calculation cannot be found by the application of simple geometric optics to the antenna. Diffraction losses around the

Table 9.4
Efficiency Calculation for a Large Cassegrain Antenna

Efficiency Factor	Symbol	Loss (%)	Loss (dB)	Efficiency (%)
Illumination efficiency	η_I	1.34	0.059	98.66
Subreflector spillover	η_{ss}	11.73	0.542	88.27
Main reflector spillover	η_{ms}	4.00	0.177	96.00
Blockage losses	η_b	7.40	0.334	92.60
Phase errors and surface errors	η_{pe}	7.56	0.340	92.44
Polarization loss	η_{xp}	1.15	0.050	98.85
Aperture efficiency, %	η			70.74
Total of losses, dB			1.502	

Source: Reproduced from N. Lockett, "The Electrical Performance of the Marconi 90 ft Space Communication Aerials," *Marconi Review*, **34**, 50–80 (1971), by permission of the editor, GEC Journal of Technology.

subreflector, phase-error losses, and polarization losses can be calculated only with a wave analysis using numerical techniques. A suite of computer programs is required to derive the factors in Table 9.4, each factor being found by the procedure described below.

The first step in the process of computing antenna gain and efficiency factors is to establish the two-dimensional pattern of the feed system. This pattern may be measured on an antenna range or generated from an algorithm that models the feed radiation. The pattern should be known in amplitude, phase, and polarization, in polar coordinates with origin at the nominal phase center of the feed. The feed pattern is used to calculate the current induced on the surface of the sub-reflector using the geometric optics approximation

$$\mathbf{K} = 2\mathbf{n} \times \mathbf{H}_i \tag{9.24}$$

where $\mathbf{K}$ is the vector surface current density, $\mathbf{n}$ is a unit normal to the reflector surface, and $\mathbf{H}_i$ is the incident magnetic field. The current must be calculated at a large number of points on the subreflector surface in order to obtain an accurate prediction of the radiation pattern of the subreflector. If the subreflector and feed are close together, the feed pattern will have to be corrected for near-field effects.

The radiation pattern from the subreflector is then calculated at a large number of points on the main reflector, so that Eq. (9.24) can be used to establish the surface current distribution on the main reflector. Finally, the radiation pattern of the main reflector can be calculated. The spillover efficiency factors in Table 9.4 are obtained by comparison of the total power radiated by the feed with the power incident on the subreflector surface and the main reflector surface, in the wanted polarization. Blockage loss is calculated by assuming an approximate blocking pattern and computing the energy lost from the main aperture by scatter from the blocked areas [9]. Surface error losses can be computed by introducing random surface errors across the reflector surface, with a given rms value, and calculating the reduction in the on-axis field strength.

As an example of what can be achieved by these techniques, Figure 9.10 shows a typical calculated pattern for the subreflector of a Cassegrain antenna and Figure 9.11 the corresponding antenna pattern assuming no surface errors. Figure 9.11 also shows the pattern calculated when surface errors on the main reflector are included—in this case, the departure from the nominal surface contour had been measured at 300 points on the main reflector surface and these data were used in the calculation. Figure 9.11 also includes the measured azimuth plane pattern of this antenna, which shows close agreement to the calculated pattern out to the seventh sidelobe peak. Note that inclusion of reflector surface errors in the calculation of the far-field pattern fills in the nulls of the theoretical pattern and modifies the sidelobe peaks away from the main beam. Null-filling is characteristic of phase errors in the antenna aperture. A null results from antiphasing of energy from different parts of the aperture to achieve total cancellation. Cancellation will occur only with a constant-phase aperture distribution. If phase errors, or curvature of the phase front, are present in the aperture field, complete cancellation of field contributions at a far field point cannot occur. This is easily

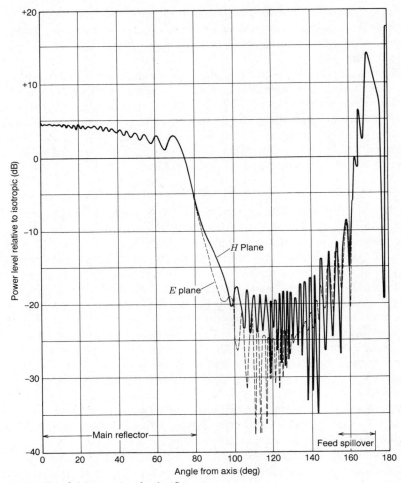

Figure 9.10 Far-field pattern of subreflector.

understood by separating the aperture field into in-phase and quadrature components. When the in-phase components cancel, there will be residual quadrature components preventing the field from falling to zero.

In the procedure described here, the surface currents in the subreflector and main reflector were calculated using the incident magnetic field. An alternative analysis procedure that leads to equivalent results uses the electric field reflected from a conducting surface according to the equation

$$\mathbf{n} \times (\mathbf{E}_i + \mathbf{E}_r) = 0 \qquad (9.25)$$

where $\mathbf{E}_r$ is the reflected electric field just above the surface and $\mathbf{E}_i$ is the incident electric field, with $\mathbf{n}$ a unit vector normal to the surface at the point of reflection.

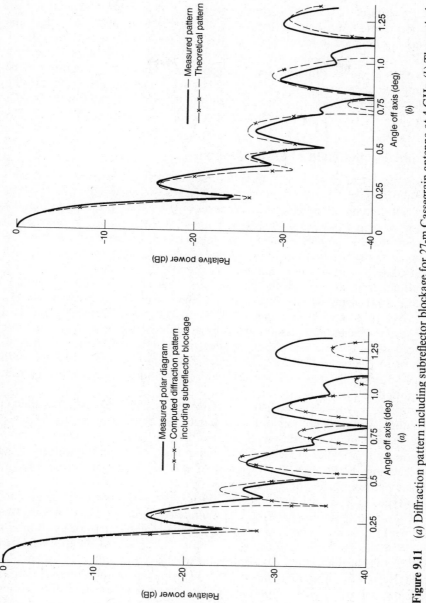

Figure 9.11 (*a*) Diffraction pattern including subreflector blockage for 27-m Cassegrain antenna at 4 GHz. (*b*) Theoretical pattern including the effects of diffraction, blockage and profile errors, and measured pattern. (Adapted from T. Pratt and B. Claydon, "The Prediction of Polar Diagrams of Large Cassegrain Antennas," *Marconi Review*, **34**, 50–80 (1971).)

Both techniques assume that the surface is flat in the region around the reflection point, and both lead to errors at the edges of the reflector where the current flow is assumed to terminate abruptly, whereas in practice it cannot do so. The errors introduced by ignoring edge currents are small for large Cassegrain antennas with low edge illumination. The geometric theory of diffraction can be used to improve the accuracy of pattern calculations by including edge effects [11]. In a large antenna, the improvement is seen mainly in the calculation of far-out sidelobes.

Optimizing the Gain of Large Antennas

The importance of achieving maximum efficiency in a large antenna has already been discussed. To achieve maximum efficiency, we must reduce all losses as far as possible and increase efficiencies whenever possible. Two powerful techniques have been used in Cassegrain antennas, both of which resulted from research aimed at improving the efficiency of large antennas used in satellite communication earth stations. The first technique is the use of feed systems with symmetrical radiation patterns; the second technique is shaping of the dual reflectors of a Cassegrain antenna to improve the illumination efficiency.

Feed-pattern symmetry is important in reflector antennas for two reasons. The spillover past the reflector cannot be kept low unless the feed radiates a circularly symmetric pattern (assuming a circular reflector). In the example in Table 9.4 the feed did not have a circularly symmetric pattern and introduced a spillover loss of 11.7 percent. If this loss had been 3 percent, which is achievable with a corrugated horn, the antenna aperture efficiency would have been 79 percent and the gain 0.48 dB greater.

Feed-pattern symmetry also helps to control cross-polarization in the antenna [12]. In most large earth station antennas, signals can be transmitted and received at the same frequency on orthogonal polarizations providing frequency reuse to increase the system traffic capacity. The antenna must maintain separation of the two polarizations by at least 27 dB, and preferably 35 dB, making control of the cross-polar pattern of the antenna an important factor.

Two types of feed have been used in earth stations to provide circular symmetry of the feed-radiation pattern. The *multimode horn*, as exemplified by the Potter horn [13], achieves circular symmetry of the radiated pattern by summing waveguide modes in the horn aperture to obtain a circularly symmetric field distribution in the horn aperture. The Potter horn adds TE_{11} and TE_{31} in circular waveguide. The TE_{11} and TE_{31} modes do not propagate at the same velocity in the horn and will arrive at the aperture in the correct relative phase only over a narrow band of frequencies. This makes the design of a Potter horn for 500-MHz bandwidths at 6 and 4 GHz simultaneously very difficult.

The corrugated horn uses hybrid modes to obtain a circularly symmetric field distribution in the horn aperture. The hybrid mode is effectively a combination of TE_{11} and TM_{11} modes in circular waveguide excited and maintained in the horn by slots cut into the internal wall of the horn. The slots are typically about

$\lambda/4$ deep and $\lambda/3$ apart. They cause the wall of the horn to present a reactive impedance to the surface field and prevent termination of the E field vector on the surface. The corrugated horn can provide excellent circular symmetry, good cross-polarization performance, and wide bandwidth. It is widely used for Cassegrain antenna feeds [14].

Shaping the reflectors of a Cassegrain or Gregorian antenna can increase the gain by 1 dB by improving the illumination efficiency [9, 15]. In a front-fed reflector antenna, or a conventional Cassegrain, there are conflicting requirements on the feed pattern. We want near-uniform illumination of the main reflector to obtain high aperture efficiency, but low edge illumination to achieve low spillover. When the horn aperture is only a few wavelengths in diameter, such precise control of its radiation pattern is not possible. However, if we change the shape of the subreflector in a dual-reflector antenna, we can redistribute the energy radiated by the feed so that more energy is directed toward the outer edge of the main reflector. Figure 9.12 illustrates the way in which this is done. A feed system that has a high taper at the subdish edge is used to give low spillover loss, edge illuminations of -15 to -20 dB being typical. Reshaping the subreflector produces a field distribution in the main reflector aperture that is uniform in the center and

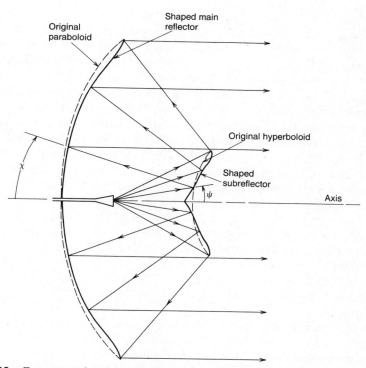

Figure 9.12 Geometry of a shaped Cassegrain antenna, illustrating redistribution of the energy radiated by the feed across the main reflector aperture.

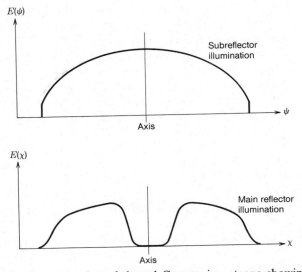

Figure 9.13   Aperture illumination of shaped Cassegrain antenna showing redistribution of energy.

heavily tapered at the periphery, as illustrated in Figure 9.13. The resulting illumination efficiency can be close to 100 percent. The figure of 98.66 percent for illumination efficiency in Table 9.4 was obtained by shaping the subdish of the antenna used in that example.

Changing the shape of the subreflector will lead to phase errors in the main aperture unless a compensatory change is made in the main reflector shape. However, we are attempting to control two parameters, the amplitude and phase of the field in the antenna aperture with two reflector profiles, so an infinite number of solutions is possible. The most widely used design approach is to start with the conventional hyperboloid-paraboloid combination of the Cassegrain antenna, and then redesign the hyperboloidal surface to redistribute the energy as required, for a given feed system radiation pattern [16]. The main reflector shape is then adjusted to correct for the phase error introduced by the change in subdish profile. The process has to be done numerically and can be iterated to yield the smallest deviation from the original paraboloid. This ensures that little variation in the amplitude distribution of the field in the antenna aperture is introduced by the change in main reflector shape. The maximum departure of the shaped main reflector from the original paraboloid can be held within $\lambda/2$ in a 25-m diameter antenna.

Note that shaping the reflectors to achieve a specified illumination pattern in the antenna aperture is based on a particular feed pattern. If the feed pattern changes, the aperture illumination will depart from the ideal and increased spillover or reduced gain may result. Most large earth station antennas must transmit and receive with the same feed, which places a stringent requirement on the feed pattern. The corrugated horn can be designed to give near-constant beamwidth over a wide frequency range by tapering the slot depth and spacing, which makes it a popular choice for shaped Cassegrains.

The choice of subreflector diameter influences the main reflector spillover loss and the blocking loss in a Cassegrain antenna. We need a large subreflector to obtain accurate illumination of the main reflector and reduce diffraction losses at the subreflector edge, but a large subreflector causes high blockage losses and needs heavy support legs. In a large Cassegrain antenna, a subreflector diameter of 20 to 40 wavelengths is commonly used. If the main reflector diameter is 25 m, the blockage introduced by a 40λ subreflector at 4 GHz is 1.44 percent of the aperture area. The subreflector diameter is 3 m in this example. Because the subreflector blocks the central region of the antenna aperture, the loss in gain is slightly greater than 1.44 percent. Some reduction in this loss can be achieved by placing a "pip" in the center of the subreflector to direct radiation away from the central part of the main aperture, as illustrated in Figure 9.12. Subreflector blocking invariably increases sidelobe levels.

Antenna Noise Temperature

Large Cassegrain antennas operating in the 6/4-GHz band must have low noise temperatures to meet stringent G/T specifications. Two items contribute most of the noise that makes up the system noise temperature: the low noise amplifier (LNA) and the antenna. Prediction of the antenna noise temperature under operating conditions is an essential part of the design process of a large earth station antenna.

The calculation of antenna noise temperature is straightforward. It involves the addition of noise temperatures, representing thermal noise contributions, for each of the losses in Table 9.4 plus a contribution from the sky. Sky noise is not constant; it depends on elevation angle and atmospheric conditions, so the antenna noise temperature must be calculated at a specified elevation angle in clear-sky conditions. An increase in antenna noise temperature will occur when attenuation in the atmosphere is present due to rain, which must be taken into account when calculating the (C/N) at the receiver for small percentages of time when rain attenuation is present.

Sky Noise

Sky noise arises because there is attenuation of a wave as it passes through the atmosphere. Causes of this attenuation include atmospheric gases and hydrometeors (rain, snow, etc.) and are discussed in detail in Chapter 8. There is also a galactic contribution to sky noise. At microwave frequencies this is 3 K, except for those directions in the sky corresponding to radio stars, but at lower frequencies (below 1 GHz) galactic noise is significant. Figure 9.14 shows the variation of "average" galactic noise with frequency. Mapping sky noise temperature at

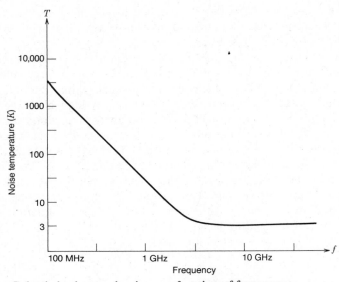

Figure 9.14 Galactic background noise as a function of frequency.

microwave frequencies is the basis of radio astronomy. The galactic background temperature of 3 K was discovered by radio astronomers when they tried to calibrate the noise temperature of their antennas. The residual 3 K is attributed to hot matter at large distances in galactic space and gave rise to the Big Bang theory [17]. For most practical purposes, galactic noise is ignored in the calculation of earth station noise temperatures.

Although sky noise can be calculated from atmospheric attenuation, it is usually determined from graphs or tables under clear-sky conditions. Figure 9.15 shows the variation in sky noise with elevation angle for two frequencies, 4 GHz and 11 GHz. The higher noise temperatures at 11 GHz are caused by the increased attenuation experienced at higher microwave frequencies, especially at low elevation angles where the path in the atmosphere is long. Sky noise increases at low elevation angles, so antenna noise temperature is calculated for the lowest working elevation angle of the antenna, usually 5° for 4-GHz systems and 10° for 11-GHz links.

Calculating Antenna Noise Temperature

Not all of the energy radiated by the atmosphere is coupled through the antenna to its output port. The losses that occur in an antenna reduce the signal delivered to the output port, relative to the total energy arriving at the antenna aperture. For a point source transmitter, the aperture efficiency η accounts for all of these losses. However, sky noise is a distributed source. To calculate the antenna noise temperature contribution due to sky noise with high accuracy we would

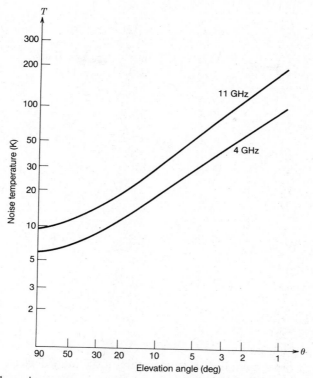

Figure 9.15 Sky noise temperature as a function of elevation angle at 4 GHz and 11 GHz. Clear-air conditions.

need to convolve the two-dimensional gain pattern of the antenna with the distribution of sky noise temperature over the upper hemisphere and earth noise temperature over the lower hemisphere. In practice we cannot perform the exact calculation because we do not have full knowledge of the two-dimensional antenna pattern and must estimate the sky and earth noise temperatures.

An approximate method that yields antenna temperature to an acceptable accuracy is described below. The procedure used is to take each loss factor in the antenna efficiency calculation, allocate a noise temperature to that source of loss, and then calculate its noise temperature contribution. We are forced to estimate the source noise temperatures for some loss factors, the antenna far-out sidelobes, for example, simply because we have no way of accurately determining them.

Table 9.4 shows the major loss factors in a typical large earth station antenna. We will allocate noise temperatures to each of these factors and calculate the noise temperature for this antenna, as illustrated in Table 9.5.

Using the noise models developed in Chapter 4, a loss is described as a gain G_l, less than unity, and the noise temperature contributed by the loss is T_l, where

$$T_l = T_p(1 - G_l) \tag{9.26}$$

Table 9.5

Example of Calculation of Antenna Noise Temperature

Source	Gain (G_l)	Loss $(1 - G_l)$	Source Temp. (T_p)	Contribution
Sky	0.707	0.293	32 K	22.6 K
Illumination efficiency	0.987	0.013	32 K	0.4 K
Subreflector spillover	0.883	0.117	90 K	10.5 K
Main reflector spillover	0.960	0.040	90 K	3.6 K
Subreflector and support blockage	0.926	0.074	90 K	6.7 K
Phase errors and surface errors	0.924	0.076	32 K	2.5 K
Total without ohmic losses				46.3 K
Feed and waveguide ohmic losses	0.959	0.041	290 K	11.9 K
Antenna noise temperature				58.2 K

and T_p is the physical or radiation temperature of the lossy device or mechanism, which is usually taken as 290 K.

The overall efficiency of the antenna used as an example in Table 9.4 was calculated to be 70.7 percent. If the antenna points at 5° elevation angle, the sky noise temperature at 4 GHz is 32 K. The antenna accepts only 70.7 percent of the energy from a distant transmitter incident on its aperture, so we can assume that only 70.7 percent of the sky noise incident on the aperture is output by the antenna, for the main beam region of the antenna pattern. Thus the sky noise contribution via the main beam is $0.707 \times 32 \text{ K} = 22.6 \text{ K}$. The figure of 70.7 percent efficiency was derived from a transmit mode analysis of the antenna. When the antenna is receiving, 70.7 percent of the antenna output comes from the main beam direction; the remaining 29.3 percent comes from other directions and sources, each of which will transmit noise to the antenna output port.

We can obtain an estimate for the noise contributed by the 29.3 percent fraction by considering the various losses that make up this figure. Referring to Table 9.4, these are illumination efficiency, subreflector and main reflector spillover, blockage, phase errors and surface errors, and polarization loss. To arrive at an estimate of the antenna temperature, we assign a source temperature to each of these losses and use Eq. (9.26) to calculate the noise temperature contributions. Table 9.5 shows an example of this calculation using the loss factors in Table 9.4.

Spillover is allocated a source noise temperature of 90 K, the average for sky at 30 K and earth at 150 K. At an elevation angle of 5°, approximately half of the subreflector and main reflector spillover is directed toward the sky and half toward the earth. The aperture efficiency is allocated a source noise temperature of 32 K because its primary influence is on the main beam region of the antenna pattern. The earth has a varying radiation temperature depending on its surface characteristics. Foliage tends to have a noise temperature close to its physical temperature, while water lying on the surface has a low noise temperature. A value of 150 K is typical.

The noise contribution from each of these sources is calculated by multiplying each loss factor by its source temperature. Addition of all the contributions, plus the sky noise at the particular elevation chosen, gives the antenna noise temperature at the output of a lossless feed. Ohmic losses in the feed and waveguide components of the antenna must be added to obtain the total antenna noise temperature at the output port.

In the example in Table 9.4, the feed and waveguide losses were 0.18 dB. These losses give a noise contribution of 11.9 K, for a physical temperature of 290 K. Summing all the contributions to antenna noise temperature at 5° elevation angle, the final figure is 58.2 K.

When we add the LNA noise temperature, we have the system noise temperature excluding any contribution from later stages in the receiver. The gain of the LNA is always made high enough to reduce the effect of later stages in the receiver to a negligible level, giving us a system noise temperature

$$T_{\text{system}} \simeq T_{\text{antenna}} + T_{\text{LNA}} \tag{9.27}$$

If we use a parametric amplifier with a 20 K noise temperature for our 4-GHz antenna example in Table 9.5, the system noise temperature will be 78.2 K. The measured system noise temperature of the large Cassegrain antenna on which this example was based was 75 K [10]. This figure will fall as the elevation angle is increased, because the contribution from sky noise will fall as shown in Figure 9.15. However, main reflector spillover "sees" a higher source temperature as the elevation angle increases and more of the spillover is directed at the earth, but at the same time subreflector spillover gives a smaller contribution. At zenith, 90° elevation angle, the sky noise temperature will be about 3 K and the antenna temperature falls to 48 K. Figure 9.16 shows an example of how G/T and T_s change with elevation angle for a large Cassegrain antenna in a Standard A INTELSAT earth station [10].

The calculated gain of the antenna in Table 9.4 is 59.5 dB at 4 GHz after inclusion of 0.2-dB waveguide loss, and the system noise temperature is 78 K from Table 9.5, giving a G/T of 40.6 K at 5° elevation angle. This is 0.1 dB below the minimum 40.7 K specified by Intelsat for a Standard A earth station. However, Figure 9.16 shows that the measured G/T of this antenna at 4 GHz was 41.2 dB, some 0.5 dB above specification, indicating a pessimistic calculation.

The system noise temperature has been calculated at the input of the low noise amplifier, following the method used in Section 4.2. When there is significant loss in the waveguide between the antenna and the LNA, the antenna noise temperature is usually quoted prior to the waveguide loss. The LNA noise temperature

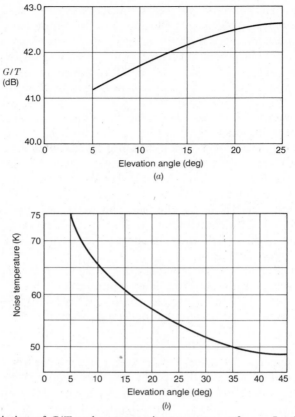

Figure 9.16 Variation of G/T and system noise temperature for an Intelsat Standard A earth station. (Reproduced by permission from N. Lockett, "The Electrical Performance of the Marconi 90 ft. Space Communication Aerials," *Marconi Review*, **34**, 1–26 (1971).) (*a*) Variation of G/T with elevation angle. (*b*) Variation of system noise temperature with elevation angle.

must be increased by a factor $1/(1 - G_w)$, where G_w is the gain (< 1) of the waveguide, and added to the antenna noise temperature to obtain the system noise temperature referred to the antenna output port.

The preceding example was for a 6/4-GHz earth station, where antenna temperature is most critical. At 14/11 GHz, the sky noise temperature rises toward 270 K during heavy rain, which causes the system noise temperature to increase correspondingly. LNAs for 11 GHz have higher noise temperatures than those for the 4-GHz band, so system noise temperatures tend to be much higher at 11 GHz than at 4 GHz. This makes antenna noise temperature much less critical in the system noise budget, although care must still be taken in the antenna design because of higher losses in waveguide at 11 GHz.

Example 9.2

A 14/11 GHz antenna has a G/T ratio of 40.3 dB at 11.2 GHz. The antenna gain is 64.0 dB and the system noise temperature at 10° elevation angle, in clear-air conditions, is 234 K. The antenna aperture efficiency and noise temperature are detailed in the list below. During heavy rain, the slant path attenuation reaches 8 dB for 0.01 percent of the year. Calculate the G/T for this fraction of the year and the corresponding reduction in (C/N) for the received signal.

> Aperture efficiency: 71.3 percent
>
> Sky noise at 10° elevation: 30 K
>
> LNA noise temperature: 150 K

The sky noise contribution to the antenna noise temperature is 30 K × 0.713 = 21.4 K. Therefore the losses in the antenna contribute the remainder of the noise, less 150 K for the LNA. Hence, all the losses add up to give a noise temperature contribution of 234 − 150 − 21.4 = 62.6 K.

When the slant path has rain attenuation of 8 dB, we can calculate the sky noise temperature from Eq. (9.26).

We will assume a medium temperature of 270 K for rain at 11 GHz. This is a typical value obtained by comparing sky noise measured with a radiometer (a noise-measuring receiver) and slant path attenuation measured with a satellite beacon for the same path and frequency. Then

$$T_{\text{sky}} = 270 \text{ K } (1 - 0.158) = 227 \text{ K}$$

The 8 dB of rain attenuation swamps the small attenuation (about 0.4 dB) that gave rise to the 30 K clear-air sky noise temperature, so we will assume that T_{sky} is 227 K when the path attenuation is 8 dB. The contribution of sky noise to antenna noise temperature will be 227 K × 0.713 = 161.8 K. Assuming that all other contributions remain unchanged, the system noise temperature is now

$$T_{\text{system}} = 161.8 + 62.6 + 150 \text{ K} \simeq 374 \text{ K}$$

$$G/T = 64.0 - 25.7 = 38.3 \text{ dBK}^{-1}$$

This represents an increase of 2.0 dB in the system noise temperature, and a corresponding reduction will occur in G/T and (C/N) when the rain attenuation on the path is 8 dB. Adding the 8 dB reduction in carrier level to the 2 dB reduction in (C/N) due to sky noise temperature increase, we have a 10-dB loss in (C/N) for 0.01 percent of the year.

Feed Systems for Large Cassegrain Antennas

The feed system in an earth station antenna may include microwave components that change the polarization of the transmitted and received signals, *diplexers* that separate signals with different frequencies or polarizations and *mode extraction*

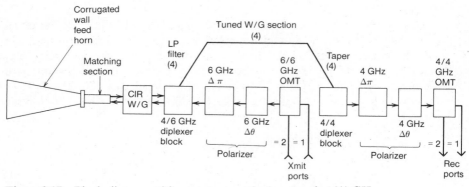

Figure 9.17 Block diagram of frequency reuse feed system for 6/4 GHz.

devices used in automatic tracking systems. As a result, such a feed can be a very complex microwave system. A block diagram for a typical frequency reuse feed for a 6/4 GHz earth station is shown in Figure 9.17.

The horn is usually a corrugated horn to provide high illumination efficiency in the antenna, with good cross-polarization characteristics.

The tracking mode extractor couples out waveguide modes generated by an off-axis transmitter (usually the satellite beacon) and provides angular error information for the autotrack system. The subject of autotrack is discussed in Section 9.4. The frequency diplexer separates the transmit and receive frequencies (6 GHz and 4 GHz) by the use of tuned sections in waveguide and must provide low loss over the 500-MHz bandwidth in both frequency bands. Loss is more of a problem at the receive frequency of 4 GHz, where the antenna gain and noise temperature are critical, so the diplexer may be optimized in the 4-GHz band at the expense of additional loss in the transmit band.

In a frequency reuse antenna, two orthogonal polarizations must be provided at both transmit and receive frequencies. The polarizations may be orthogonal circular (as in the Intelsat system) or orthogonal linear, as in most U.S. domestic systems. *An orthogonal mode transducer (OMT)* is used to separate two polarizations in one waveguide into separate signals in two waveguides. The OMT will generate or separate two orthogonal linearly polarized waves, which must be converted to orthogonal circularly polarized waves in an antenna using circular polarization. A device to do this is called a *polarizer*, and it works by shifting the phase of one of the linearly polarized waves by $\pm 90°$. There are many designs of polarizer: some use dielectric vanes within the waveguide, some have waveguides of elliptical or other asymmetric cross-section, and some add together two linearly polarized waves, one of which has first been phase shifted by $\pm 90°$. The last design is sometimes known as a *turnstile junction*.

Large earth station antennas are often equipped with motor-driven polarizers so that the antenna polarization can be changed from linear to circular. This allows the antenna to be used with more than one design of satellite. It also allows the reception of elliptical polarization when the satellite does not radiate true circular polarization, and can thus improve the isolation between the orthogonal polarization channels of a frequency reuse transmission.

The Cassegrain antenna is the preferred choice for large earth stations because of the extra gain that can be obtained by reflector shaping. However, the conventional Cassegrain antenna illustrated in Figure 9.8 places the feed output port behind the center of the main reflector. To minimize the length of waveguide between the feed output and the receiver low noise amplifier (and thus minimize the system noise temperature), the LNA must be placed immediately behind the feed. (The transmitter can be coupled to the feed via waveguide and rotating joints, since loss in the waveguide is less important on transmission.) Locating the LNA behind the center of a 25-m reflector makes maintenance very difficult and inconvenient, especially when the LNA has cryogenic cooling. To overcome this problem, the *beam waveguide feed* was developed [18, 19].

Figure 9.18a shows an illustration of a beam waveguide feed. The beam waveguide feed reflects the incoming plane wave through a series of paths, as illustrated in Figure 9.18b, to a feed system placed at the base of the antenna. This

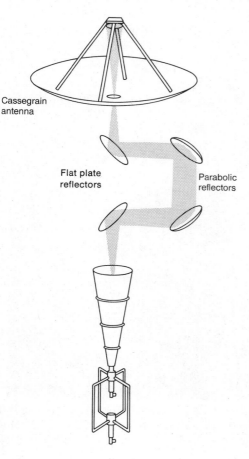

Cassegrain
antenna

Flat plate
reflectors

Parabolic
reflectors

Figure 9.18a Illustration of a beam waveguide feed. (Reprinted by permission from *Microwave Journal*.)

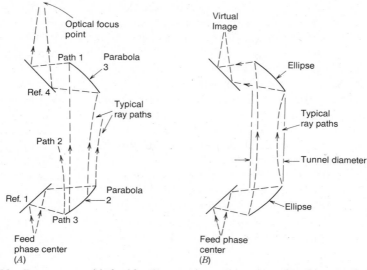

Figure 9.18b Beam waveguide feed for Cassegrain antennas. (Reprinted by permission from D. J. Sommers, L. J. Parad, and J. G. DiTullio, "Beam Waveguide Feed with Frequency Reuse Diplexer," *Microwave Journal*, **18**, 51–59 (1975).)

Figure 9.18c Beam waveguide for a 19-m 14/11 GHz Cassegrain antenna, under test at a ground site. This feed system is used in the 19-m antenna shown in Figure 9.1. (Photograph courtesy of Marconi Research Centre, U.K., and Marconi Communication Systems Ltd., U.K.)

overcomes the problem of feed and LNA accessibility. The center line of path 1 is arranged to lie along the elevation axis of the antenna so that elevation angle motion of the antenna simply rotates the planar reflector 4 and does not affect the signal transmission via this reflector. Path 2 is arranged to lie along the azimuth axis of the antenna, and the feed system is also aligned with this axis. The planar reflector 1 rotates with azimuth angle motion of the antenna, but signal transmission is unaffected.

The two concave reflectors 2 and 3 in Figure 9.18 focus the waves reflected from reflector 4 down to reflector 1. The entire beam waveguide is usually enclosed within a large cylindrical or conical tube to prevent scatter of energy on transmit or unwanted interference on receive. Typically, the loss in a beam waveguide can be kept below 0.15 dB, with cross-polar isolation of better than 40 dB [19]. Figure 9.18c shows a beam waveguide feed for a 19-m 14/11 GHz Cassegrain antenna under test at a ground site.

A variation on the designs shown in Figure 9.18b does not have a virtual image at the focus of the Cassegrain antenna's subreflector. The hyperboloidal subreflector of the conventional Cassegrain can be replaced by a paraboloid that reflects a plane wave toward the vertex of the main reflector (on reception). A beam waveguide is then used that accepts the plane wave and transfers it to a point focus at the base of the antenna. This configuration is known as a *near-field Cassegrain* antenna, because the subreflector lies in the near field of the beam waveguide aperture (and vice versa). Some modification of the subreflector surface profile may be needed in the near-field Cassegrain because the wave reflected by a paraboloidal subreflector is not truly plane at the aperture of the beam waveguide.

9.4 TRACKING

The 3-dB beamwidth of a 25-m antenna operating at 4 GHz is typically 10 min of arc, so the antenna must be pointed with an accuracy of ± 1 min of arc if pointing loss is to be avoided. If a satellite moves at all, a very accurate program-track facility is required, or autotrack must be used. There are three main systems for tracking satellites with a large antenna. They are illustrated in Figure 9.19.

1. *Monopulse* The antenna generates difference patterns with nulls on the axis, in azimuth and elevation planes. Separate tracking receivers, or one receiver time shared, are used to detect error outputs from the *Az/El* channels, and thus to drive the antenna servos. Complete separation of the tracking and communication systems is possible, but the monopulse feed and receivers tend to be expensive.

2. *Conical Scan* The receive frequency beam is rotated about the axis by the small angle, typically 1.5 min. A constant signal indicates that the satellite is on axis, whereas a rise and fall per revolution shows a pointing error. By correlating the maximum signal position with the feed position, error signals for driving the servos can be generated.

3. *Hill Climbing* The antenna beam is moved about in a predetermined fashion, and the signal amplitude noted. Maximum signal indicates the best

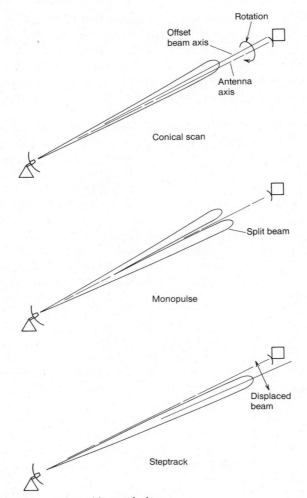

Figure 9.19 Illustration of tracking techniques.

beam position. The beam must continually be moved to check that it is in the right position. This is also known as *step track*. In a hill-climbing system the antenna is pointed away from the nominal position of the satellite by a fraction of a degree, in several directions. From a knowledge of the main beam shape, the true direction of the satellite is estimated and the antenna is then pointed in that direction.

Tracking Feeds

Both hill-climbing and step-tracking systems can be affected by scintillation or fading of the received signal. If the signal amplitude changes during a measurement period, which typically lasts a few seconds while the antenna is repositioned several times, the tracking system cannot distinguish between these changes due to propagation effects and the intended changes due to antenna movement. The result is that the antenna will be repositioned to a false angle, possibly causing a drop in signal level instead of an increase. Most step-track systems monitor scintillation of the received signal and switch to program track if the scintillation amplitude becomes large enough to cause tracking errors.

The problem of making a tracking feed can best be understood by considering the field in the focal region of a paraboloid when a satellite beacon transmitter is slightly off axis. (All satellites carry beacons. For example, INTELSAT V satellites carry two beacons on spot frequencies in the guardband 3.93–3.97 GHz and two additional beacons at 11.19 and 11.45 GHz.) The focal plane distribution (*FPD*) will be unchanged in form, but displaced from the horn axis, and the direction of the displacement in angle corresponds to the position of the satellite. This is illustrated in Figure 9.20. The displacement gives rise to three effect:

1. Additional asymmetrical waveguide modes are generated in the horn. These can be detected and used to extract tracking information (mode extraction).
2. The energy contained within the main lobe of the FPD that fails to enter the horn can be detected by additional horns outside the main horn (monopulse).
3. The reduction in gain due to poorer correlation between the feed aperture field and the FPD can be detected, and the feed moved (conical scan) or the

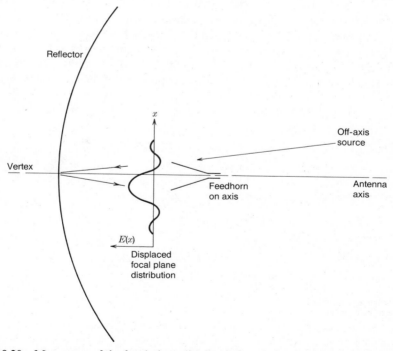

Figure 9.20 Movement of the focal plane distribution away from antenna axis when transmitting source is off-axis.

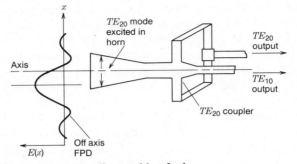

Figure 9.21 TE_{20} mode extractor for tracking feed.

FPD moved (step track) to find the position of the satellite. These schemes have the advantage that the main LNA can be used to amplify the tracking signal, and only a single feed system is required.

Mode Extraction Schemes

Mode extraction is widely used for tracking. The easiest to understand is TE_{20} in rectangular or square horns: TE_{20} is the first asymmetrical mode to be generated in a TE_{10} horn when the FPD is off axis. With linear polarization there is one TE_{20} mode; with circular polarization there will be two orthogonal modes. Figure 9.21 illustrates a TE_{20} mode extraction system.

In circular waveguide using TE_{11} as the dominant mode, a circularly polarized signal generates TM_{01} when off axis. This mode can be detected directly; the amplitude is proportional to the off axis angle, and the phase, relative to the TE_{11} mode, gives information on the angular position of the satellite. With linear

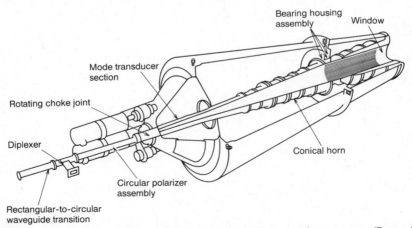

Figure 9.22 Conical scan tracking feed using tuned TM_{01} mode extractor. (Reproduced by permission from N. Lockett, "The Electrical Performance of the Marconi 90 ft. Space Communication Aerials," *Marconi Review*, **34**, 50–80 (1971).)

polarization, a number of tracking modes are generated, of which TE_{21} is the most useful, but the others can be a problem if they are not cut off in waveguide large enough to support TE_{21}.

An example of a conical scan tracking system is that used on the Goonhilly II antenna of British Telecom, as illustrated in Figure 9.22 [10]. A TM_{01} waveguide mode is coupled into a pair of resonant cavities near the throat of the horn, tuned to the satellite beacon frequency. The TM_{01} is converted to TE_{11} and reradiated into the horn with a phase shift relative to the main TE_{11} mode. The phase shift depends on the position of the cavity relative to the peak of the FPD, so by rotating the horn and cavities about the feed axis, TE_{11} mode fed to the receiver is amplitude modulated at the beacon frequency only. Modulation of the main receive channel and transmit frequencies is highly undesirable and is avoided in this system. By correlating the cavity position and the beacon output, the satellite position can be found.

Tracking Geostationary Satellites

Most geostationary satellites are not truly geostationary; they drift along the geostationary orbit in an east–west direction and usually have a small orbital inclination angle. When observed from an earth station, the motion of the satellite in the sky tends to be around a narrow ellipse, which moves slowly east–west. An example is shown in Figure 9.23 for the ATS-6 satellite at a time when its orbital

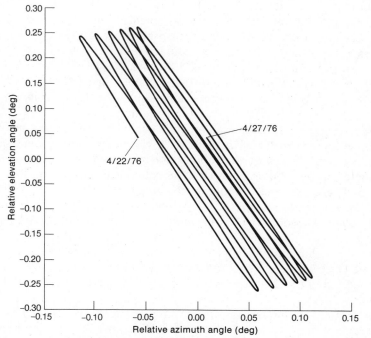

Figure 9.23   Drift in ATS-6 satellite position during period April 22 to 27, 1976.

inclination was 0.5°. The satellite moves around the ellipse once every 24 hours. The peak-to-peak amplitude of the ellipse is approximately equal to twice the orbital inclination angle when this is small.

Satellites that are held to very small inclination angles, such as $\pm 0.1°$, move very slowly in azimuth and elevation when viewed from an earth station. The orbit of the satellite can be predicted very accurately using the techniques discussed in Chapter 2, making accurate program track of the satellite from look angle predictions a straightforward operation. This is the preferred tracking method for most satellites, and only when a satellite has a large inclination angle is autotracking really needed. Nevertheless, it is provided on many large earth station antennas, at considerable expense, so that the capability is available when needed.

9.5 SMALL EARTH STATION ANTENNAS

Small earth stations escape many of the problems encountered with large antennas and cost a small fraction of their larger counterparts. The broader beamwidth of a small antenna allows fixed pointing of the antenna when the satellite is held within $\pm 0.1°$ of its nominal position, eliminating the need for expensive autotrack equipment, dual-axis drives, and servo systems. Thus, the fixed-pointing, small-aperture antenna is an attractive design where the lower G/T can be compensated for by an increase in satellite transmitted power or a reduction in RF bandwidth.

Small earth stations have been constructed in large numbers for domestic *TVRO* reception of cable TV signals distributed at 4 GHz by U.S. domestic satellites. In 1984, these earth stations could be purchased for $1300, with a 10-ft diameter antenna and feed costing as little as $300. The introduction of direct broadcast satellites (DBS) will see a further reduction in earth station size and cost as satellite power is increased and production quantities move from thousands to millions.

The Cassegrain and Gregorian antenna configurations cannot be used when the main reflector diameter is less than 50 wavelengths without a considerable loss in efficiency. This is because the subreflector must have a diameter of at least 8 wavelengths to prevent diffraction of waves round it. Smaller subreflectors do not act as good reflectors and cannot control the illumination of the main reflector adequately. If a large subreflector is used, blockage loss becomes severe and the antenna pattern sidelobes rise significantly due to blocking of the aperture by the subreflector.

Below a main reflector diameter of 50 wavelengths, front-fed paraboloid antennas are used. The *scalar feed* has been developed to provide good illumination of front-fed reflectors [20]. Aperture efficiency with a scalar feed can be up to 65 percent in a typical antenna, but cannot be as high as in a shaped Cassegrain antenna. Figure 9.24 shows an example of a scalar feed, and its radiation pattern.

Control of the antenna pattern sidelobes becomes increasingly important as antenna aperture size is reduced and the pattern broadens. Satellite spacing in

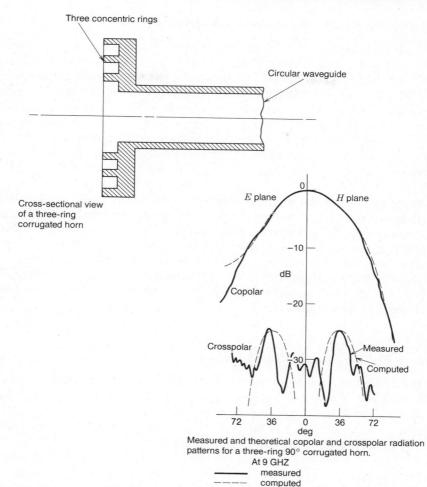

Figure 9.24   Illustration of a 90° corrugated horn and measured and theoretical far field patterns. (Patterns reprinted by permission from G. A. Hockham, "Investigation of a 90° Corrugated Horn," *Electronics Letters*, **12**, 199–201 (1976). Published by the Institution of Electrical Engineers.)

geostationary orbit is being reduced from 3° to 2°, which will increase interference from adjacent satellites on reception and also lead to increased uplink interference when a small earth station transmits. The latter problem places a lower limit on the antenna size used by transmitting earth stations. However, for small diameter reflectors, the cost of a larger reflector is not very great (cf Eq. 9.20), especially when compared to high-power microwave transmitters. It is the transmitter that is often the most costly item in small earth stations that transmit and receive.

Design of Small Earth Station Antennas

Small earth stations are defined here as those using antennas less than 60 wavelengths in diameter, corresponding to 5-m dishes at 4 GHz and 1.6-m dishes at 11 GHz, with gains below about 44 dB. Most of these antennas are symmetrical, front-fed paraboloidal reflectors with scalar feeds. Blockage losses become excessive in symmetrical dual-reflector antennas when the main reflector diameter is below 60λ, because the subreflector diameter must be 8λ or 10λ to obtain good control of the aperture illumination. Blockage of the main aperture also causes high sidelobes in the far-field pattern, making it difficult for the antenna designer to meet the stringent sidelobe envelope specifications that are mandated with 2° satellite spacing.

Although symmetrical front-fed paraboloids are widely used for small earth station antennas, *offset-fed antennas* employing single- and dual-reflector configurations are also used. The offset feed lies above or below the ray paths from the main reflector, which eliminates the blockage problem. We will discuss offset-fed antennas later in this section. Multiple beams can be obtained with one reflector if a *spherical reflector* or *parabolic torus reflector* is used, allowing one antenna to be used with several closely spaced satellites simultaneously. This topic is considered in more detail later.

Front-Fed Paraboloidal Reflector Antennas

The conventional circular paraboloidal dish with a scalar feed represents the simplest and lowest cost antenna for a small earth station. The scalar feed produces a symmetrical radiation pattern that is broad, flat-topped, and close to Gaussian in shape (see Figure 9.24). It is also frequency independent over a wide bandwidth, making it an excellent choice for an earth station antenna feed. The design of front-fed parabolic antennas is straightforward. The half angle of the reflector is chosen to obtain a particular edge illumination with a given feed, or a scalar feed can be selected to provide a required edge illumination with a given reflector.

The geometry of the paraboloidal reflector is shown in Figure 9.25. The distance OF is called the *focal length* of the reflector, denoted by F. The curvature of the reflector can be specified by the *half-angle* θ_0, or by the F/D ratio. F/D and θ_0 are related by

$$\tan \theta_0 = \tfrac{1}{2} \frac{\dfrac{D}{F}}{1 - \left(\dfrac{D^2}{16F^2}\right)} \tag{9.28}$$

The surface of the reflector is given by

$$x^2 + y^2 = 4Fz \tag{9.29}$$

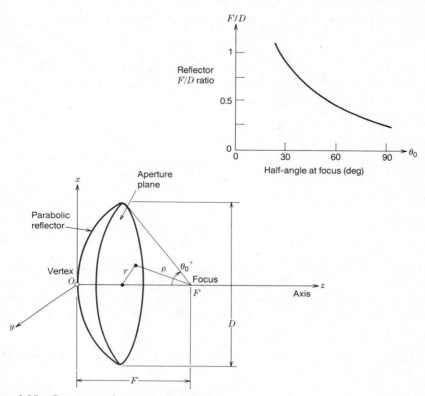

Figure 9.25 Geometry of a parabolic reflector.

or

$$r^2 = 4Fz \qquad\qquad (9.30)$$

where r is the radial distance from the axis OF to the point (x, y) on the reflector.

Most front-fed paraboloid antennas have reflectors with F/D in the range 0.25 to 0.4 corresponding to half-angles of 90° to 64°. The $F/D = 0.25$ reflector is known as a *focal plane reflector*, since the focus of the reflector lies in the aperture plane. Feed design for a focal plane reflector is difficult because the feed must illuminate a full angle of 180°. Scalar feeds tend to have rather narrower beams and are best suited to reflectors with F/D ratios in the 0.3 to 0.4 range.

Maximum antenna gain results when the illumination amplitude taper across the dish aperture gives an edge illumination level 10 to 12 dB below the axis level [2]. A smaller edge taper increases the illumination efficiency by making the illumination closer to uniform, but results in increased spillover. Lower edge illumination values reduce spillover, but also reduce the illumination efficiency. However, lowering the edge illumination reduces the sidelobe levels, whereas high edge illuminations lead to high sidelobe levels. Figure 9.26 shows the way in which

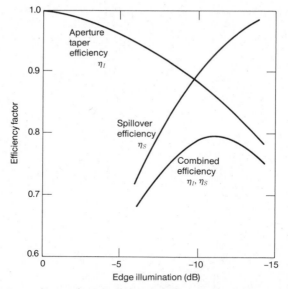

Figure 9.26   Antenna gain as a function of main reflector edge illumination level. (Adapted from W. L. Stutzman and G. A. Thiele, *Antenna Theory and Design*, John Wiley & Sons, New York, 1981.)

antenna gain is related to reflector edge illumination level for a typical feed system [2].

Small antennas have relatively broad antenna patterns. Satellites spaced close to the wanted satellite will radiate interfering signals into the sidelobes of the earth station antenna when it is receiving and will be affected by interference from the antenna sidelobes when the earth station is transmitting. The smaller the antenna, the worse this problem becomes. The Regional Administrative Radio Conference (*RARC*) in 1983 laid down a tightened specification for the sidelobe envelope of earth stations receiving from direct broadcast satellites, and the FCC drew up new regulations for U.S. domestic satellite system earth station antenna sidelobe envelopes [21, 22]. Previously, the ITU had set standards for earth station sidelobe performance through the CCIR, with a sidelobe envelope requirement $G(\theta) = (32 - 25 \log \theta)$ dB, θ in degrees, shown in Figure 9.27. The FCC's requirement for satellite spacing of 2° is $(32 - 25 \log \theta)$ dB beyond 9° and $(29 - 25 \log \theta)$ dB within 7° of the beam axis. Figure 9.27 illustrates the requirements. The earlier specification referred to an average sidelobe envelope, whereas the new specification requires that no sidelobe peak should exceed the specified envelope curve.

Offset-Fed Antennas

Blockage of the aperture by the feed and its supports can be avoided in an offset-fed antenna. Figure 9.28 shows the blockage shadow pattern for a Cassegrain antenna with a tripod leg arrangement to support the subreflector. There is shadowing of both the main reflector aperture and also the subreflector radiation

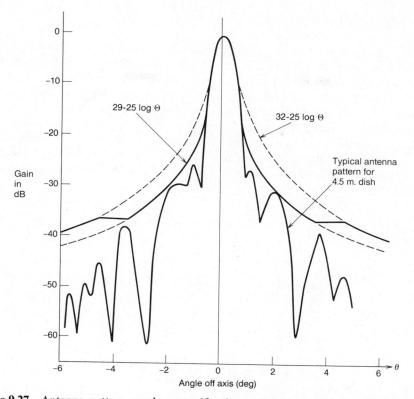

Figure 9.27 Antenna pattern envelope specifications. $32 - 25 \log \theta$ was set by CCIR Recommendation No. 465 for large earth stations. The more stringent requirement of $29 - 25 \log \theta$ close to the antenna axis was set by the FCC in Regulation No. 25.209 for new U.S. domestic satellite system earth stations in 1983.

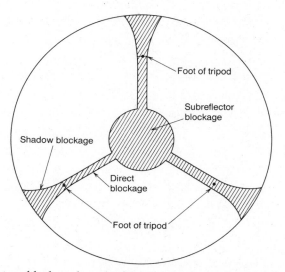

Figure 9.28 Aperture blockage by subreflector and tripod support legs in a large Cassegrain antenna. (Reprinted with permission from T. Pratt and B. Claydon, "The Prediction of Polar Diagrams of Large Cassegrain Antennas," *Marconi Review*, **34**, 1–26 (1971).)

pattern, leading to significant loss of gain and high near-in sidelobes. Front feeds also need supports, but because of the smaller size and weight of their feeds the supports can be much thinner than those needed for a subreflector.

Figure 9.29 shows two offset-fed antenna configurations. The main reflector is a section of a paraboloid, cut off above the axis. The feed (or a feed–subreflector combination) is located below the axis in Figure 9.29a, giving a completely unblocked aperture. In the *open Cassegrain* arrangement [23], shown in Figure 9.29b, the feed is located within the main reflector aperture, leading to slight blockage losses. Figure 9.30 shows photographs of two offset antennas of the type illustrated in Figure 9.29.

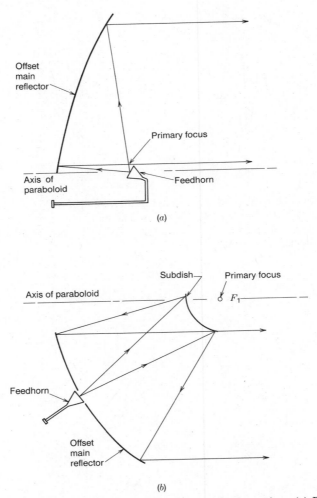

Figure 9.29 Geometry of two offset reflector antenna configurations (a) Front fed offset antenna. (b) "Open Cassegrain" design of dual reflector offset antenna.

Figure 9.30a A small offset-fed reflector antenna for the reception of direct broadcast satellite television signals. The reflector diameter is 0.6 m, and the operating frequency 11.7 to 12.1 GHz, with a gain of 34.7 dB. (Photograph courtesy of ERA Technology, Ltd., U.K., c ERA Technology, 1985.)

Figure 9.30b The 6-m Open Cassegrain antenna at the University of Birmingham, U.K. (Photograph by Tim Pratt.)

Although offset antennas can be designed to produce superior performance compared to symmetrical antennas, their construction tends to be more expensive, at least in the larger diameters. Symmetrical reflectors can be made up of a large number of identical "petals," which have the same curvature. An offset reflector possesses symmetry only around the axis of the generating paraboloid and requires a number of differently shaped sections, with a corresponding increase in tooling costs. Smaller reflectors, of about 1-m diameter, can be formed from a single sheet of metal drawn into a mold; the exact shape of the mold does not affect the cost of its production very much, so there is little difference in cost for small antennas between symmetrical and offset types.

Dual-reflector offset Cassegrains can use shaped reflectors to obtain increased gain by improved illumination efficiency. The mathematics used to derive the reflector shapes is particularly difficult and requires numerical solution of a set of partial differential equations [24].

Beam Steering by Feed Movement

The beam of a parabolic reflector antenna can be steered over a few beam-widths by moving the feed transversely away from the axis of the paraboloid. The sidelobe structure of the beam deteriorates as the beam scan angle is increased, due to *coma*. Coma is an optical term for the effect of beam scanning in lenses, which have similar properties to reflectors. In a parabolic reflector, scanning the feed away from the axis causes the phase front in the reflector aperture to distort into an S-shape. For small scan angles, up to one beamwidth typically, the distortion is small; at larger scan angles scanning results in reduced antenna gain and an increase in the levels of sidelobes on the axis side of the displaced beam, called *coma lobes*. Coma lobes can be reduced and beam-scanning performance improved by the use of an array feed with multiple feed elements each separately fed with the appropriate amplitude and phase. This technique is widely used in satellite antennas to provide shaped beams for zone coverage (see Chapter 3), and also to provide several simultaneous beams with one reflector.

Beam steering by feed movement, and also multiple beams, are possible over a wider angular range when a *spherical reflector* is used [25, 26]. The spherical reflector is symmetrical about its center of curvature but does not focus incoming rays to a single point as does the paraboloid. Instead, it generates a *caustic*, a region in which rays intersect. This is illustrated in Figure 9.31. If we restrict the angular range over which a feed receives or transmits and locate it at the point F in Figure 9.31 (called the *paraxial focus*), we can collect energy from the central region of the reflector. Scanning can then be achieved by rotating the feed system about the sphere's center of curvature, O, without any change in beam shape.

When a spherical reflector is illuminated over a narrow angular range by a feed at the paraxial focus, it is being used as an approximation to a paraboloid. The surface of a paraboloidal reflector differs very little from that of a spherical reflector for angles close to the reflector axis. This technique has been successfully applied to the production of multiple beam antennas, which use one reflector

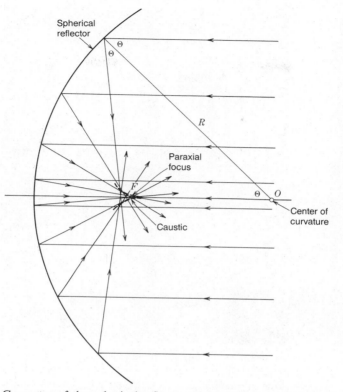

Figure 9.31 Geometry of the spherical reflector showing the caustic created by crossing rays.

and several feeds to simultaneously provide communications with a number of closely space satellites in geostationary orbit [27, 28]. In this application, the reflector has spherical curvature in one plane and a parabolic curve in the orthogonal plane; it is known as a *parabolic torus reflector*. The parabolic torus reflector is mounted so that the reflector axis is perpendicular to the earth's north–south axis, to provide a *polar mount*. Rotation of a feed about the center of curvature of the torus scans an arc in the sky very close to the geostationary orbit. Figure 9.32 shows a multiple-beam parabolic torus antenna designed to provide simultaneous reception from a number of TV relay domestic satellites.

The failure of a wide-angle spherical reflector to produce a point focus results in *spherical aberration*. Spherical aberration can be corrected by collecting energy with a suitably phased line feed, since all incident parallel rays are reflected to cross the reflector axis, or with a dual-reflector system in which the subreflector is shaped to correct the spherical aberration [25]. The subreflector shape is determined by path length considerations and cannot be used to control the aperture illumination in a dual-reflector spherical reflector antenna. The aperture amplitude distribution has an inverse taper (more energy at the outer edge than in the center),

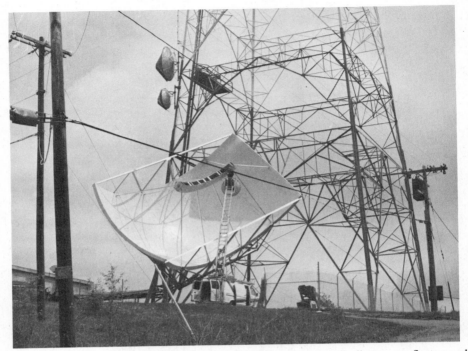

Figure 9.32 A multiple-beam earth station antenna using a parabolic torus reflector and several feeds. The antenna in this photograph is located at a terrestrial TV transmitting station and receives program material from several satellites simultaneously. (Photograph courtesy of Antenna Technology Corporation.)

leading to very high first sidelobe levels, typically 11 dB, and poor efficiency. A three-reflector system allows beam steering and independent control of the aperture amplitude and phase distributions. An offset configuration can be used to reduce blockage effects, and a high efficiency can be achieved. However, the necessary 10λ diameter of the subreflectors prevents close spacing of multiple beams in a two- or three-reflector antenna.

9.6 EQUIPMENT FOR EARTH STATIONS

Figure 9.33 shows a simplified diagram of the major items of equipment needed at an earth station that receives and transmits. In a large earth station there will be many receivers and transmitters multiplexed together onto one antenna to provide channelized communication through separate transponders. In a TVRO earth station, for example, there will be only one receive channel and no transmitting equipment. Transmitters are very much more expensive than receivers, and the cost increases rapidly as the transmitter power increases; this is partly because receivers are made in much larger quantities than transmitters, leading to economies of scale, and partly because of the tight specifications on out-of-band emission,

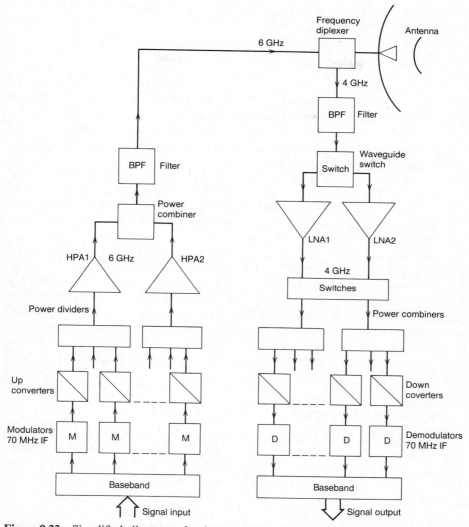

Figure 9.33 Simplified diagram of a large earth station's equipment using FDM/FM/FDMA technology.

frequency stability, and power control that are necessary to avoid interference with other channels and satellites. Microwave transmitters are expensive devices that employ costly high-power amplifiers such as traveling wave tubes and multicavity klystrons.

Baseband equipment (modulators, demodulators, and multiplex equipment) is considered in the next section. The major RF components in an earth station are the low noise amplifier (LNA) of the receiver and the high-power amplifier (HPA) of the transmitter. Also required are up and down converters to translate signals from (or to) VHF intermediate frequencies (usually 70 or 140 MHz) to (or from) microwave carriers.

Low Noise Amplifiers

Large earth stations need very low noise amplifiers. Cryogenically cooled parametric amplifiers are widely used, with liquid helium cooling at 4° K above absolute zero to achieve noise temperatures of 20 to 40 K at 4 GHz. Medium and small earth stations use GaAsFET amplifiers with no cooling or electrothermal (Peltier) cooling. These achieve noise temperatures in the range 50 to 120 K at 4 GHz and 120 to 300 K at 11 GHz.

The FET amplifier is much simpler than the cooled parametric amplifier and is particularly attractive for unattended and TVRO earth stations, especially where cost is an important factor. Development of 20- and 44-GHz GaAsFET amplifiers is in progress, although noise temperatures tend to be much higher than those achieved at 4 and 11 GHz.

The LNAs used in earth stations usually cover the 500-MHz fixed service band at 4 GHz and 750 MHz at 11 GHz. In large earth stations a one-for-one redundancy arrangement such as that shown in Figure 9.33 is widely used. Failure of one LNA, indicated by loss of a pilot signal at the receiver output, results in immediate switchover to the second LNA. The spare (unused) LNA is often kept on test with a pilot signal or noise source input so that its state of readiness can be monitored continuously. Dual polarization earth stations need two RF receive channels and may use one-for-two redundancy (one spare for two operational) in their LNAs.

High-Power Amplifiers

Large earth stations frequently use large numbers of high-power amplifiers (HPA) with output power levels up to 8.5 kW. The configuration employed depends on the number of carriers to be transmitted and whether these are FDM or TDM signals. The most common configuration employs one HPA for each transponder to be used. At 6 GHz, HPAs having bandwidths of 40 or 80 MHz are used in large earth stations, using either air-cooled TWT (traveling wave tube) amplifiers or water-cooled klystrons. TWTAs have wider operating bandwidths than klystrons and can cover the full 500-MHz bandwidth at 6 GHz [29], allowing the TWT to be tuned to any transponder band.

FDM transmission of several carriers to one or more transponders requires a linear high-power amplifier if intermodulation is to be avoided. At an earth station neither input power nor efficiency are prime concerns, so considerable input back-off can be used with the HPA to achieve near-linear operation and low intermodulation. Typically, a 3-kW HPA will be operated with 12 or 14 dB input back-off giving an output power in the 300 to 500 W range.

Although QPSK and other digital modulations are often represented as constant-envelope signals, this condition is present only in an infinite bandwidth

QPSK signal. Out-of-band emissions and intersymbol interference are controlled by careful filtering of the PSK signal with Nyquist-type filters (see Chapter 5). The filtering of the signal results in a nonconstant envelope signal with amplitude as well as phase variation. When the filtered signal is hard limited by a saturating amplifier, the signal is returned to a constant envelope form, which broadens its spectrum considerably, increasing the level of out-of-band emissions. Consequently, HPAs used at earth stations carrying digital traffic are often run with input backoff of 10 to 14 dB.

When several HPAs are used with one antenna, a combining network is needed to sum their outputs into a single transmit waveguide. Frequency-selective networks and waveguide hybrid junctions are used to couple the HPAs, with a typical loss of 4 dB per HPA [29]. As a result, a 3-kW HPA run with an output backoff of 10 dB might actually deliver only 120 W to the earth station antenna.

Single-channel-per-carrier (SCPC) systems allocate a separate transmitter frequency to each channel (usually FM or PSK voice). The channel frequencies are generated by a programmable synthesizer and then used to upconvert the modulated signals to the transmit frequency. In a large earth station, the outputs of the upconverter are then summed with hybrid couplers and a single broadband FDM signal is applied to the HPA. In small earth stations carrying only a few voice channels, it is possible to use solid-state amplifiers for the HPAs, by having one amplifier for each voice channel. The transmitted power of a single SCPC voice channel is typically below 1 W for medium-sized earth stations; GaAsFET amplifiers at 6 GHz can produce up to 20 W output power in saturation. At the time of writing (1984), the cost of these amplifiers and the combining losses for a large number of HPAs make it uneconomical to use solid-state HPAs when more than a few voice channels are to be combined.

Solid-state HPAs have the additional benefit that they do not require the very high voltages needed by TWT and klystron amplifiers (typically 10–50 kV). High voltage equipment tends to suffer more failures than lower voltage equipment, and is also bulky and heavy. In transportable earth stations, or mobile systems, solid-state HPAs are attractive. However, reducing the size of the antenna demands more transmitter power for a given EIRP, and small earth stations often require relatively high transmitter power levels.

Bipolar transistor HPAs with up to 50-W saturated output are widely used in maritime satellite communication systems for shipboard transmitters operating at 1.6 GHz.

An earth station serves as an interface between the RF carriers sent to and from a satellite and the baseband voice, data, or video signals sent via the terrestrial network or provided by the user. A great deal of signal processing is required when many voice or data channels are multiplexed together into single carriers. The multiplexing and modulation–demodulation operations are almost always carried out at baseband and intermediate frequencies, which can be handled more easily than microwave frequencies. The up- and downconverters form an interface between the RF and IF portions of transmitters and receivers; the only operations that are normally carried out on RF signals are amplification and filtering, with minimal combining and splitting. This part of the earth station is known as *ground control equipment (GCE)*.

There are significant differences between equipment designed for FDM operation and that designed for TDM, so we will discuss them separately. The upconverters and downconverters are similar in each case, although bandwidths may differ.

FDM Systems

FDM systems transmit and receive many voice or data signals by allocating separate frequencies to each signal, at either baseband or RF, or both. The assembly of FDM voice and data channels into groups was discussed in Chapter 5. At an earth station that operates in an FDM mode, a terrestrial link is used to send and receive FDM or TDM groups of channels. The link may be microwave, coax cable, or an optical fiber. The terrestrial interface may separate incoming channels back to baseband for reassembly as new FDM groups for transmission via the satellite link, or it may assemble incoming FDM groups. When the terrestrial link uses TDM (e.g., optical fiber links), the channels must be converted back to baseband and then reassembled in FDM groups for satellite transmission. The reverse process must be performed on received FDM groups.

Large earth stations using FDM/FDMA operation, such as Standard A stations in the Intelsat network and many domestic satellite system stations, may carry several thousand 4-kHz voice channels. The voice channels are collected from a wide geographic area and sent via a *gateway*. In international systems, the gateway is usually located in a major city, and all international traffic from the whole country may be routed through this one point. Voice channels that are to be routed to each country, or each earth station, are assembled into groups at the international exchange and then sent to the earth station. Similar receive paths are established to complete the two-way link.

It is not essential that both the go and return paths be via satellite. Undersea or terrestrial cable or terrestrial microwave links can be used for one path, and a satellite link for the other. This reduces the round-trip delay in a long telephone circuit and also allows maintenance of parts of the earth station equipment. An advantage of the international gateway arrangement is that all international circuits—satellite, cable, and optical fiber—terminate at one point, allowing traffic to be distributed between the available transmission paths.

Figure 9.34 shows a typical arrangement for a FDM/FDMA earth station's ground control equipment, corresponding to the IF and baseband equipment in Figure 9.33. The transmitting section, shown in Figure 9.34a, accepts baseband signals from the terrestrial interface and assembles these into FDM groups for different destinations. In an FDM/FDMA system, each route from one earth station to another earth station has an allocated frequency. Destinations are associated with RF frequencies at a transmitting station, so baseband channels must be translated to specific RF frequencies determined by their eventual destination.

The system shown in Figure 9.34a uses a double-frequency conversion with two IF frequencies, 70 MHz and 770 MHz. Each 70-MHz channel leads to a separate RF carrier in the transmitted spectrum. The FDM signal, consisting of as few as 12 telephone channels or as many as 1872, is frequency modulated onto a 70-MHz IF carrier.

The 70-MHz IF filter defines the bandwidth of the FM signal very accurately and is the major bandwidth control element in the transmit chain. Its bandwidth lies between 1.25 and 36 MHz depending on the carrier size, and a variable group-delay equalizer may be included to compensate for group delay in the uplink. The 70-MHz IF carrier is then upconverted to a 770-MHz IF band, where other carriers are added to form a composite FM/FDMA signal. The 770-MHz broad-band signal is finally upconverted to 6 GHz for amplification by the HPA.

Not shown in Figure 9.34a are power control devices needed to set the level of the 6-GHz carrier. FDMA requires very accurate power sharing when more than one carrier accesses a single transponder, to avoid excessive intermodulation on any one downlink carrier. Closed-loop control of the uplink carrier can be used; the downlink carrier is monitored at the transmitting station using a spectrum analyzer, and the level of the uplink carrier is set accordingly. This cannot be done when a satellite such as INTELSAT V is used and the transmitting and receiving stations are in separate zone beams. In this case, the absolute level of the transmitted carrier must be correctly set at the transmitter. Monitoring of the transponder carrier power distribution is then carried out by a station within the receive zone beam region, which reports back to each transmitting earth station via a telex or voice channel allocated for this purpose. A considerable amount of monitoring and supervisory equipment is also required at each stage of the GCE, and energy dispersal waveform generators are needed in the uplink equipment to add triangular waveforms to the transmitted signals when traffic volume on any carrier is low.

The downlink GCE complements that of the uplink and contains almost the same equipment except that frequency conversion is down rather than up. Figure 9.34b shows an arrangement to complement the transmit equipment of Figure 9.34a. Carriers at 4 GHz are amplified in a broadband LNA and then downconverted to 770-MHz IF using appropriate local oscillators. The local oscillators separate the received 4-GHz carriers into individual 770-MHz receivers with 40-MHz bandwidth so that each receiver corresponds to one transponder. A second downconversion to 70 MHz allows accurate filtering and group delay equalization to be performed before demodulation. The bandwidth of the 70-MHz IF stages is set between 1.25 and 36 MHz depending on the number of channels carried by the transponder. The FM demodulators used in the GCE usually employ threshold extension techniques to achieve a low threshold in (C/N) for carriers with up to 252 channels. Threshold extension may not be needed with the higher density carriers; these usually have much higher (C/N) and originate only in large earth stations served by higher gain zone beams. After demodulation, the voice channel signals are translated to baseband and bandlimited to 3.1 or 3.4 kHz, continuity pilots are extracted, and the signal is passed to the terrestrial

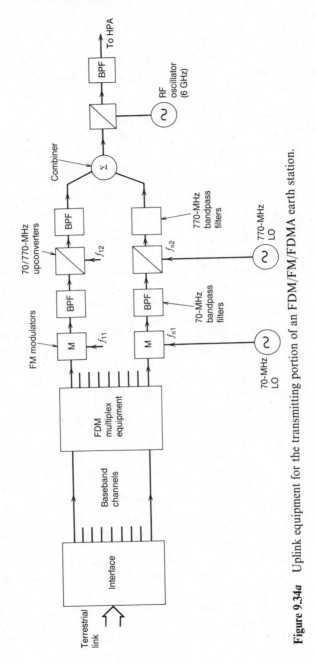

Figure 9.34a Uplink equipment for the transmitting portion of an FDM/FM/FDMA earth station.

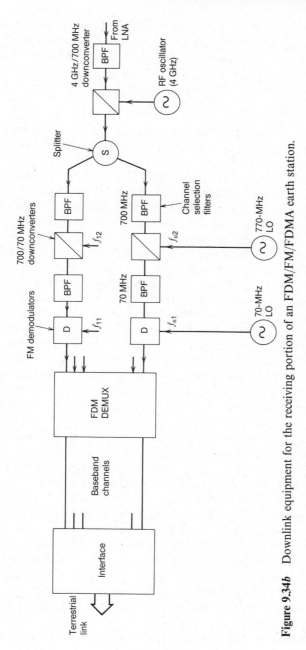

Figure 9.34b Downlink equipment for the receiving portion of an FDM/FM/FDMA earth station.

interface for onward transmission. Deemphasis filters are required in the baseband equipment, as well as noise and pilot level monitoring equipment. The 70-MHz IF incorporates automatic gain control (AGC) to ensure that a constant level signal is applied to the demodulator.

TDM Systems

TDM systems are time division to interlace digital signals into frames, which are transmitted sequentially through separate transponders on a satellite (see Chapters 5 and 6 for details of TDM/TDMA systems). The equipment requirements for TDM digital systems are quite different from those of FDM/FDMA earth stations. Terrestrial interconnection can be either FDM or TDM, the latter being more convenient for high-speed data transmission. Figure 9.35 shows a simplified diagram of the major baseband and IF elements in a TDM digital earth station. The 70-MHz IF used for FDM systems must be replaced by a 140-MHz IF when 120-Mbps data is sent by QPSK using an 80-MHz bandwidth. The 770-MHz IF may be replaced by a 1.2-GHz IF in earth stations using the 14/11-GHz (Ku) band, where the RF bandwidth is 750 MHz.

Signals are transferred by the terrestrial interface to four phase PSK modulators driven by 140-MHz IF carriers. The voice and data signals supplied by the interface are simply binary bit streams, in TDM format. These data must be formatted into frames according to the system requirements, with one frame per transmission. Normally one earth station does not fill a complete frame in a TDMA system; it transmits a burst of QPSK signal at the appropriate time to partially fill a frame at the satellite. Very accurate timing is required at the earth station to ensure that when the burst of RF energy arrives at the satellite it is in exactly the correct position (in time) to interlace between similar bursts arriving from other stations. Thus a considerable quantity of synchronization and timing equipment is needed in the transmit portion of the GCE of a TDMA earth station.

Digital speech transmission efficiency can be increased by using *digital speech interpolation* (DSI). With DSI, any speech channel (a time slot) that is not being used can be seized by another active channel for transmission of its data. Since telephone circuits have a typical activity of only 40 percent, at least half of the circuits are inactive (on average) at any instant. In systems handling very large numbers of telephone channels simultaneously, this averaging works well, and DSI can be used to double the number of speech circuits. DSI equipment is required on both transmit and receive sides of the GCE to insert and separate the individual speech channels. A common *mapping channel* is transmitted alongside DSI channels to inform the receive end of the position of each channel within the time frame sent by the transmitter.

Chapter 6 reviews the basic ideas of DSI, and a good discussion of various DSI techniques can be found in Feher [30, pp. 65–72], and an analysis of DSI performance and system capacity in Bhargava et al. [31, Appendix E, pp. 544–545].

The receiving equipment needed at a large TDM/TDMA earth station is shown in Figure 9.36b; it complements the transmitting equipment shown in

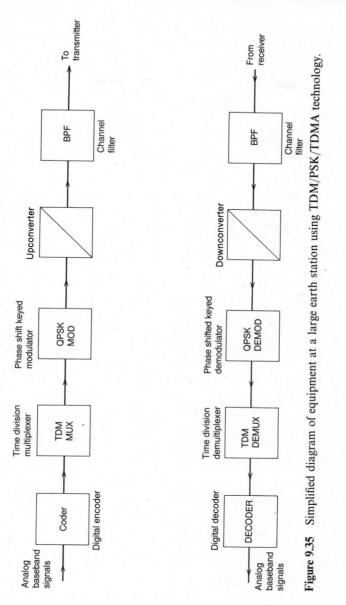

Figure 9.35 Simplified diagram of equipment at a large earth station using TDM/PSK/TDMA technology.

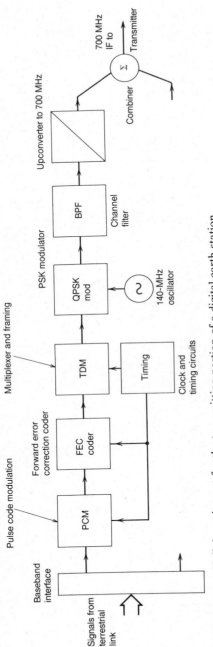

Figure 9.36a Uplink equipment for the transmitting portion of a digital earth station.

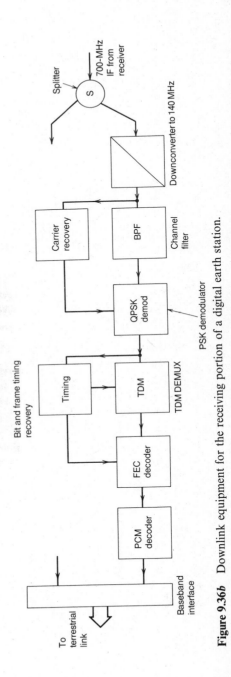

Figure 9.36b Downlink equipment for the receiving portion of a digital earth station.

Figure 9.36*a*. The most critical item in a digital receiver is the demodulator, usually for QPSK in satellite links. QPSK demodulation was discussed in Chapter 5, where the techniques for carrier and bit timing recovery were also discussed. The bit error rate (BER) from a QPSK demodulator is critically dependent on the accurate recovery of the carrier with correct phase and on bit sampling instant as determined by the bit clock recovery circuits in the demodulator. Most satellite systems achieve (C/N) of between 10 and 25 dB at the demodulator input. A BER of 10^{-6} can theoretically be obtained with a (C/N) of 10.6 dB at the demodulator input, but most demodulators require 1.5 to 3 dB greater (C/N) to achieve this BER.

In Chapter 7 the benefits of forward error correction (FEC) coding of digital signals were discussed. FEC coding equipment can be placed at the input to the modulator and the corresponding decoder placed at the output of the demodulator, when coding is applied to the satellite link alone by the earth station operator. Since additional benefits to the user can be obtained by using FEC on the terrstrial section of the link, individual data signals can be encoded at any point between the user and the earth station, and the decision to use FEC coding is normally left to the user.

Speech signals sent by satellite incur a one way transmission delay of up to 240 ms. Any mismatch at the receiving end of a link can result in an echo of the speaker being heard at the transmitting end with a delay of 450 to 500 ms. This is sufficiently disturbing to many people that *echo suppressors* or *cancelers* are employed at each end of a satellite link, either at the earth station or at the international gateway exchange. The simplest type, the echo suppressor, inserts attenuation in the return path when speech is detected in the forward path. Effectively, the speaker seizes the channel and makes it virtually one way. When the speaker stops talking, the attenuation on the return path is removed, and the other person can be heard. This type of echo suppression is effective, but makes it difficult for one person to break into the conversation when the other is talking. More sophisticated versions remove the suppression when both people talk at the same time. The echo canceler detects the presence of a delayed version of the forward signal in the return path and dynamically cancels out the unwanted echo using a transversal filter. This is much more complex, and generally a little less effective, than the first type, but allows the second party in a conversation to be heard at full volume at any time. Feher [30] describes echo suppressors and echo cancelers in some detail.

9.7 VIDEO RECEIVE-ONLY SYSTEMS

In terms of the number of earth stations manufactured, the television receive-only (TVRO) and direct broadcast system (DBS) terminal is likely to remain the most common. By comparison with the large earth stations used in the Intelsat Standard A system, these earth stations are very simple. However, considerable sophistication must be achieved at a very low price if a manufacturer is to sell his product in the competitive TVRO and DBS earth station market.

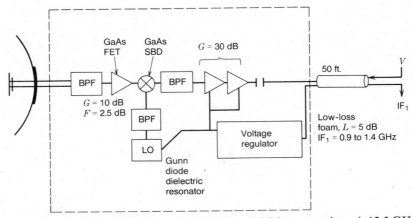

Block diagram of an outdoor unit (ODU) for a DBS home receiver. A 12.5 GHz FM signal from the DBS satellite is picked up by the antenna on the left and is amplified, downconverted to approximately 1 GHz, and fed through a 50-ft low-loss cable to the indoor unit (IDU) in the home. The line marked V at the right represents the power from the home to the ODU mounted on the back of the antenna. Specific symbols are: G, gain of the amplifiers; F, noise figure; BPF, bandpass filter; FET, field-effect transistor; LO, local oscillator; and SBD, Schottky-barrier diode.

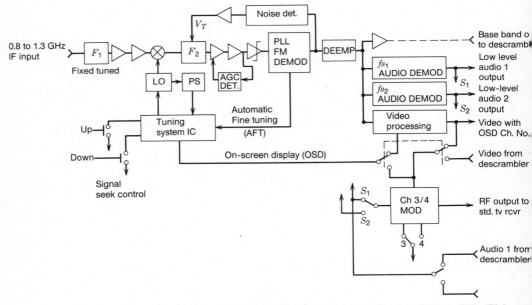

Block diagram of the indoor unit (IDU) for a DBS home receiver. The 1-GHz FM signal from the ODU is processed in this unit and put into a format that can be accepted by the TV set in the home. The symbols in this diagram are: AGC, automatic gain control; DEMOD, demodulator; DEEMP, deemphasis network; DET, detector; IC, integrated circuit; LO, local oscillator; MOD, modulator; PLL, phase-locked loop; and PS, prescaler.

Figure 9.37 Receiving system for a DBS (TV receive-only) earth station for home reception at 12 GHz. (Reprinted with permission from Klensch et al., "Critical System Parameters in the DBS Service," *RCA Engineer*, **28**, 58–63 (1983).)

Figure 9.37 shows the general layout of a DBS earth station [32]. There are two units: the *outdoor unit* (ODU), which mounts on the antenna feed, and the *indoor unit* (IDU), which is fed by coaxial cable from the ODU and contains the interface to a domestic VHF/UHF TV receiver. A high gain GaAsFET amplifier is commonly used for the RF section of the receiver, mounted immediately behind the feed to minimize waveguide losses. At 12 GHz, the noise temperature of a GaAsFET amplifier is typically 220 to 300 K. At 4 GHz, noise temperatures down to 70 K can be achieved. Image enhancement mixers with noise temperatures down to 340 K have been developed for DBS service and may offer lower cost ODUs than FET amplifier systems, which also require a mixer.

A fixed frequency Gunn diode local oscillator (LO) is normally used, with a frequency around 11 GHz. This gives a first IF around 1.2 GHz for the 12-GHz DBS band. The 11 GHz LO is in the ODU, which sends the 1.2-GHz IF signal to the IDU for downconversion to 70 MHz. Selection of channel frequency is achieved with a second local oscillator at 1.1 GHz to obtain a 70-MHz IF signal with 30-MHz bandwidth.

Further processing of the 70-MHz IF signal allows separation of the FM sound carrier, which is transmitted separately from the FM video signal. A threshold extension demodulator is used to extract the video signal. The threshold is at approximately 7.5 dB in demodulators developed specially for DBS and TVRO use, a significantly lower value than in conventional FM demodulators. Figure 9.38 shows the characteristic of one such demodulator. Note the gradual degradation in video (S/N) as the input (C/N) is reduced. This gradual deterioration ("graceful degradation") in video (S/N) is very valuable in satellite television because it prevents sudden loss of the TV picture. Instead, an increasing amount of *snow* or

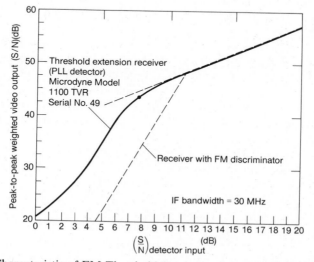

Figure 9.38 Characteristic of FM Threshold Extension demodulator for TV receive only systems. (Courtesy Microdyne Corporation, Ocala, Florida.)

sparklies (black and white specks due to noise transients from the demodulator) is seen on top of an otherwise acceptable picture.

Comparison of the curve in Figure 9.38 with the BER curve for a QPSK system shows one reason why FM is preferred over digital techniques for television transmission in a DBS system; degradation of the TV picture is much more gradual with FM. (A second reason is that the bandwidth of a QPSK TV signal with 4.2-MHz video bandwidth is well in excess of 30 MHz unless video bandwidth compression techniques are employed.)

After demodulation, video and sound signals are remodulated onto a VHF or UHF carrier using vestigial sideband (VSB) AM for the video and FM for sound to produce a combined signal compatible with domestic TV receivers.

9.8 FREQUENCY COORDINATION

Satellite communications systems and terrestrial microwave links share common frequency allocations in a number of bands. This means that when a new earth station or terrestrial link is planned, care must be taken to ensure that the new installation will not interfere with or be subject to interference from existing systems. This process is called *frequency coordination*, and it is particularly important in the crowded 4- and 6-GHz bands. The area around an earth station site within which interference with another link must be considered is called the *coordination area*. It is bounded by a curve called the *coordination contour*; the distance along any given azimuth from the earth station to the coordination contour is called the *coordination distance*.

Coordination is effected by picking a possible earth station location and determining its coordination distances in various directions and thus its coordination contour. All transmitting and receiving stations that share the earth station's frequencies and lie within the coordination contour are identified, and the probable interfering signal levels are calculated. If these are below the allowed maxima, the actual interference environment at the site may be measured using a transportable terminal. While computer programs are available commercially to make interference calculations, these do not usually resolve the terrain in sufficiently fine detail to account for site shielding by buildings, small depressions, and so on. Thus on-site measurements are usually necessary, particularly in an urban area with a lot of 4- and 6-GHz terrestrial traffic.

The process by which coordination distance should be calculated is discussed in several CCIR reports, particularly 724-1 [33] and 382 [34]. At any given time the current CCIR recommendation may not be the same as that specified in a particular country's radio regulations. Thus, U.S. readers involved in frequency coordination calculations should consult the *FCC Rules and Regulations* before they begin.

The first step in the process is to determine the *basic transmission loss* L_b in decibels where [34]

$$L_b = P_T + G_T + G_R - A_S - P_R \text{ dB} \tag{9.31}$$

and P_T is the transmitter output power in dBW, G_T and G_R are the transmitter and receiver antenna gains in the appropriate directions, and P_R is the maximum allowed received power from the interferer. A_S is called the site shielding factor. The second step is to determine the distance that corresponds to L_b. CCIR provides equations and curves to calculate this distance for two modes of propagation; Mode A corresponds to great-circle path mechanisms and Mode B corresponds to scattering by rain. (Rain can cause interference by scattering energy from the beam of a transmitting antenna into the aperture of a receiving antenna.)

The CCIR curves for Mode A (see Figure 9.39) display coordination distance versus frequency and $L_b - A_S$. To use them, one may estimate A_S from the frequency f in GHz and the elevation angle θ of the terrain above the horizon by [35]

$$A_S = 20 \log_{10}[1 + 4.5\theta f^{0.5}] + \theta f^{0.33} \qquad \theta \geq 0°$$
$$= 0 \qquad\qquad\qquad\qquad\qquad\qquad \theta < 0° \qquad (9.32)$$

In this equation θ is in degrees.

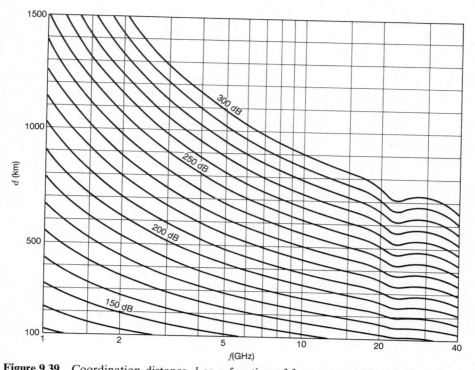

Figure 9.39   Coordination distance d as a function of frequency and basic transmission loss L_b — site shielding factor A_s for great-circle path propagation mechanisms (CCIR Mode A) over land. (Reprinted with permission from International Radio Consultative Committee (CCIR), *Recommendations and Reports of the CCIR, 1982, Volume V, Propagation in Non-Ionized Media,* International Telecommunication Union, Geneva, Switzerland, 1982.)

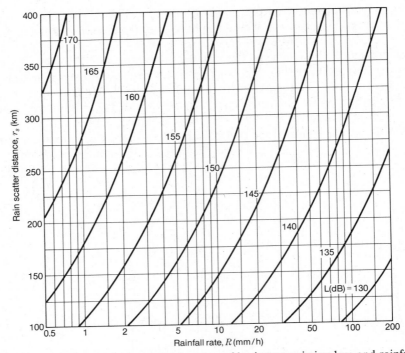

Figure 9.40 Rain scatter distance as a function of basic transmission loss and rainfall rate at a frequency of 4 GHz. (Reprinted with permission from International Radio Consultative Committee (CCIR), *Recommendations and Reports of the CCIR, 1982, Volume V, Propagation in Non-Ionized Media*, International Telecommunication Union, Geneva, Switzerland, 1982.)

Site shielding is not a factor in rain scatter interference and the CCIR curves for rain scatter coordination distance involve rainfall rate and frequency directly; see Figure 9.40. Rainfall rates exceeded for percentages of time of interest may be obtained from the CCIR rain climate regions described in the previous chapter.

9.9 SUMMARY

Earth stations for use in satellite communication systems are characterized by their G/T ratio, which combines antenna gain with system noise temperature. Because the received signals are weak, the G/T ratio must be maximized within constraints of antenna size and receiver cost.

Earth stations carrying many telephone signals, high bit rate data links, or several television channels simultaneously use large antennas, with diameters up to 30 m. These cost millions of dollars and are used to link major cities in domestic and international satellite communication systems. Small earth stations that carry only one voice channel or receive one television signal have lower G/T ratios and can use much smaller, lower cost antennas.

In large antennas, the gain is maximized by using the shaped Cassegrain design, in which near-uniform illumination of the antenna is achieved, giving high illumination efficiency. Corrugated horns, often combined with a beam waveguide feed, provide low antenna noise temperature and good cross-polarization characteristics. Transmitting earth stations must have transmit antenna patterns that conform to published specifications so that they do not cause interference to adjacent satellites. Careful design and analysis using diffraction theory and numerical integration techniques are needed to ensure that large antennas achieve the required G/T ratio.

Small antennas often use a front-fed configuration with a scalar feed. The scalar feed provides a high illumination efficiency with low spillover by radiating a flat-topped beam with steep sides. Offset configurations can be used to avoid blockage of the antenna aperture by the feed system and its supports.

Large antennas have narrow beams and frequently are equipped with facilities to automatically track movement of a satellite. The autotrack system derives an error voltage as the satellite moves away from the beam axis, which is used to drive the antenna servos so that the beam is recentered on the satellite. Small antennas have broader beams that can encompass the full range of movement of geostationary satellites that are held within $\pm 0.1°$ of the nominal orbital position. Most receive-only earth stations do not have autotrack, at a considerable cost saving.

Antennas with spherical or torus reflectors can track a satellite over a limited angular range by movement of the feed. Multiple beams with one fixed reflector can be obtained by using several feeds and a spherical reflector. Receive-only earth stations can be very simple, requiring only the antenna, a low noise RF amplifier and downconverter, and an IF receiver. Some DBSTV receiver designs use a low noise mixer without an RF amplifier. Earth stations that transmit are very much more expensive than receive-only stations. Generation of microwave power with high frequency stability is expensive. The large earth stations in the Intelsat network have large numbers of transmitters and receivers, and associated multiplexing and demultiplexing equipment. The transmitters are usually high-power TWTAs or klystrons, with up to 5-kW saturated output. The HPAs are usually run with considerable backoff to obtain quasilinear operation.

It is important that earth stations do not suffer interference from other communication systems operating in the same frequency band, and that transmitting stations do not cause interference to other systems. Frequency coordination is carried out using procedures specified by CCIR to determine the likelihood of interference occurring.

REFERENCES

1. *Standard A Performance Characteristics of Earth Stations in the Intelsat IV, IV-A, and V Systems having a G/T of 40.7 dB/K*, (BG-28-72E Rev. 1), Intelsat, Washington, D.C., December 15, 1982.

2. W. L. Stutzman and G. A. Thiele, *Antenna Theory and Design*, John Wiley & Sons, New York, 1981.

3. S. Silver, Ed., *Microwave Antenna Theory and Design* Vol. 12, M.I.T. Radiation Lab. Series, McGraw-Hill, 1949, pp. 122–125. (Republished as Vol. 19 of the *IEE Electromagnetic Wave Series*, Peter Perigrinus Ltd., Stevenage, Herts, UK, 1984.)

4. S. Silver, Ed., (Reference 3) Tables 6.1 and 6.2, pp. 187, 195.

5. N. E. Feldman, "The Link from a Communications Satellite to a Small Ground Terminal," *Microwave Journal*, 7, 39–44 (1964).

6. C. M. Abrahmanson, and G. P. Petrick, "Economic Considerations for Low Capacity SHF Satellite Communications Earth Terminals," *AIAA 5th Communications Satellite Systems Conference Proceedings*, p. 175 (1974).

7. J. N. Hines, T. Li, and R. H. Turrin, "The Electrical Characteristics of the Horn Reflector Antenna," *Bell Systems Technical Journal*, 42, Part 2, 1187–1211 (1963).

8. F. J. D. Taylor, Ed., *The Goonhilly Project*, Institution of Electrical Engineers, London, 1964.

9. T. Pratt and B. Claydon, "The Prediction of Polar Diagrams of Large Cassegrain Antennas," *Marconi Review*, 34, 1–26 (1971).

10. N. Lockett, "The Electrical Performance of the Marconi 90 ft. Space Communication Aerials," *Marconi Review*, 34, 50–80 (1971).

11. R. G. Kouyoumjian, and P. H. Pathak, "A Uniform Geometrical Theory of Diffraction for an Edge in a Perfectly Conducting Surface," *Proceedings of the IEEE*, 62, 1448–1461 (1974).

12. A. W. Rudge and N. A. Adatia, " Offset Parabolic Reflector Antennas—a Review," *Proceedings of the IEEE*, 66, 1592–1618 (December 1978).

13. P. D. Potter, "A New Horn Antenna with Suppressed Sidelobes and Equal Beamwidths," *Microwave Journal*, 6, 71 (1963).

14. P. B. Clarricoats and G. T. Poulton, "High-Efficiency Microwave Reflector Antennas—A Review," *Proceedings of the IEEE*, 65, 1470–1504 (1977).

15. V. Galindo, "Design of Dual-reflector Antennas with Arbitrary Phase and Amplitude Distributions," *IEEE Transactions on Antennas and Propagation*, AP-12, 403–408 (1964).

16. W. F. Williams, "High Efficiency Antenna Reflector," *Microwave Journal*, 8, 79–82 (July 1965).

17. J. D. Kraus, *Radio Astronomy*, McGraw-Hill, New York, 1966. (Republished by Cygnus Quasar Publishing Co., Powell, Ohio, 1983.)

18. H. R. Guy and J. R. Brain, "Waveguide Feed for an 11/14 GHz Antenna," *Proceedings of IEE Conference on Antennas and Propagation*, York, England, pp. 223–226 (1977).

19. D. J. Sommers, L. J. Parad, and J. G. DiTullio, "Beam Waveguide Feed with Frequency Reuse Diplexer," *Microwave Journal*, 18, 51–59 (1975).

20. P. J. B. Clarricoats and P. K. Saha, "Propagation and Radiation Behavior of Corrugated Feeds. Part I—Corrugated Waveguide Feed. Part II—Corrugated Conical Horn Feed." *Proceedings of the IEE*, London, 118, 1167–1186. (1971).

21. FCC Requirement No. 25.209, revised August 1983.
22. *Actions of the Regional Administrative Radio Conference (RARC '83)*, 1983.
23. J. S. Cook, E. M. Elam, and H. Zucker, "The Open Cassegrain Antenna: Part 1, Electromagnetic Design and Analysis," *Bell System Technical Journal*, **44**, 1255–1300 (September 1965).
24. V. Galindo-Israel, R. Mittra, and A. Cha, "Aperture Amplitude and Phase Control in Offset Dual Reflectors," *USNC/URSI International Symposium on Antennas and Propagation*, May 1978.
25. T. Pratt, "Offset Spherical Reflector Aerial with a Line Feed," *Proceedings of the IEE*, London, **115**, 633–641 (1968).
26. J. Ashmead and A. P. Pippard, "The Use of Spherical Reflectors as Microwave Scanning Aerials," *Journal IEE*, London, **93**. Part IIIA, 627 (1946).
27. E. A. Ohm. "A Proposed Multiple Beam Microwave Antenna for Earth Stations and Satellites," *Bell System Technical Journal*, **53**, 1657–1665 (October 1974).
28. G. Hyde, R. W. Kreutel, and L. V. Smith, "The Unattended Earth Terminal Multiple Beam Torus Antenna," *COMSAT Technical Review*, **4**, 231–264 (1974).
29. C. B. Wooster, "Engineering Satellite Ground Stations," *Electronics and Power (JIEE)*, **27**, 618–622 (September 1981).
30. K. Feher, *Digital Communications: Satellite/Earth Station Engineering*, Prentice-Hall, Englewood Cliffs, NJ, 1983.
31. V. K. Bhargava, D. Haccoun, R. Matyas, and N. Nuspl, *Digital Communications by Satellite*, John Wiley & Sons, New York, 1981.
32. R. J. Klensch, et al., "Critical System Parameters in the DBS Service," *RCA Engineer*, **28**, 58–63 (1983).
33. International Radio Consultative Committee (CCIR), *Recommendations and Reports of the CCIR, 1982, Volume V, Propagation in Non-Ionized Media*, International Telecommunication Union, Geneva, Switzerland, 1982.
34. International Radio Consultative Committee (CCIR), *Recommendations and Reports of the CCIR, Volume IV/IX-2, Frequency Sharing and Coordination Between Systems in the Fixed-Satellite Service and Radio-Relay Systems*, International Telecommunication Union, Geneva, Switzerland, 1982.
35. K. Miya, Ed., *Satellite Communications Technology*, KDD Engineering and Consulting, Tokyo, Japan, 1981.

PROBLEMS

1. A uniformly illuminated rectangular aperture is 1 m high and 2 m wide, and transmits at 4 GHz. Find
   a. The vertical plane 3-dB beamwidth and first sidelobe level.
   b. The horizontal plane 3-dB beamwidth and first sidelobe level.
   c. The gain of the antenna.
   d. The position of the first null in the vertical plane.

2. An antenna has a rectangular aperture 0.6 m high and 1.5 m wide, and operates at 10 GHz. The illumination of the aperture is uniform in the vertical

plane and cosine squared in the horizontal plane. Find
   a.   The gain of the antenna.
   b.   The 3-dB beamwidths in the vertical and horizontal planes.
   c.   The first sidelobe levels in these two planes.

   3.   A geostationary satellite transmits at 4 GHz from a rectangular horn operated in the TE_{10} mode, polarized in the N–S direction. The horn provides service to an elliptical zone 8° by 12° at the 3-dB contour of the antenna beam.
   a.   Calculate the horn aperture dimensions in the N–S and E–W planes.
   b.   Find the first sidelobe levels in the principal planes.
   c.   Calculate the gain of the antenna, assuming a plane wavefront in the aperture of the horn.

   4.   An earth station antenna has a circular aperture 2 m in diameter and transmits at 14 GHz. The aperture efficiency of the antenna is 60 percent. Find
   a.   The antenna gain.
   b.   The first sidelobe level and 3-dB beamwidth, if the illumination function is $\left(1 - \left(\dfrac{r}{r_0}\right)^2\right)$.

   5.   A large earth station operates in the 6/4-GHz band and has an antenna with a circular reflector 30 m in diameter. The antenna uses the shaped Cassegrain configuration to achieve an illumination efficiency of 96 percent at 4 GHz. At this frequency, blockage losses are 4 percent and feed spillover is 5 percent. Main reflector spillover is 2 percent and other diffraction losses total 6 percent. The feed system uses a beam waveguide feed and corrugated horn with an ohmic loss of 0.3 dB.
   a.   Calculate the aperture efficiency of the antenna, without ohmic losses.
   b.   Calculate the gain of the antenna referred to the feed output flange. Hence calculate the antenna efficiency including feed losses.
   c.   Estimate the first sidelobe level and the 3-dB beamwidth of the antenna.

   6.   A 15-m antenna is used to receive transmissions from an 11-GHz satellite with an elevation angle of 10°. When the antenna is pointed at zenith, the measured noise temperature at the feed output flange is 55 K. The aperture efficiency of the antenna is 65 percent.
   a.   Assume that the noise temperature contribution from the antenna sidelobes is constant at all angles. Estimate the antenna noise temperature at an elevation angle of 10°. [*Hint*: Use Figure 9.15 to find the sky noise temperature at zenith, and calculate its contribution to the antenna noise temperature. Hence find the sidelobe contribution.]
   b.   Heavy rain in the antenna beam increases the sky noise temperature to 180 K, at 10° elevation angle. Find the antenna noise temperature under these conditions.
   c.   A LNA with a noise temperature of 160 K is connected to the antenna by a waveguide with a loss of 0.8 dB and a physical temperature of 290 K.

Calculate the system noise temperature
  (i)   under clear-sky conditions
  (ii) in the heavy rain described in Part b.
d. Find the earth station G/T in clear air.

7. A 4-GHz TVRO antenna has a front-fed reflector with a diameter of 3 m. Using Figure 9.26, estimate the optimum edge illumination of the reflector. If losses other than illumination efficiency and spillover are 12 percent, what is the overall efficiency of the antenna? What is its gain at 4.2 GHz?

8. The antenna in Problem 7 could be used to transmit to the satellite, at a frequency of 5.9 GHz. However, the antenna efficiency will be lower because the feed pattern will be narrower at the higher frequency. Suppose the antenna feed is a circular waveguide horn with a pattern given by $G(\theta) = -kf\theta^2$ dB, where k is a constant, θ is the pattern angle in degrees, and f is the frequency in GHz. The constant k is set such that the edge illumination of the reflector is -10 dB at a frequency of 4.2 GHz. Find the edge illumination of the reflector at a frequency of 5.9 GHz, and the overall efficiency of the antenna if other losses total 14 percent.

10

INTELSAT AND INMARSAT

At the present time, late 1984, Intelsat and Inmarsat provide the bulk of international satellite communications. Their published standards are well known in the industry, and we have used them for many of the examples in this book. Although their monopolies are currently being challenged, it is likely that the majority of our readers who work in satellite communications will have some interaction with the Intelsat and Inmarsat systems. For that reason we devote this chapter to their structure and organization.

Partly for reasons of time and space and partly because of a lack of publicly available technical standards, we have not provided similar information and examples for the growing number of domestic and regional satellite communications systems. For information about these systems we refer the reader to the November 1984 issue of *Proceedings of the IEEE* and to *Satellite Communications* magazine.

10.1 HISTORY AND STRUCTURE OF INTELSAT

Intelsat began on August 20, 1964, as an interim organization of 11 countries with the initial name of International Telecommunications Satellite Consortium. The interim structure continued until it was replaced by two new international agreements on February 12, 1973. At that time 80 countries belonged. With the new agreements the name was changed to International Telecommunications Satellite Organization [1].

Currently Intelsat has 109 members and provides service to over 600 earth stations in more than 149 countries, territories, and dependencies [2]. Member nations range in size and wealth from the United States, Japan, the United Kingdom, through the Peoples Republic of China and India, to the Vatican City State. Each member country owns a share of Intelsat proportional to its use of the system; the minimum investment is 0.05 percent [3]. As of 1982, the total investment in the space segment of the Intelsat system was estimated at $1.076 billion.

Technical details of Intelsat spacecraft and member country earth station operation are adopted by a Board of Governors, which follows general rules established by the Meeting of Signatories [4]. The latter convenes once a year and discusses general operational and financial matters. The Board of Governors includes representatives of the 20 largest investors plus as many as five regional representatives [5]. The Board of Governors publishes mandatory technical and performance characteristics that earth stations must meet in order to participate in the Intelsat system.

10.2 THE INTELSAT NETWORK

Currently Intelsat operates 16 spacecraft. The model we have stressed in this text is INTELSAT V; INTELSAT VI is scheduled for launch in 1986 or 1987 [1]. See Table 10.1 for details of all the Intelsat spacecraft, and also Figure 3.9 in Chapter 3.

Earth stations in the Intelsat network are classified as Standard A, B, C, D, and E. The last two are small installations for regional business service. Standard A and B stations both work at 4 and 6 GHz and have minimum G/T values of 40.7 dBK^{-1} and 31.7 dBK^{-1}, respectively. Typically these numbers correspond to 30-m antennas for Standard A and 11-m antennas for Standard B. Standard C earth stations work in the 11 and 14 GHz bands; their G/T requirements are spelled out in terms of frequency and rain attenuation as follows [6]. At any frequency f (GHz) in the 10.95 to 11.20 and 11.45 to 11.70 GHz downlink bands,

$$G/T \geq 39 + 20 \log_{10} (f/11.2) + L_1 \text{ dBK}^{-1} \tag{10.1}$$

where L_1 is the attenuation exceeded 10 percent of the time, and

$$G/T \geq B + 20 \log_{10} (f/11.2) + L_2 \text{ dBK}^{-1} \tag{10.2}$$

where L_2 is the attenuation exceeded 0.017 percent of the time. B is 29.5 for stations receiving the west spot beam and 32.5 for stations receiving the east spot beam.

The Intelsat network is controlled from the Intelsat Spacecraft Technical Control Center in Washington, D.C.

10.3 INMARSAT

Inmarsat is the short form name of the International Maritime Satellite Organization. Generally patterned after Intelsat, it was formally organized on July 16, 1979, and began full operation on February 1, 1982 [7]. Currently (1984) it has 38 member nations. The largest investment shares are held by the U.S. (23.33 percent) and the U.S.S.R. (14.07 percent). The U.S. signatory is Comsat. As of January 1984, the Inmarsat network included 2124 ship earth stations, 7 coast

Table 10.1
Intelsat Satellites

Satellite type	Date of first launch in series	Type vehicle/ Number of satellites launched	Satellite capacity (telephone & television)	Basic communications characteristics and improvements	Basic spacecraft characteristics	Average cost per satellite ($ US.)	Cost of launch ($ US.)	Total cost satellite and launch ($ US.)	Design satellite lifetime (yr)
INTELSAT I	4/6/65	Thrust Augmented Delta/there was only 1 satellite launched in this series.	240 voice circuits or TV	Omni-antenna with squinted beam. Coverage to North Atlantic Region only. Restricted to point-to-point communication.	Spin stabilization in geosynchronous orbit. *Weight*: 68 kg at launch; 39 kg in orbit.	7,000,000	4,700,000	11,700,000	1.5
INTELSAT II	10/26/66	Improved Thrust Augmented Delta/3 satellites successfully launched and 1 launch failure.	240 voice circuits or TV	Global beam coverage. Multipoint communication among earth stations in region.	Spin stabilization in geosynchronous orbit. *Weight*: 162 kg at launch; 86 kg in orbit.	3,600,000	4,600,000	8,200,000	3.0
INTELSAT III	9/18/68	Improved Thrust Augmented Long-Tank Delta/5 satellites successfully launched and 3 launch failures.	1200 voice circuits plus 2 TV channels	Fivefold increase in communications capacity as a result of efficient new antenna. New capability of providing TV and voice simultaneously.	Spin stabilization in geosynchronous orbit. Mechanically despun antenna. *Weight*: 294 kg at launch; 152 kg in orbit.	6,250,000	5,750,000	12,000,000	5.0
INTELSAT IV	1/25/71	Atlas-Centaur launch vehicle/7 satellites successfully launched and 1 launch failure.	4000 voice circuits plus 2 TV channels	More than twofold increase in capacity over INTELSAT III as a result of increased power and efficient new spot beam antennas.	Spin stabilization in geosynchronous orbit. Mechanically despun platform, including antennas. Double the power of INTELSAT III. *Weight*: 1418 kg at launch; 732 kg in orbit.	14,000,000	18,500,000	32,500,000	7.0

INTELSAT IV-A	9/25/75	Atlas-Centaur launch vehicle/5 out of the 6 satellites in this series successfully launched.	6000 voice circuits plus 2 TV channels	50 percent increase in capacity over INTELSAT IV as a result of frequency reuse by hemispheric beam isolation.	Spin stabilization in geosynchronous orbit. Complex antenna farm, which is mechanically despun. *Weight:* 1516 kg at launch; 863 kg in orbit.	21,500,000	26,000,000	47,500,000	7.0
INTELSAT V	12/6/81	Both Atlas-Centaur and ESA's Ariane to be used for 9 satellites in this series.	12,000 voice circuits plus 2 TV channels	Double the capacity of INTELSAT IV-A by use of new frequency at 14/11 GHz and by frequency reuse both through dual polarization and hemispheric beam isolation. Maritime mobile communications capacity on INTELSAT V (F-5) to (F-9).	Three-axis body stabilization through use of momentum wheel. Deployable solar array. *Weight:* 1946 kg at launch; 1012 kg in orbit.	30,000,000	50,000,000	80,000,000	7.0
INTELSAT V-A	1983	These 6 satellites will be launched by the Ariane, Atlas-Centaur, or the STS with perigee engine.	15,000 voice circuits plus 2 TV channels	25 percent increase in capacity over INTELSAT V. Higher power EIRP for domestic service	Three-axis body stabilization through use of momentum wheel. Deployable solar array. *Weight:* 2141 kg at launch; 1159 kg in orbit.	35,000,000	50,000,000 to 60,000,000	85,000,000 to 95,000,000	7.0
INTELSAT VI	1986/87	To be decided between Ariane 4 and STS with perigee engine.	36,000 voice circuits plus 2 TV channels	150 times the capacity of Early Bird. Multiple beam antennas with complex feed system. Sixfold frequency reuse. Space switched TDMA/DSI operation. 38 C-band and 10 Ku-band transponders	Spin stabilized power: Solar cell drum. 2 kW end of life. *Weight:* 3600 kg at launch; 1800 kg in orbit.	N.A.	N.A.	N.A.	10

Source: Reprinted with permission of the North-Holland Publishing Company from Joseph N. Pelton, "INTELSAT: Making the Future Happen," *Space Communication and Broadcasting,* **1**, 36–37 (April 1983).

earth stations, and 6 spacecraft. By 1987 the number of coast earth stations is expected to grow to 29 [8]. Inmarsat headquarters are in London.

Inmarsat leases transponders from Marisat, ESA, and Intelsat. Marisat is a U.S. system that serves the U.S. Navy and merchant marine; Comsat General is the system manager and majority stockholder [3]. The ESA spacecraft that Inmarsat leases is MARECS A; MARECS B was lost during launch and will be replaced in 1985 by MARECS B2. The INTELSAT V spacecraft that carry Inmarsat traffic are equipped with a maritime communications subsystem (MCS).

REFERENCES

1. Joseph N. Pelton, "INTELSAT: Making the Future Happen," *Space Communication and Broadcasting*, **1**, 33–52 (April 1983).
2. Jose L. Alegrett, "U.S. Role in International Satellite Communications II: Leadership Continues," *Satellite Communications*, **8**, 20–23 (March 1984).
3. *COMSAT Guide to the INTELSAT, MARISAT, and COMSTAR Satellite Systems*, Communications Satellite Corporation, Washington, DC.
4. *Standard A Performance Characteristics of Earth Stations in the Intelsat IV, IVA, and V Systems Having a G/T of 40.7 dB/K* (BG-28-72E Rev. 1), Intelsat, Washington, DC, December 15, 1982.
5. K. Miya, Ed., *Satellite Communications Engineering*, Lattice Publishing Co., Tokyo, Japan, 1975.
6. *Standard "C" Performance Characteristics of Earth Stations in the INTELSAT V System (14 and 11 GHz Frequency Bands)*, (BG-28-73E Rev. 1), Intelsat, Washington, DC, December 15, 1982.
7. Paul Branch and Alex Da Silva Curiel, "Inmarsat and Mobile Satellite Communications," Part I, *Telecommunications*, pp. 30–32, 38, 40, 65 (March 1984).
8. "Inmarsat Data," *Ocean Voice*, **4**, 14–15 (January 1984).

11

SATELLITE TELEVISION: NETWORK DISTRIBUTION AND DIRECT BROADCASTING

Television is probably more strongly associated by the public with satellite communications than any other aspect of the industry. Quite rare in 1980, home satellite TV receivers are now common in the United States, and a whole new industry has grown up to sell and service them. Originally intended as relays for network TV and cable TV systems, current communications satellites are becoming de facto direct broadcast satellites (DBS) with their signals received by large numbers of private individuals. This has accelerated movement toward the construction and launch of the first true DBSs, intended to broadcast TV and possibly audio programs to the general population.

Some authorities feel that TV broadcasting will be one of the most important future uses for communications satellites in that it is difficult to foresee how any competing technologies (e.g., optical fiber cables) can offer the accessibility of satellite transmissions. This is particularly important to those users who want their television receivers to be easily portable. Satellites also offer the possibility of broadcasting high-definition TV, a service for which terrestrial television stations lack the necessary bandwidth.

In this chapter we will attempt to describe the current state of satellite TV distribution and home satellite TV reception. We will also summarize the current plans for DBS systems. The reader who would like more details on how to construct or assemble his or her own home earth station should consult reference 1 or 2 and *Satellite TV Magazine*.

11.1 TRANSPONDER FREQUENCIES AND DESIGNATIONS

The cable TV industry has developed a standard numbering system for satellite TV channels that has become widely accepted and that is used to label the tuning controls of most commercially available satellite TV receivers. As summarized in Table 11.1, it is based on 24 downlink center frequencies spaced 20 MHz apart with channel 1 centered at 3720 MHz and channel 24 centered at 4180 MHz. The channels themselves are 40-MHz wide. The spectra of adjacent channels radiated by a "24-channel satellite" thus overlap; interference is avoided by using alternating horizontal and vertical polarization on adjacent channels. The common practice is to use horizontal polarization for the odd-numbered channels and vertical polarization for the even-numbered channels. But this scheme is not universal, and WESTAR IV, for example, reverses it. Some satellites lack dual polarization and transmit only 12 40-MHz channels corresponding to the odd-numbered channels of the standard numbering system. The "12-channel satellites" are usually horizontally polarized [3]. In a 24-channel satellite the transponder numbers correspond to the channel numbers, while in a 12-channel satellite transponder numbers 1 through 12 correspond to odd-numbered channels 1 through 23. Reflecting this, some home satellite receivers have a tuning control with 12 positions. The first is marked with something like 1/2 for channels 1 and 2; the second is marked with 3/4 for channels 3 and 4, and so on. A toggle switch or pushbutton controls the antenna polarization and provides the center frequency offset necessary to receive the desired channel.

Published directories provide detailed listings of what is carried on each channel of each commercial satellite. One of the most complete appears in reference 4.

Table 11.1
Satellite TV Downlink Channel Numbering System

Channel	Center Frequency (MHz)	Channel	Center Frequency (MHz)
1	3720	13	3960
2	3740	14	3980
3	3760	15	4000
4	3780	16	4020
5	3800	17	4040
6	3820	18	4060
7	3840	19	4080
8	3860	20	4100
9	3880	21	4120
10	3900	22	4140
11	3920	23	4160
12	3940	24	4180

11.2 SATELLITE TELEVISION RECEIVERS

While there is considerable variation between manufacturers, the typical TV receive-only (TVRO) earth terminal consists of a small antenna (typically 8 to 15 ft in diameter) a low noise amplifier (LNA), a downconverter, and an IF receiver (simply called a receiver). The market closely resembles that for audio equipment, with some vendors offering integrated systems and others specializing in one or two components. In this section we will briefly survey the characteristics of the equipment that was available at the time of writing. Some of the specifications quoted will quickly become outdated, and the reader who is interested in assembling a personal TVRO terminal should consult the popular electronics press. We will begin with the receiver and work our way back up the chain to the antenna.

In Chapter 5 we summarized the modulation standards for television signals distributed by satellite. The composite video signal that frequency modulates the uplink carrier consists of a baseband video waveform extending from 0 to 4.2 MHz plus one or more frequency-modulated audio subcarriers, which are typically at 6.8 or 6.2 MHz. The first is now the most common, and the less-expensive receivers typically provide selection between only these two. More expensive receivers offer continuously tunable subcarrier reception at frequencies as high as 8 MHz.

While as much as 40 MHz of transponder bandwidth may be available, typical satellite TV signals require 23 to 30 MHz IF bandwidth in the receiver. The industry standard is supposed to be 30 MHz, but a random survey of current receiver specifications shows that 28 and 27 MHz are common, and at least one unit offers selectable IF bandwidths of 23, 25, and 27 MHz. There is an obvious advantage to being able to minimize noise by choosing the smallest bandwidth consistent with desired video quality. As we indicated in Chapter 5, satellite TV signals are frequently overdeviated (sent through a transponder bandwidth narrower than their Carson's rule bandwidth) to enhance video (S/N) in exchange for slightly lower picture quality.

The standard input frequency for TVRO IF receivers is 70 MHz. Most units have a threshold (C/N) at the IF input of 7 or 8 dB, and an FM improvement of about 34 dB can be expected. There is considerable difference between receivers in the detailed video and audio specifications for such things as distortion, differential gain, and the like; these may be important to an electronic hobbyist and unimportant to someone whose main interest in a TVRO terminal is to watch TV. Noise figures are usually not quoted for IF receivers since these make a negligible contribution to the overall TVRO terminal noise performance.

Downconverters translate a selected channel in the 3.7 to 4.2 GHz downlink band to IF. The downconverter local oscillator must be controlled by the IF receiver so that the local oscillator frequency corresponds to the channel for which the receiver is tuned. Typical noise figures for downconverters range from 15 dB down to a nominal 12 dB typical, 10 dB minimum, depending on the manufacturer and cost. Some manufacturers integrate the downconverter and the low noise amplifier into one unit, while others provide separate assemblies.

Low noise amplifiers are available at noise temperatures that range upward from 30 K, but those priced for home TVRO stations typically offer values between 90 and 120 K. Gains range from 30 to 50 dB with the higher figure being typical of units that combine an LNA and a downconverter in one package.

Feed systems for home TVRO antennas are primarily circular horns with rectangular waveguide outputs. The simplest provide only a single polarization and must be physically rotated to change between the horizontal and vertical polarizations of a 24-channel satellite. More sophisticated feeds provide remote polarization adjustment by means of a servo motor or a switch. The feed polarization control system must be compatible with the receiver, since some receivers combine polarization selection with channel selection.

Reflectors range upward in size from 7.5-ft diameters. With present transponders even a 9-ft antenna is marginal for most users, and 10- to 12-ft reflectors are more common. Except for zoning laws, the only limits to home TVRO antenna size are economic, and 15- and 16-footers are available.

Antenna positioners range from the "armstrong rotator" (an old radio amateur term for a hand-operated crank) to sophisticated servo systems that will point the antenna to the desired satellite at the touch of a pushbutton. The owner of a home TVRO system can have the degree of convenience that he or she is willing to pay for!

As in all problems of earth satellite design, it is not possible to give a simple answer to the question "Which system should I buy?" often asked of satellite communications engineers. The problem is exactly like that of assembling a home audio system in that it involves trading dollars for performance. One user may be very happy with a snowy picture while another may require network quality. See reference 5 for a discussion of the relationship between perceived picture quality and weighted (S/N) values. According to the results presented there, about 90 percent of the population considers a 35-dB signal-to-noise ratio "passable" and a 50-dB signal-to-noise ratio "excellent." Rounding the 36.5 dB minimum improvement factor of Chapter 5 up to 37 dB, these numbers correspond to (C/N) values ranging from −2 to 13 dB. The first is below threshold and unrealistically low, but these numbers indicate that (C/N) values only a few decibels above the demodulator threshold will be acceptable to most viewers. A recent article for prospective TVRO owners describes (C/N) values in the range 8 to 10 dB as offering good video quality and 10 to 14 dB as providing excellent quality [6]. These numbers are consistent with the foregoing discussion.

Construction details for those readers who want to build their own TVRO system from scratch are beyond the scope of this text. The primary problems are the reflector, feed, and LNA, three areas in which it is difficult to come close to the price and performance offered by industry. Do-it-yourself articles are available from mail order sources and in the popular electronics press; see the April 1980 *Radio Electronics* magazine for a particularly good set.

11.3 LEGAL MATTERS

At the time this book was written, many legal questions about home satellite TV reception were unresolved—at least in the minds of the interested parties. The originators of most of the television programs currently distributed by satellite (Home Box Office, for example) earn their revenues by selling their programs to cable TV systems. They believe that people who receive the programs directly from a satellite without permission or payment are violating their property rights. At the other extreme are electronic experimenters who believe that U.S. law has always permitted anyone to receive any available signal so long as (1) it is for a noncommercial purpose and (2) it violates no one's rights of privacy. The situation is exactly analogous to that involving copying of books and records. When technology makes it easy to copy something and when copying is cheaper than buying the original, many people feel that they have a right to copy. When it is as cheap to receive TV signals on your own terminal as it is to join the local CATV system, most people will want their own terminals. How they can have them without violating the property rights of the program originators is still unresolved. At the present time it is legal to own and operate a TVRO terminal, and no license is required. Whether permission to receive a particular program is required from the program originator is, in our opinion, still unresolved. See reference 2 for further discussion.

11.4 DIRECT BROADCAST SATELLITES

The satellite systems described in the previous sections of this chapter were designed to distribute television programs to TV broadcast stations and to CATV systems. Their widespread reception by private individuals and the current popularity of home TVRO terminals were probably not foreseen. But while most distributors of satellite TV may be looking for ways to discourage home TVRO viewers, a large number of companies are preparing to launch direct broadcast satellites (DBS) designed for convenient home reception. Some will probably provide encoded transmissions, which only paying subscribers can unscramble, while others will presumably be paid for by advertisers in the same way that broadcast TV is now.

This book was written before widespread DBS service began, and the material we will present about it may be outdated in a short time. Most of it was taken from the special satellite communications issue of *IEEE Communications Magazine* published in March 1984.

As approved by the FCC in 1983, the primary allocation for the so-called broadcast satellite service (*BSS*) is an uplink band from 17.3 to 17.8 GHz and a downlink from 12.2 to 12.7 GHz. In addition, that portion of the "normal" fixed satellite service (*FSS*) allocation from 11.7 to 12.2 GHz may be used for direct broadcast satellites on a noninterfering basis [7].

Currently there are 15 potential operators of DBS systems. Of these, Satellite Television Corporation (STC) is probably closest to operation. As an interim measure it expects to provide DBS service to a limited area of the northeast United States beginning in 1984 with several transponders of the Satellite Business Systems SBS IV spacecraft. This will offer a 53-dBW EIRP [8]. The FCC has authorized STC to launch two DBS satellites to share an orbital location of

100.8° W and broadcast a total of six TV channels to the eastern United States [9]. Dedicated DBS spacecraft will be launched by a number of operators in the 1985–86 time frame.

Proposed DBS transmitters will offer EIRP values in the 54 to 60 dBW range—10 to 20 dB better than what is now available from U.S. domestic satellites. These will provide clear-weather (C/N) values on the order of 14 dB in a 24 MHz bandwidth using home TVRO terminals with G/T values on the order of 12 dBK^{-1}. These may be achieved, for example, by employing 0.9-m dishes and receivers with overall noise temperatures of 480 K [10]. This is possible because small offset-fed antennas are available with aperture efficiencies of 75 percent and because GaAsFET amplifiers are in production with noise figures in the 2.8 to 3.2 dB range [8].

The high transmitter powers required by DBS spacecraft—typically 200 W per transponder—will limit the number of channels that a single satellite can carry to something between three and six [8]. Exactly how many DBS channels will ultimately be available is not yet clear, since it will depend on the bandwidths, polarizations, and frequency plans used. The most commonly accepted numbers are in the 32 to 40 channel range [11].

11.5 SUMMARY

Communications satellites are widely used to distribute television programming to terrestrial TV stations and to CATV systems. Many individuals receive these signals as if they were broadcast to the general public, and a number of companies are seeking to provide true direct broadcast satellite (DBS) service.

Most satellites currently used for TV distribution offer either 12 or 24 40-MHz bandwidth transponders. If only 12 are used, they are all on one polarization (usually horizontal). If 24 are used, they overlap in frequency and avoid interference by putting overlapping channels on orthogonal polarizations.

Home TVRO (television receive-only) terminals are available with antenna diameters from 7.5 ft up. For satisfactory performance, they must deliver (C/N) values of at least 8 dB to their FM demodulators. If proposed high-power DBS spacecraft are launched, terminals to receive them are expected to be significantly smaller and cheaper than those now required for conventional satellites.

REFERENCES

1. Anthony T. Easton, *The Home Satellite TV Book*, Putman Publishing Group, New York, 1982.
2. Robert J. Traister, *Build a Personal Earth Station for Worldwide Satellite TV Reception*, TAB Books, Blue Ridge Summit, PA, 1982.
3. C. J. Schultheiss, "The Ultimate Receiver," *Satellite TV Magazine*, **1**, 30–34 (March–April 1983).
4. Cheryl R. Carpinello, Lucinda Weindling, and Pat Hoyos, "Satellite Tran-

sponder Chart: Who's Carrying What on Which Satellite," *Satellite Communications*, **7**, (un-numbered foldout section) (June 1983).

5. *Reference Data for Radio Engineers* (6th ed.), Howard W. Sams and Co., Indianapolis, 1975.

6. C. J. Schultheiss, "Understanding TVRO," *Satellite TV Magazine*, **1**, 8–13, 84–85 (March–April 1983).

7. Trudy E. Bell, "Technology '84 Communications," *IEEE Spectrum*, **21**, 53–57 (January 1984).

8. D. K. Dement, "United States Direct Broadcast Satellite System Development," *IEEE Communications Magazine*, **22**, 6–10 (March 1984).

9. Hale Montgomery, "First In, First Out," *Satellite Communications*, **8**, 12 (July 1984).

10. Wilbur L. Pritchard and Harley W. Radin, "Direct Broadcast Satellite Service by Direct Broadcast Satellite Corporation," *IEEE Communications Magazine*, **22**, 19–25 (March 1984).

11. Richard G. Gould, "Transmission Standards for Direct Broadcast Satellites," *IEEE Communications Magazine*, **22**, 26–34 (March 1984).

PROBLEMS

1. Calculate the video-weighted (S/N) that can be expected from a "super-cheap" home TVRO earth terminal that uses a 7.5-ft diameter dish with a noise temperature of 50 K and an LNA with a noise temperature of 120 K. Assume that all of the receiver noise is contributed by the LNA, the operating frequency is 4 GHz, the satellite EIRP is 34 dBW, the distance from the satellite to the earth station is 38,000 km, the earth station antenna's aperture efficiency is 0.55, and the FM demodulator has an 8-dB (C/N) threshold.

2. Direct broadcast satellites (DBS) will probably be launched in 1985 to provide low-cost home TV via inexpensive terminals. The allocated frequencies are 17.3 to 17.8 GHz uplink and 12.2 to 12.7 GHz downlink; these provide a total bandwidth of 500 MHz. Transmission standards for DBS are not yet fully developed; obviously every prospective operator wants to maximize the number of channels that can be carried in the available bandwidth, minimize the number of satellites required to cover this bandwidth, and minimize the cost of the required earth terminals.

In this problem you will examine some of the trade-offs involved in planning a DBS system. Your goal is to design a system that meets the following specifications so that (1) the number of channels (transponders) carried by an individual satellite is as large as you can make it but (2) if more than one satellite is required to cover the allocated 500-MHz band, each satellite carries the same number of transponders.

a. Spacecraft Specifications:
   (i) Receive G/T: $+1$ dBK^{-1}
   (ii) Saturation uplink flux density: -90 dBW/m^2

(iii) Total RF power: 1400-W end of life. The sum of the total transponder output powers on any one spacecraft cannot exceed 1400 W.

(iv) Downlink antenna pattern: Circularly symmetric with 3° minimum half-power beamwidth across downlink band.

(v) Number of transponders: One for each TV channel carried.

(vi) Transponder bandwidth: TV channel bandwidth multiplied by 1.10

(vii) Total bandwidth (sum of transponder bandwidths): 500 MHz maximum.

(viii) Backoff required: None. Only one carrier is present and the transponder is assumed to be linear.

(ix) Orbital location: 101° W.

b. Uplink Earth Station Specifications:

(i) Location: 80.438° W, 37.229° N.

(ii) Transmitting antenna: 20 ft in diameter with a 55 percent aperture efficiency.

c. Downlink Earth Station Specifications:

(i) Location: Anywhere in eastern half of the United States.

(ii) Antenna: 1 m in diameter with 75 percent aperture efficiency.

(iii) Antenna noise temperature: 90 K in clear weather.

(iv) Receiver U.S. standard noise figure: 3 dB (includes all noise contributions except those from the antenna).

(v) Demodulator threshold: 7.5 dB.

(vi) Receiver noise bandwidth: Negotiable. It can be made equal to the specified TV channel bandwidth.

(vii) Video (S/N) required: 45 dB at beam edge, calculated by $(S/N) = (C/N) + 18.8 + 10 \log_{10} [3\, m^2(m + 1)]$ dB. This includes all preemphasis and weighting factors. The modulation index m is given by $m = f_p/f_v$ where f_p is the peak deviation (you choose its value) and f_v is 4.2 MHz.

In your design do not worry about rain margin. Do not assume overdeviation; calculate occupied bandwidths from Carson's rule.

APPENDIX

A.1 DECIBELS IN COMMUNICATIONS ENGINEERING

Most readers of this book will be familiar with the practice of expressing power ratios in decibels, abbreviated dB. The dB ratio A of two power levels p_1 and p_2 is given by

$$A = 10 \log_{10}\left(\frac{p_1}{p_2}\right) \text{dB} \tag{A.1}$$

provided that p_1 and p_2 are expressed in the same units. Although the decibel is formally defined only for a power ratio, p_1 and p_2 and A can also be expressed in terms of many combinations of voltage, current, resistance, electric field strength, magnetic field strength, and so on.

It is common practice in communications engineering to use decibels and the mathematical properties of the logarithm to transform multiplicative equations to additive equations, to manipulate the additive equations into particularly convenient forms, and to define new logarithmic units with dB in their names for some of the quantities that appear. When first presented, this practice is confusing to many people, and we hope to clarify it here.

Consider the simple voltage divider circuit with resistors R_S and R_L shown in Figure A.1. The rms voltages across the source and the load resistance are V_S and V_L, respectively; the rms power supplied by the source is p_S W and the rms power delivered to the load is p_L W. From elementary circuit theory, these quantities are related by

$$p_S = \frac{V_S^2}{R_S + R_L} \tag{A.2}$$

$$p_L = \frac{V_L^2}{R_L} \tag{A.3}$$

439

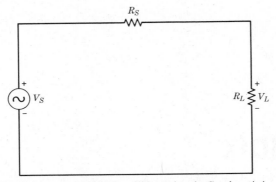

Figure A.1 Voltage divider circuit used as illustration in Section A.1.

$$V_L = \frac{V_S R_L}{R_L + R_S} \tag{A.4}$$

$$p_L = \frac{p_S R_L}{R_L + R_S} \tag{A.5}$$

letting

$$R_L + R_S = R_T \tag{A.6}$$

then

$$p_L = \frac{p_S R_L}{R_T} \tag{A.7}$$

Equation (A.7) is a multiplicative equation, and p_S and p_L must have the same units. Expressed another way, whatever units we substitute in for p_S, p_L will have the same units. The reader should keep this in mind for what comes next.

Solving for the ratio p_L/p_S and expressing the result in decibels, we have

$$10 \log_{10}\left(\frac{p_L}{p_S}\right) = 10 \log_{10}\left(\frac{R_L}{R_T}\right) \tag{A.8}$$

Invoking the properties of logarithms we may rewrite this as

$$10 \log_{10}(p_L) - 10 \log_{10}(p_S) = 10 \log_{10}\left(\frac{R_L}{R_T}\right) \tag{A.9}$$

Without affecting the correctness of this equation or of its predecessor we may divide p_L and p_S by 1 W. Expressed in the form of Eq. (A.9), the result is

$$10 \log_{10}\left(\frac{p_L}{1\text{ W}}\right) - 10 \log_{10}\left(\frac{p_S}{1\text{ W}}\right) = 10 \log_{10}\left(\frac{R_L}{R_T}\right) \tag{A.10}$$

The first term above is the decibel ratio of p_L to 1 W. This is defined as p_L expressed in units of decibels above 1 W, or p_L in dBW. If we represent this quantity as P_L then

$$P_L(\text{dBW}) = 10 \log_{10}\left(\frac{p_L}{1\text{ W}}\right) \tag{A.11}$$

Likewise the source power in dBW is P_S where

$$P_S(\text{dBW}) = 10 \log_{10}\left(\frac{p_S}{1\text{ W}}\right) \tag{A.12}$$

Substituting P_L and P_S into (A.10) yields

$$P_L(\text{dBW}) - P_S(\text{dBW}) = 10 \log_{10}\left(\frac{R_L}{R_T}\right) \tag{A.13}$$

If we had expressed the powers in Eq. (A.9) in milliwatts and then divided both of them by 1 mW, the only effect would have been to express P_L and P_S in decibels above 1 mW or dBm

$$P_L(\text{dBm}) = 10 \log_{10}\left(\frac{p_L}{1\text{ mW}}\right) \tag{A.14}$$

$$P_S(\text{dBm}) = 10 \log_{10}\left(\frac{p_S}{1\text{ mW}}\right) \tag{A.15}$$

and Eq. (A.9) would have become

$$P_S(\text{dBm}) - P_L(\text{dBm}) = 10 \log_{10}\left(\frac{R_L}{R_T}\right) \tag{A.16}$$

Equations (A.13) and (A.16) differ only in the logarithmic power units that appear on their left-hand sides. These equations would be true so long as P_L and P_S were both expressed in the same logarithmic units. This happens because units that cancel by division in multiplicative equations like Eq. (A.7) cancel by addition or subtraction in additive decibel equations like Eqs. (A.13) and (A.16). We can use this to write a general form for both these equations:

$$P_S - P_L = 10 \log_{10}\left(\frac{R_L}{R_T}\right)\text{dB} \tag{A.17}$$

The "dB" after the equation means that the quantities involved must be expressed in a consistent set of logarithmic units.

Equation (A.17) is typical of many of the equations used in this text that contain a mixture of decibel and nondecibel units. We can change it to one involving only decibel quantities if we divide both the resistances by 1 ohm (Ω) and transform the ratio on the right-hand side to a difference

$$P_S - P_L = 10 \log_{10}\left(\frac{R_L}{1\,\Omega}\right) - 10 \log_{10}\left(\frac{R_T}{1\,\Omega}\right) \tag{A.18}$$

Now let us define a new unit for our own use called the dBΩ for decibels above one ohm. This is a dubious but expedient use of the term decibel! Thus

$$R_L(\text{in dB}\Omega) = 10 \log_{10}\left(\frac{R_L}{1\,\Omega}\right) \tag{A.19}$$

We can also express R_T in dBΩ by taking its log. Thus in these new units Eq. (A.16) becomes

$$P_L(\text{dBm}) - P_S(\text{dBm}) = R_L(\text{dB}\Omega) - R_T(\text{dB}\Omega) \qquad (\text{A.20})$$

Likewise we could have expressed the resistance in kilohms (kΩ) and then divided all of the resistance terms by 1 kΩ, inventing a new unit that we will call the dBkΩ. The result could have been either

$$P_L(\text{dBm}) - P_S(\text{dBm}) = R_L(\text{dBk}\Omega) - R_T(\text{dBk}\Omega) \qquad (\text{A.21})$$

or

$$P_L(\text{dBW}) - P_S(\text{dBW}) = R_L(\text{dBk}\Omega) - R_T(\text{dBk}\Omega) \qquad (\text{A.22})$$

The units in these two equations cancel by addition and subtraction rather than by multiplication and division. Hence so long as both powers are in the one common decibel unit and so long as both resistances are in another common decibel unit, we may write a general form of these equations

$$P_L - P_S = R_L - R_T \text{ dB} \qquad (\text{A.23})$$

A common alternative is to rearrange Eqs. (A.21) and (A.22) so that input quantities and output quantities are on opposite sides of the equation. For example, if the usual problem is to find the load power, then Eq. (A.21) would most usefully be expressed

$$P_L(\text{dBW}) = P_S(\text{dBW}) + R_L(\text{dB}\Omega) - R_T(\text{dB}\Omega) \qquad (\text{A.24})$$

Readers seeing expressions like Eq. (A.24) for the first time object to its apparent addition of dBW and dBΩ, since they have learned by sad experience that quantities with different units should not be added. But logarithmic units are different: quantities with different logarithmic or decibel units *are* added; the test of correctness is whether or not the units cancel by addition or subtraction. Thus in Eq. (A.24) the dBΩ units cancel in the subtraction of R_T in dBΩ from R_L in dBΩ and the dBWs cancel because they appear on both sides of the equal sign.

The approach we have followed in this text is to use decibel units where the practice is common (principally for power) and to call the reader's attention to other common dB units at the point of first introduction. Besides the dBW and the dBm, a common power unit is the dBp, which is power expressed in dB above 1 picowatt.

We must emphasize again that decibel units of resistance are irregular and that we have introduced them here only as a teaching tool.

A.2 RF FILTERING OF PSK CARRIERS

BPSK and QPSK signals have broad spectra that cause interference in adjacent channels unless filtered. The ideal Nyquist filter is unrealizable since its transfer function becomes zero beyond a certain frequency; real filters never have infinite attenuation and can only approximate the ideal case.

Butterworth and Chebyshev filters are widely used to limit the bandwidth of PSK signals, and standard designs are readily available. However, since they only approximate the desired Nyquist characteristic, some ISI will result from their use. Figure A.2, from reference 1, shows the spectrum of QPSK signals without filtering and after filtering by a Butterworth filter with bandwidth B. The product BT_b relates filter bandwidth to bit rate. A BT_b product of one corresponds to a filter with a bandwidth in hertz equal to the bit rate in bits per second. Butterworth filters always have a gain of -3 dB at their cutoff frequencies, and therefore approximate the square-root raised cosine filter in the bandpass region.

Figure A.3 [1] shows the impulse response $v_r(t)$ of several Butterworth filters with BT_b ratio 0.5. The parameter n is the order of the filter; higher order filters have steeper roll-off characteristics. Note that the zero crossings of the pulses do not occur at exactly $2T_b$, $3T_b$, and so on, indicating that ISI will occur.

The overall effect of filtering the QPSK signal with nonideal filters is to introduce ISI, which increases the bit error rate when the E_b/N_0 ratio is near threshold. Figures A.4 and A.5 [1] show results for the degradation of BER for a BPSK wave with Chebyshev filters and an integrate and dump receiver. Figure A.4 gives the probability of bit error as a function of E_b/N_0, for three BT values. $BT = \infty$ corresponds to the unfiltered case and is the ideal result usually quoted for BPSK and QPSK. $BT = 1.5$ and $BT = 1$ produce much narrower PSK spectra, but the probability of error increases at a given E_b/N_0. Figure A.5 shows the increase

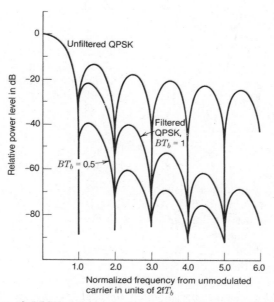

Figure A.2   Spectra of QPSK carriers; unfiltered, and filtered by Butterworth bandpass filters with $BT_b = 1$ and 0.5. (*Source:* Robert M. Gagliardi, *Satellite Communications*, Lifetime Learning Publications, Belmont, CA, 1983, p. 148. Copyright © 1984 Wadsworth, Inc. Reprinted by permission of Van Nostrand Reinhold Co., Inc.)

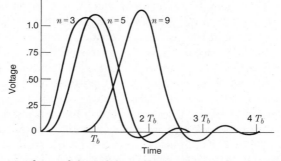

Figure A.3 Time waveform of demodulated QPSK pulses after bandpass filtering with a Chebyschev filter of order n with $BT_b = 0.5$. Note that the zero crossings do not coincide with integer values of T_b. (*Source:* Robert M. Gagliardi, *Satellite Communications*, Lifetime Learning Publications, Belmont, CA, 1983, p. 150. Copyright © 1984 Wadsworth, Inc. Reprinted by permission of Van Nostrand Reinhold Co., Inc.)

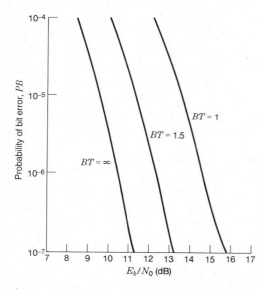

Figure A.4 Probability of error for a BPSK signal after bandpass filtering. B is the 3-dB bandwidth of the filter. $BT = \infty$ is the unfiltered case. (*Source:* Robert M. Gagliardi, *Satellite Communications*, Lifetime Learning Publications, Belmont, CA, 1983, p. 151. Copyright © 1984 Wadsworth, Inc. Reprinted by permission of Van Nostrand Reinhold Co., Inc.)

in E_b required to maintain a probability of error of 10^{-6}, for several orders of Chebyschev filter with given BT_b value. Here, B is the 3-dB bandwidth of the Chebyschev filter. In a typical application with filtering having $BT_b = 1.25$, a fifth-order filter ($N = 5$) requires an additional 2.2 dB of carrier power to achieve a 10^{-6} probability of bit error. Thus the E_b/N_0 ratio needed for this error rate is 12.7 dB, rather than the ideal 10.5 dB.

Designing PSK links requires careful trade-off of bit error rate and filter characteristics. RF filtering must be provided to meet out-of-band spectral emission requirements and to minimize noise, but results in increased error rates and a demand for more carrier power for a given BER. The effect of practical filters and equalizers can be modeled by simulation of the link using a computer. The

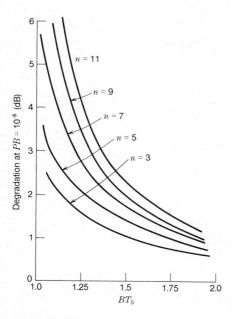

Figure A.5 Degradation of probability of error due to bandpass filtering of a BPSK carrier. The ordinate is the increase in carrier power in decibels to maintain a BER of 10^{-6} as the bandwidth of the filter is narrowed. Chebyshev filter, order n, integrate, and dump receiver. (*Source:* Robert M. Gagliardi, *Satellite Communications*, Lifetime Learning Publications, Belmont, CA, 1983, p. 152. Copyright © 1984 Wadsworth, Inc. Reprinted by permission of Van Nostrand Reinhold Co., Inc.)

simulation models each element of the link, in either the time domain or the frequency domain, and uses fast Fourier transform (FFT) algorithms to convert the signal from a time waveform to a frequency spectrum, as it progresses through the link. Calculation of RF spectra and receiver BER for a given transmitted power and receiver noise power is possible with this approach, for any combination of filters. Nonlinearities in the earth station HPA and satellite transponder can also be modeled. Reference 2 contains a good survey of communication link simulation programs and gives an example of the design of a PSK link using one such program.

Feher [3] gives a detailed survey of the effects of filtering for several forms of PSK signal, as well as techniques that keep BER degradation small while controlling RF bandwidth.

REFERENCES

1. R. M. Gagliardi, *Satellite Communications*, Lifetime Learning Publications, Belmont, CA, 1984.
2. L. C. Palmer, "Computer Modeling and Simulation of Communication Satellite Channels," *IEEE Journal on Selected Areas in Communications*, **SAC-2**, 89–102 (1984).
3. K. Feher, *Digital Communications: Satellite/Earth Station Engineering*, Prentice Hall, Englewood Cliffs, NJ, 1983.

GLOSSARY

AC Assignment channel. A channel carrying assignment information in the Intelsat TDMA system.

ACK Acknowledge signal. This is sent when data are received correctly in an ARQ system.

ADM Adaptive delta modulation. Delta modulation that varies its step size in response to the behavior of the input signal.

AKM Apogee kick motor. A rocket engine that boosts a satellite from the Space Shuttle orbit into a transfer orbit.

ALOHA A random access system.

Antenna gain An increase in flux density in a given direction produced by an antenna relative to flux density produced by an isotropic antenna transmitting the same total power.

Aperture The area across which an antenna radiates or receives energy.

Aperture antenna A microwave antenna using a horn or a feed and reflector.

Aperture efficiency The ratio of effective aperture of an antenna to its physical area, typically in range 50 to 75 percent. A lossless, uniformly illuminated aperture has an aperture efficiency of 100 percent.

Apogee The point in an orbit of greatest distance from the earth.

APT Automatic picture transmission. A format for sending weather data from a satellite.

ARQ Automatic repeat request.

ASCII code A standardized binary code to represent alphabetic letters, numbers, and symbols. ASCII stands for American Society for Computer Information Interchange.

AT&T American Telephone and Telegraph Company.

Attenuation Time-varying fading.

Attitude and Orbit Control System (AOCS) A system that keeps a satellite in the correct orbit and pointing in the correct direction. Made up of gas jets and/or momentum wheels.

Audio subcarrier A carrier that is modulated by the audio component of a television signal.

Autotrack Automatic tracking of satellite motion by earth station antenna.

Azimuth The horizontal angle measured east from north to the line from an observer to a satellite.

Backoff The process of reducing the input power level of a traveling wave tube to obtain more linear operation.

Baseband The frequency band that a signal occupies when initially generated.

BCH codes Bose-Chaudhuri-Hocquenghem codes, named after discoverers. These are linear block codes.

BER Bit error rate. The fraction of a sequence of message bits that are in error.

Binary cyclic codes Block codes with cyclic properties. These can be generated with shift registers.

Block code A binary data transmission format in which message bits and parity check bits are formed into blocks, each with a predetermined number of bits.

Blockage The loss of energy in a reflector antenna caused by the presence of obstacles in the aperture (feed, feed supports, etc.).

Boresight The center of an antenna beam, usually the direction of maximum gain.

Boltzmann's constant The constant used to obtain noise power from noise temperature. Value is 1.38×10^{-23} J/K, or -228.6 dBW/K/Hz.

BPSK Binary (or bipolar) phase shift keying. A digital modulation technique in which the carrier phase takes on one of two possible values.

BSS Broadcast satellite service.

Burst errors Errors that occur in adjacent bits.

C band Frequency range 4 GHz to 8 GHz.

Carson's rule bandwidth The theoretical bandwidth needed to transmit an FM signal without distortion.

Cassegrain antenna A two-reflector antenna with a configuration originally used by William Cassegrain for an optical telescope.

CATV Originally, this stood for community antenna television, but now it is used as a general abbreviation for cable television.

CCIR International Radio Consultative Committee.

CCITT International Telegraph and Telephone Consultative Committee.

CDMA Code division multiple access. A multiple access scheme in which stations use spread-spectrum modulation and orthogonal codes to avoid interfering with one another.

CES Coastal earth station in the Inmarsat system.

Circumscribed circle A circle that is tangential to an orbit and that encloses the orbit.

Clear sky Conditions under which atmosphere does not cause excess attenuation or depolarization of RF signals.

Clip The time delay between the onset of sound and the completion of a channel assignment and connection in a demand-assignment system.

(C/N) Carrier-to-noise ratio.

Code vector A codeword.

Codec A coder-decoder.

Codeword A combination of message bits and parity check bits into a word of fixed length.

Coma Distortion of plane wave in antenna aperture caused by movement of the feed away from the focus in the focal plane.

Companding The process by which the dynamic range of speech is compressed before transmission and expanded after detection. The name is a contraction of compression and expanding.

Comsat The Communications Satellite Corporation.

Conical scan Autotrack technique that derives angular error of a satellite in two planes simultaneously by rotation of the antenna beam around the boresight axis.

Convolutional code   A code in which there is not a one-to-one correspondence between data bits and coded bits.

Corrugated horn   A microwave antenna with high efficiency. It is used as a feed for reflector antennas because of its symmetrical beam shape and low cross-polarization.

CR/BTR   Carrier recovery/bit timing recovery.

Crosstalk   Transfer of signal between two separate channels or circuits.

CSC   A common signaling channel in the SPADE system.

CSSB   Companded single sideband. An analog FDM technique using companding and single sideband amplitude modulation.

dBK   Noise temperature unit expressed in decibels greater than 1 K.

dBK^{-1}   Units of G/T ratio, decibels per Kelvin.

dBm   Decibels relative to one milliwatt.

dBp   Power in dB above 1 picowatt. It may be weighted or unweighted, depending on context.

DBS   Direct broadcast satellite.

dBW   Decibels relative to one watt.

Deemphasis   The removal of some high-frequency noise from a demodulated FM signal. See *preemphasis*.

Delta modulation   A process for digitally encoding and transmitting an analog waveform that avoids the use of analog-to-digital and digital-to-analog converters.

Demultiplexer   A device that recovers individual signals that had been multiplexed for transmission.

Despun antenna   An antenna, mounted on a satellite with a spinning body, which is rotated in the opposite direction to the body rotation so that the antenna beam points in a fixed direction.

Differential modulation   Digital modulation in which the change in the carrier phase or amplitude from one symbol to the next is determined by the modulating signal. This is in contrast to direct modulation, in which the state of the carrier phase (or amplitude) is determined by the modulating signal.

Direct modulation   A digital modulation scheme in which the phase (or amplitude) of the carrier is determined by the modulating signal. This is in contrast to differential modulation, where the change in carrier amplitude (or phase) is determined by the modulating signal.

DM   Delta modulation.

Doppler shift   The change in radio frequency that results from motion of the transmitter or receiver.

Downlink   The communications channel from a satellite to an earth station.

DSI   Digital speech interpolation. The digital implementation of TASI, a technique for reassigning channels during speech pauses.

Duplex link   A link capable of simultaneous two-way transmission of data.

Eccentricity   A measure of the ellipticity of an orbit. If the eccentricity is zero, the orbit is circular.

Eclipse   A satellite in the shadow of the earth.

EIRP   Effective isotropically radiated power. This is equal to antenna gain multiplied by transmitted power.

Elevation   The angle measured up from local horizontal to the line from an observer to a satellite.

ELV Expendable launch vehicle.

Encryption Systematic modification of a signal to prevent unauthorized use.

Energy dispersal The process of modulating a lightly loaded link to maintain a low power spectral density and avoid interference with terrestrial links.

Equalizer A filter with an amplitude or phase characteristic designed to correct for amplitude or phase distortion in another part of a communication channel.

Erlang The unit of traffic flow in a communication system.

Erlang B model The preferred equation for calculating the number of channels required to carry a given amount of traffic.

Error correction The correction of bits in a digital data stream that have been corrupted during transmission.

Error detection The detection of bits in a digital data stream that have been corrupted during transmission.

Error syndrome A word that indicates whether a codeword is correct or in error.

ESA European Space Agency.

Even (odd) parity Addition of one or more bits to a data signal to permit error detection or correction. In even parity, the binary sum of the bits in a character or block is even. In odd parity, the binary sum is odd.

Far field A region sufficiently distant from a transmitting antenna that the observed wave is locally plane.

FAW Frame alignment word. A pattern of bits that identifies the start of a frame.

FCC Federal Communications Commission. The government agency in the United States that regulates radio communications and allocates frequencies.

FDM Frequency division multiplexing. A technique whereby several signals from the same earth station share a transponder by using different frequencies.

FDMA Frequency division multiple access. A technique whereby signals from several earth stations share a satellite or transponder by using different frequencies.

FEC Forward error correction.

FH Frequency hopping. A spread-spectrum technique in which the transmitter frequency is "hopped" over a wide bandwidth.

Flux density Power per unit area when an electromagnetic wave is incident on a surface.

FM Frequency modulation.

FM improvement Increase in (S/N) at the output of an FM demodulator relative to the (C/N) at its input.

FPD Focal plane distribution. Spatial variation of energy in the focal plane of an antenna.

Frame (TDM) A portion of a digital transmission that contains one word from each channel and sufficient synchronization information to identify the start of the frame.

Frame (TDMA) A portion of a digital transmission that contains one burst from each earth station plus synchronization and housekeeping information.

Frequency coordination A process by which interference is avoided when planning an earth station installation.

Frequency reuse Transmitting to or receiving from a satellite two independent data channels at the same radio frequency. The channels are separated by the spacecraft antenna using directional beams in spatial frequency reuse and by orthogonal polarizations in polarization frequency reuse.

Front-fed antenna An antenna with a single reflector and a radiating feed at the focus of the reflector.

FSK Frequency shift keying.

FSS   Fixed satellite service. An FCC term describing satellite communications with fixed (i.e., nonmobile) earth stations.

GaAsFET   Gallium Arsenide Field Effect Transistor.

GCE   Ground control equipment (at earth station).

Geostationary arc   A circle of radius 42,242 km lying in the plane of the equator and having its center at the center of the earth.

Geostationary orbit   A circular orbit of radius 42,242 km lying in the plane of the equator. To an observer on the ground, a satellite in geostationary orbit remains at the same fixed position in the sky.

Geosynchronous orbit   A circular orbit of radius 42,242 km that is not in the earth's equatorial plane. A satellite in geosynchronous orbit has an orbital period equal to the earth's rotational period, but its inclination with respect to the equatorial plane makes its position with respect to an observer on the ground change with time.

Global beam   A satellite antenna beam covering the whole of earth, as seen from the satellite.

Grating lobe   An unwanted sidelobe caused by two radiating or receiving antennas or elements that are separated by more than one half wavelength.

Gregorian antenna   A two-reflector antenna with a configuration originally used by James Gregory for an optical telescope.

G/T Antenna gain to noise temperature ratio. Used to characterize earth stations.

Guardband   An empty frequency band used to separate FDM signals.

Guard time   An empty time interval used to separate TDMA bursts.

Hamming code   A simple block code that is useful for correcting or detecting a small number of errors per block.

Housekeeping   This term refers to systems used on a satellite to keep it operating, but that do not form part of the payload.

HPA   High-power amplifier.

I channel   The bit stream that modulates the in-phase carrier of a QPSK system.

IF   Intermediate frequency.

Illumination efficiency   A measure of the efficiency with which an aperture collects energy from an incident plane wave. Uniform illumination of the aperture gives 100 percent efficiency when the antenna transmits or receives.

Inmarsat   The International Maritime Satellite Organization.

Intelsat   The International Telecommunications Satellite Organization.

Intermodulation noise   Noise generated by the interaction of signals in a nonlinear device (usually a TWT in the context of this text).

ISI   Intersymbol interference. This is interference between different symbols in a serial digital transmission.

Isotropic source (antenna)   An antenna that radiates equal power in all directions.

Julian date   The number assigned to a day in a standard astronomical dating system.

Kelvin   Unit of noise temperature.

Key   The information needed to successfully decode an encrypted data sequence.

Ku-band   Frequency range of 12 GHz to 18 GHz

L-band   Frequency range of 1 GHz to 2 GHz

LDM   Linear delta modulation. Delta modulation that uses a fixed step size.

LHCP   Left-hand circular polarization.

Link power budget   Method for calculation of received power in a satellite link.

LNA   Low noise amplifier.

Loading   A measure of the traffic carried by a multiplex telephone link.

Loading factor   A factor that is multiplied by the rms test-tone deviation to yield the rms deviation of an FDM/FM multiplexed telephone signal.

Look angles   The coordinates to which an earth station antenna must be pointed to communicate with a satellite.

Luminance   The amount of white light at a point in a television picture.

Margin   The amount (usually in decibels) by which a received signal exceeds a predetermined lower limit.

Modem   A modulator/demodulator.

Momentum wheel   A heavy wheel, mounted within a satellite, which rotates at high speed to provide angular momentum in one direction. It is used in the attitude control system.

Monopulse   An autotrack technique that derives angular error of a satellite in two planes simultaneously.

MTBF   Mean time before failure.

Multiple access   The sharing of a transponder or satellite by signals from several earth stations.

Multiplexer   A device that combines signals for transmission.

Multiplexing   The sharing of a transponder by several signals from the same earth station.

(n, k)code   An arrangement of binary data such that in a block of n bits, k bits are the message and $(n - k)$ are parity check bits.

NAK   Not acknowledge signal. It is sent when data are received incorrectly in an ARQ system.

NASA   The National Aeronautics and Space Administration.

NOAA   National Oceanographic and Atmospheric Agency. U.S. Government body that operates weather satellites.

Noise figure   A measure of the noise power generated by a device or system.

Noise power budget   A method for calculation of total noise power in a receiving system.

Noise temperature   A measure of noise power that is independent of measurement bandwidth.

North–South, East–West, stationkeeping maneuver   Movement of the satellite perpendicular to (N–S) and along (E–W) the geostationary orbit to correct orbital errors.

NPR   Noise power ratio. A measure of intermodulation noise in a multiplexed telephone channel.

NRZ   Non-return-to-zero. A baseband digital transmission scheme in which the logical 1 and 0 signals correspond to equal-amplitude positive and negative voltages.

NTSC   The 525-line/60-Hz television transmission standard used in North America and Japan.

Nyquist filter   A filter that produces zero ISI waveforms when driven by an impulse.

Offset reflector   A reflector that is not a rotationally symmetric conic section.

OMT   Orthogonal mode transducer. This separates polarizations in an antenna feed.

OQPSK   Offset QPSK modulation.

Orbital elements   A set of six constants that are sufficient to specify an orbit.

Osculating orbit   The orbit that a spacecraft would follow if all perturbing forces were removed.

OTS   Orbital Test Satellite, built by the European Space Agency.

Outage   Loss of communication.

Overdeviation   Intentionally modulating an FM transmitter so that the Carson's rule bandwidth of the uplink signal will be significantly greater than the transponder (and possibly the IF receiver) bandwidth.

P channel   The bit stream that modulates the in-phase carrier of a QPSK system.

PAL   The European 625-line/50-Hz television transmission system.

PAM   Payload assist module. A rocket engine used in Space Shuttle launches to move a satellite from low orbit to geostationary orbit.

Parabolic torus reflector   A reflector that is circular in one plane and parabolic in the orthogonal plane. Beam can be scanned by feed movement in the plane of circular curvature.

Paramp   Parametric amplifier.

Parity check matrix   A method for calculating the error syndrome in a block codeword.

Path loss   The apparent loss of power caused by spherical expansion of a transmitted wave. Equal to $(\lambda/4\pi R)^2$.

PCM   Pulse code modulation. Any modulation scheme that transmits digitally encoded quantized samples of an incoming signal.

Peak factor   A factor that is multiplied by the rms deviation to yield the true deviation of an FDM/FM multiplexed telephone signal.

Perigee   The point in the orbit where a satellite is closest to the earth.

PN   Pseudonoise. This is also the abbreviation for a spread-spectrum scheme in which a transmitted signal is spread through multiplying it by a PN sequence.

PN sequence   A sequence of apparently random ones and zeros that actually repeats after a finite time. The spectrum of a PN sequence is similar to that of white noise.

Polarizer   A device to generate wanted polarization in an antenna feed.

Preamble   The part of a TDMA burst that contains synchronization and other housekeeping information.

Preemphasis   A process for increasing the signal-to-noise ratio of a demodulated FM signal by filtering out some of the high-frequency noise. The signal is distorted before modulation by a preemphasis filter, and the distortion is removed along with some of the noise by a deemphasis filter.

Principal planes   The planes in an antenna that align with reference axes such as the direction of linear polarization and its orthogonal plane.

PSK   Phase shift keying. A modulation technique in which the phase of the RF carrier is varied.

Psophometric weighting   A correction factor used in signal-to-noise ratio calculation that accounts for the nonuniform response of the human ear to white noise.

Pulse stuffing   Insertion of dummy words into a multiplexed digital transmission. It is done to allow an input channel with a low transmission rate to catch up with the rest of the system.

pWp   Picowatts psophometrically weighted. Power in picowatts corrected for the nonuniform response of the human ear to white noise for use in signal-to-noise ratio calculation.

Q channel   The bit stream that modulates the quadrature channel of a QPSK system.

QPSK   Quadrature phase shift keying. A digital modulation technique in which the carrier phase takes on one of four possible values.

Quantization   The process of resolving analog samples of a signal into one of a finite number of possible values.

RARC   Regional Administrative Radio Conference. Sets frequency allocations for a region of the world.

RCA   Radio Corporation of America.

Redundancy   The provision of spare components in a satellite or earth station to ensure continuous operation when one component fails.

Redundant bits   Extra bits inserted in a digital data stream to enable error detection or correction to be carried out.

Reference burst   The burst that marks the start of a TDMA frame.

Reliability   Measure of the likelihood of a component or system continuing to operate correctly after a given time.

RF   Radio frequency.

RHCP   Right-hand circular polarization.

rms multicarrier deviation   The rms deviation of an FDM/FM telephone signal.

Roll, Pitch, Yaw   Rotational motion about three Cartesian axes centered in the spacecraft.

S band   Frequency range of 2 GHz to 4 GHz.

SAM   Simple attenuation model.

SBS   Satellite Business Systems. A U.S. company that operates satellite networks, primarily digital.

SC   Satellite channel. The bits representing 16 terrestrial channels (TC) in the Intelsat TDMA system.

Scalar feed   Feed for front-fed reflector antennas with high efficiency and symmetry.

Scintillation   Rapid fluctuations in signal strength.

SCPC   Single channel per carrier. An FDM technique in which each signal is modulated onto its own carrier for transmission.

SER   Symbol error rate. The fraction of a sequence of message symbols that are in error.

SES   Ship earth station in the Inmarsat system.

Shannon capacity   The maximum possible data rate of a communication channel for a given signal-to-noise ratio, with no bit errors.

Shaped reflector   Antenna reflector with modified profile for improved efficiency.

SIC   Station identification code.

Simplex link   A link that is capable of only one-way transmission of data.

Site-diversity   A scheme in which two or more redundant earth stations located a few kilometers apart communicate with the same satellite.

Sky noise   Electromagnetic radiation from galactic sources and thermal agitation of atmospheric gases and particles.

Slot   A grouping of bits from one channel (in TDM) or from one earth station (in TDMA).

(S/N)   Signal-to-noise ratio.

SOF   Start of frame in a TDMA system.

Solar sails   Long flat panels covered with solar cells, projecting from a satellite. The panels are rotated to face the sun and generate the spacecraft's electrical power from incident sunlight.

SORF   Start of receive frame. The reference time for an earth station's TDMA downlink.

SOTF   Start of transmit frame. The reference time for an earth station's TDMA uplink.

Space qualification   Special test procedures to ensure high reliability of spacecraft parts.

Spacecraft   Any vehicle in orbit. All communication satellites are spacecraft.

SPADE   Single channel per carrier PCM multiple-access demand assignment equipment. An Intelsat demand-access system.

Sparklies   White flashes or dots appearing on a TV screen in FM TV systems. They are caused by noise impulses from the FM demodulator.

Spherical reflector   A reflector having a spherical surface, which can employ a movable feed to scan the beam.

Spillover   Energy lost in an antenna that does not contribute to the illumination efficiency.

Spinner   A satellite with a body that rotates to provide gyroscopic stabilization of attitude.

Spot beam   Very narrow beam covering a small part of the earth's surface.

SQPSK   Staggered QPSK modulation.

SS   Spread-spectrum. A modulation technique in which a signal's energy is spread over a bandwidth much wider than that required for transmission.

SS-TDMA   Satellite switched TDMA. A TDMA system in which the satellite can interconnect different uplink and downlink spot beams.

SSPA   Solid-state power amplifier.

STS   Space Transportation System. The Space Shuttle.

Subdish   Secondary reflector in a Cassegrain or Gregorian antenna.

Subsatellite point   The place where a line drawn from the center of the earth to a satellite passes through the earth's surface.

Superframe   The number of TDM or TDMA frames required to transmit all housekeeping and synchronization information once.

Surface errors   Departure of reflector surface from ideal curvature due to manufacturing tolerance, alignment errors, etc.

Symbol   One of the unique states of the carrier in digital modulation. A symbol may carry one or more bits.

System noise temperature   The total apparent noise temperature at a reference point in a receiving system.

Taper   Distribution of field strength across an aperture or reflector.

TASI   Time-assigned speech interpolation. A technique for reassigning channels during speech pauses.

TC   Terrestrial channel. The bits containing 16 samples of one voice channel carried in the Intelsat TDMA system.

TDM   Time division multiplexing. A technique whereby several signals from the same earth station share a transponder by using it at different times.

TDMA   Time division multiple access. A technique whereby signals from several earth stations share a satellite or transponder by using it at different times.

Telemetry, Tracking, and Command (TT&C)   A system that provides two-way communication between an earth station and the satellite to monitor spacecraft systems and to send instructions for changes.

Test tone   A 1-mW (0-dBm) 1-kHz signal used to adjust telephone equipment.

Three-axis stabilization (body stabilization)   A technique to maintain the body of a satellite in the same orientation relative to earth at all times. The body of the satellite does not rotate.

3-dB beamwidth   A measure of the width of an antenna beam. The power level observed by a receiving antenna placed at the 3-dB point of an antenna pattern is one-half of the power received when the antenna is at boresight.

Threshold   A level at which the operation of a system changes suddenly.

Thruster   A small rocket used to provide fine control of the velocity or attitude of a spacecraft.

T1 carrier   A 24-channel TDM system developed by the Bell System for telephony.

Transfer orbit   An intermediate orbit used in the process of launching a geostationary satellite.

Transparent transponder   A transponder that does not modify the signal, other than by amplification and frequency conversion.

Transponder   Basically a receiver followed by a transmitter. A transponder receives an uplink signal at one frequency, amplifies it, and retransmits it on another frequency.

Transponder hopping   One TDMA earth station working several transponders.

Trellis diagram   A technique for locating errors in convolutional code codewords.

TRT   The timing reference transponder for an Intelsat TDMA network.

TVRO   Television receive-only terminal.

TWT   Traveling wave tube.

TWTA   Traveling wave tube amplifier.

Universal time (UT)   Standard time for space operations and scientific observations.

Uplink   The communications channel from an earth station to a satellite.

UW   Unique word. A sequence of bits used in TDMA to mark each frame or superframe.

Viterbi codes   Convolution codes named for A. J. Viterbi.

WARC   World Administrative Radio Conference. This sets frequency allocations for the whole world.

Windowing   A process in which a TDMA earth station searches for the unique word only during some time window.

X band   Frequency range of 8 GHz to 12 GHz.

XPD   Cross-polarization discrimination. A measure of the conversion of power from one wave polarization to the orthogonal polarization.

XPI   Cross-polarization isolation. A measure of the coupling between nominally orthogonally polarized channels.

Z-axis intercept   Intersection of the satellite Z axis and Earth's surface. Defines the antenna pointing direction.

Zone beam   Satellite antenna beam covering a part of the earth's surface.

LIST OF SYMBOLS

The following is a list of the symbols used in most of the equations of this text. We have tried to adopt the symbols used in the relevant literature, recognizing that in some cases this causes the same symbol to be used for more than one quantity, while in other cases the same quantity is represented by more than one symbol.

a	Arc length used in look angle calculation.
a	Height of antenna aperture.
a	Semimajor axis of orbit.
$\mathbf{a}$	Electric field vector of a vertically polarized transmission.
$\mathbf{a}_c$	Co-polarized electric field vector that is received when $\mathbf{a}$ is transmitted.
a_n	Coefficients used in the infinite-series representation of a general nonlinear device.
$\mathbf{a}_x$	Cross-polarized electric field vector that is received when $\mathbf{a}$ is transmitted.
aR^b	Specific attenuation at rain rate R.
A	Amplitude of a general sinewave.
A	Traffic intensity in erlangs.
A	Attenuation, dB.
A	Subjective signal-to-noise ratio improvement provided by a compandor.
A	Area of an antenna aperture.
A_c	Total attenuation due to the clear atmosphere on a vertical path.
A_e	Effective area of an antenna aperture.
A_j	Joint attenuation in a diversity reception system.
$A(P)$	Attenuation exceeded for P percent of the time.
A_s	Average single-site attenuation in a diversity system.
Az	Azimuth angle.
$A_{0.01}$	Attenuation value exceeded for 0.01 percent of a year.
b	Width of an antenna aperture.
b	Semiminor axis of orbit.

b	Voice channel bandwidth in hertz.
$\mathbf{b}$	Electric field vector of a horizontally polarized transmission.
$\mathbf{b}_c$	Co-polarized electric field vector that is received when $\mathbf{b}$ is transmitted.
$\mathbf{b}_x$	Cross-polarized electric field vector that is received when $\mathbf{b}$ is transmitted.
B	Bandwidth in hertz.
B_0	Geomagnetic flux density in Teslas.
B_{IF}	IF bandwidth in hertz.
BO_i	Input backoff in decibels.
BO_o	Output backoff in decibels.
c	Arc length used in look angle calculation.
C	Codeword
C	Polar angle between earth station and subsatellite point.
C	Received carrier power.
C	Undetermined constant in differential equation solution.
C	Total number of available channels.
$(C/N)_i$	Carrier-to-noise ratio at input of an FM demodulator.
$(C/N)_i$	Overall carrier-to-noise ratio on a satellite link.
(C/N_0)	Carrier-to-noise power density ratio.
$(C/N)_D$	The value that the downlink carrier-to-noise ratio would have if the transponder did not retransmit incoming noise.
$(C/N)_I$	Term introduced into (C/N) equations to account for the effects of intermodulation noise.
$(C/N)_{\text{IS}}$	The value of $(C/N)_I$ when a TWT is operating at saturation.
$(C/N)_U$	Uplink carrier-to-noise ratio.
dD	Incremental raindrop diameter.
d_m	rms multichannel deviation at full loading.
dt	Differential time.
D	Antenna aperture diameter.
D	Data message
D_m	Median raindrop diameter.
e	Eccentricity of an orbit.
e_{in}	Input voltage of a general nonlinear device.
e_{out}	Output voltage of a general nonlinear device.
E	Electric field.
E	Orbital eccentric anomaly.
E	Number of bits by which a received sequence can differ from the unique word and still be counted as correct.
E_b	Energy per bit.
El	Elevation angle.
E_s	Energy per symbol.
EIRP_S	Satellite effective isotropic radiated power.
f_c	Carrier frequency in hertz.
f_{max}	Maximum modulating frequency in hertz.
f_{mod}	Modulating frequency in hertz.
f_r	frequency of minimum attenuation in a preemphasis filter.
f_N	IF carrier frequency of Nth channel in SPADE.
f_R	Received frequency.
f_T	Frequency that a moving transmitter would radiate when at rest.
f_V	Maximum video-modulating frequency.

F	Focal length of a parabolic reflector.
F	Probability of a false alarm (confusing an extraneous bit sequence with a unique word).
F	Flux density.
F^{-1}	Inverse Fourier transform.
F_{max}	Deviation in hertz that the maximum dispersal waveform causes.
F_S	Single-carrier saturation flux density at a satellite in dBW/m^2.
g	Peak factor for a multiplexed telephone signal.
$g(f)$	Factor used in scaling attenuation values with frequency.
G	Antenna gain.
G	Amplifier gain.
G	Green signal in color TV transmission.
G	Universal gravitational constant (6.67×10^{-11} Nm^2/kg).
G	Channel utilization in ALOHA.
G_l	Gain (less than unity) of a lossy device.
G_p	Spread-spectrum processing gain.
G_r	Receiving antenna gain.
G_t	Transmitting antenna gain.
$G_v(f)$	Power spectral density of random sequence of rectangular pulses.
G_D	Diversity gain.
$(G/T)_E$	Figure of merit of an earth station.
$(G/T)_S$	Figure of merit of a transponder's uplink antenna and receiver.
$G(\theta)$	Antenna gain as a function of angle θ.
h	Quantity of traffic (total holding time).
$\mathbf{h}$	Orbital angular momentum.
h_0	Earth station elevation above sea level in kilometers.
h_R	Rain height used in CCIR rain-attenuation model.
H	Parity check matrix.
H	Magnetic field.
H_e	Effective height of the top of a rainstorm in kilometers.
H_i	Height of zero-degree isotherm in kilometers.
H_0	Earth station elevation above sea level in kilometers.
H^T	Transpose of parity check matrix.
i	Orbital inclination.
I	Identity matrix
I	In-phase component of a TV chrominance signal.
I	Number of received incorrect bits.
I_D	Diversity improvement.
J_n	Bessel function of order n.
J_n	Legendre polynomial of order n.
k	Boltzmann's constant, 1.38×10^{-23} JK^{-1}.
k	Propagation constant, $2\pi/\lambda$.
k	Number of message bits in a codeword.
k	Number of stages in a shift register.
$\mathbf{K}$	Surface current density.
K	Transfer characteristic of FM demodulator.
l	Loading factor.
l_A	West longitude of point A.
l_B	West longitude of point B.
l_e	West longitude of earth station.

l_s	West longitude of the subsatellite point.
L	Mean time duration of a speech spurt.
L	Path length in kilometers.
L_a	Loss due to atmospheric effects.
L_A	North latitude of point A.
L_B	North latitude of point B.
L_e	North latitude of earth station.
L_{eff}	Effective path length in kilometers.
L_G	Horizontal projection of the portion of a satellite path length that is in rain.
L_p	Total downlink path loss.
L_{ra}	Loss in receiving antenna system.
L_s	North latitude of the subsatellite point.
L_S	Path length in rain in CCIR attenuation model.
m	FM modulation index.
m	Mean time before failure.
M	Orbital mean anomaly.
M_E	Mass of the earth.
M_r	Required fade margin for link reliability r.
n	Number of bits in a codeword.
n	The number of PN sequences that a shift register can generate.
$\mathbf{n}$	Unit vector normal to reflector surface.
N	Number of channels carried by a multiplexed telephone system.
N	Received noise power.
N	Channel number in SPADE-unprimed numbering system.
N	Mean number of times that ALOHA packets must be retransmitted.
N	Electron density in electrons/m^3.
N	Number of chips used to spread each bit in a PN spread-spectrum system.
N'	Channel number in SPADE-primed numbering system.
N_b	Number of bits required to transmit an alphabet.
$N(D)$	Number of raindrops.
N_f	Number of failed components in a life test.
N_s	Number of components at start of life test.
N_0	Noise power density.
N_0	A constant describing the size distribution of raindrops.
N_0	Number of surviving components in a life test.
p	Psophometric weighting factor expressed as a ratio.
p	Semilatus rectum of an orbit.
p	Power level in watts.
p	Bit error probability.
$p_r(t)$	Output voltage of a Nyquist filter when excited by a unit impulse.
$p(v)$	Probability that an instantaneous voltage has value v.
$p(x)$	Probability that a new call will be made in time interval dt.
p_L	Power delivered to a load.
p_S	Power supplied by a source.
P	Psophometric weighting factor, 2.5 dB.
P	Percent of time.
P	Number of bits in a preamble.
P	Probability of all combinations of I errors and $N - I$ errors in N consecutive bits.

P	Probability of failure.
P	Probability that a unique word will be received correctly.
P_n	Noise power.
P_r	Received power.
$P_{r_{\text{clear\_weather}}}$	Received power under clear-weather conditions.
$P_r(f)$	Amplitude response of a Nyquist filter.
P_t	Transmitted power.
P_0	Total power.
PB	Bit error probability.
PE	Probability of a symbol error.
$P(C, A)$	Probability that the number of required channels will exceed the number of available channels when the traffic intensity is A.
$P_J(A)$	Probability that the joint attenuation exceeds A.
P_S	Average power in a voice channel before compression.
$P_S(A)$	Probability that single-site attenuation exceeds A.
$P(\theta)$	Power in a direction θ.
Q	Weighting factor in video signal-to-noise ratio calculation.
Q	Probability of a miss (failure to recognize a correct unique word).
Q	Probability of failure.
r	Link reliability, the percentage of time that a link performs satisfactorily.
r	Distance between two points.
r	Number of parity bits in a codeword.
r	Radial distance in a circular aperture.
r_e	Radius of earth, 6370 km.
(r_o, ϕ_o)	Satellite polar coordinates in the orbital plane.
r_p	Factor used in CCIR rain-attenuation model to account for spatial non-uniformity of rain.
r_s	Orbital radius.
r_s	Symbol rate.
r_0	Radius of a circular aperture.
R	Distance from a satellite to an earth station.
R	Red signal in color TV transmission.
R	Rainfall rate in millimeters per hour.
R	Transmission bit rate.
R	Distance from the center of an antenna aperture to an observation point.
R	Reliability.
RA	Right ascension.
R_b	Bit rate.
R_c	Chip rate of a PN spread-spectrum system.
R_L	Load resistance.
$R(P)$	Rainfall rate exceeded for P percent of the time.
R_s	Symbol rate.
R_S	Internal resistance of source.
R_T	Total resistance in circuit.
s	Half-perimeter of spherical triangle used in look angle calculations.
S	Demodulated signal power.
S	Error syndrome of a block code.
$S(f)$	Spectrum of a general pulse.
(S/I)	Voice channel signal to co-polarized interference ratio.

$(S/N)_D$	The value that the downlink signal-to-noise ratio would have if the transponder did not retransmit incoming noise.
$(S/N)_i$	Overall signal-to-noise ratio of a CSSB link.
$(S/N)_I$	Term introduced into (S/N) equations to account for the effects of intermodulation noise.
$(S/N)_o$	Signal-to-noise ratio at output of FM demodulator.
$(S/N)_U$	Uplink signal-to-noise ratio.
$(S/N)_V$	Weighted video signal-to-noise ratio.
$(S/N)_{wc}$	Signal-to-noise ratio in worst channel at demultiplexer output.
(S/X)	Signal to cross-polarized interference ratio.
t	Time.
t_1	Time at which satellite position is to be determined.
t_o	Time at which satellite position is known.
t_p	Time of perigee.
t_x	The total time that x out of N channels are in use during time interval T.
T	Noise temperature.
T	Orbital period.
T	Reference time period for traffic calculations.
T_A	Antenna noise temperature.
T_b	Duration of a rectangular baseband pulse.
T_b	Increase in antenna noise temperature in Kelvins caused by rain.
T_c	Elapsed time in Julian centuries between 0 hours UT Julian day JD and noon UT on January 1, 1900.
T_e	Elapsed time since the x_r axis coincided with the x_i axis.
T_F	Frame period.
T_l	Noise temperature of a lossy device.
T_n	Noise temperature of a device.
T_p	Physical temperature.
T_s	Duration of one symbol.
T_s	System noise temperature.
(T_T/N)	Test-tone-to-noise ratio.
u	$1/r_o$ where r_o is the satellite radial coordinate in the orbital plane.
U	Coefficient used in calculating XPD from attenuation.
v	Instantaneous value of voltage.
v	Orbital linear velocity.
V	Coefficient used in calculating XPD from attenuation.
V	Bit rate for a single voice channel.
$V(D)$	Terminal velocity of raindrop with diameter D.
$V(f)$	Voltage spectral density.
V_P	Phase velocity of light in free space.
V_T	Component of transmitter velocity directed toward the receiver.
w	Preemphasis improvement factor expressed as a ratio.
w_b	Bandwidth required to transmit symbols at bit rate R_b.
W	Preemphasis improvement factor in decibels.
W	Bandwidth required for a spread-spectrum system with chip rate R_c.
$W(f)$	Power spectral density of an FDM/FM signal at RF.
x	Distance in aperture plane.
x	Number of channels in use.
(x_i, y_i, z_i)	Satellite rectangular coordinates in the geocentric equatorial system.

(x_o, y_o, z_o)	Satellite rectangular coordinates in the orbital plane.
xpd_V	Numerical cross-polarization discrimination on V-polarization channel.
xpi_V	Numerical cross-polarization isolation on V-polarization channel.
(x_r, y_r, z_r)	Rotating rectangular coordinate system.
X	Vertex angle in look angle calculation.
X	Decibel increase in average power provided by a compressor.
XP	Logical variable used in unique word identification.
XPD_V	Decibel cross-polarization discrimination when V-polarized waves are transmitted.
XPI_V	Decibel cross-polarization isolation on V-polarized channel.
XQ	Logical variable used in unique word identification.
y	Distance in aperture plane.
Y	Luminance signal of a television transmission.
Y	Vertex angle in look angle calculation.
Z	Path length in meters.
α	Vertex angle used in look angle calculation.
α	Rolloff factor of a Nyquist filter.
$\alpha_{g,o}$	Right ascension of the Greenwich meridian at 0 hours UT.
γ	Central angle between subsatellite point and earth station.
γ	Empirical constant used in the simple attenuation model.
$\delta(t)$	Unit impulse function.
Δf	Frequency deviation in hertz.
Δf	Frequency shift.
Δf_{rms}	rms test-tone deviation.
ΔN	Increase in received thermal noise power resulting from rain.
$\Delta \phi$	Change in polarization angle of a linearly polarized wave.
η	Average orbital angular velocity.
η	Single-sided rms noise power spectral density.
η	Aperture efficiency.
θ	Angle between antenna axis and a line to an observation point.
λ	Wavelength in meters.
λ	Packet transmission rate in ALOHA.
λ	Reciprocal of MTBF.
λ'	Packet reception rate in ALOHA.
Λ_e	Earth station latitude.
μ	Kepler's constant ($3.9861352 \times 10^5 \text{ km}^3/\text{s}^2$).
σ	rms value of a signal.
τ_L	Telephone load activity factor.
ϕ	Arbitrary phase angle.
ϕ	Angle between received electric field and local horizontal.
ϕ	Rotation angle in Faraday rotation.
ϕ	Angle between geomagnetic field and direction of propagation.
ω	Argument of perigee.
ω	Radian frequency.
ω_c	Radian carrier frequency.
ω_{mod}	Radian modulating frequency.
Ω	Right ascension of the ascending node.
Ω_e	Orbital angular velocity of the rotating coordinate system.

INDEX